About Island Press

Island Press is the only nonprofit organization in the United States whose principal purpose is the publication of books on environmental issues and natural resource management. We provide solutions-oriented information to professionals, public officials, business and community leaders, and concerned citizens who are shaping responses to environmental problems.

In 1994, Island Press celebrates its tenth anniversary as the leading provider of timely and practical books that take a multidisciplinary approach to critical environmental concerns. Our growing list of titles reflects our commitment to bringing the best of an expanding body of literature to the environmental community throughout North America and the world.

Support for Island Press is provided by The Geraldine R. Dodge Foundation, The Energy Foundation, The Ford Foundation, The George Gund Foundation, William and Flora Hewlett Foundation, The James Irvine Foundation, The John D. and Catherine T. MacArthur Foundation, The Andrew W. Mellon Foundation, The Joyce Mertz-Gilmore Foundation, The New-Land Foundation, The Pew Charitable Trusts, The Rockefeller Brothers Fund, The Tides Foundation, Turner Foundation, Inc., The Rockefeller Philanthropic Collaborative, Inc., and individual donors.

Endangered
Species
Recovery

Endangered Species Recovery

Finding the Lessons, Improving the Process

Edited by
Tim W. Clark,
Richard P. Reading,
Alice L. Clarke

ISLAND PRESS

Washington, D.C. • Covelo, California

Library of Congress Cataloging-in-Publication Data

Endangered species recovery: finding the lessons, improving the
  process / edited by Tim W. Clark, Richard P. Reading, and Alice L.
  Clarke.
    p. cm.
    Outgrowth of a conference held at the University of Michigan, Jan.
  8–9, 1993.
    Includes bibliographical references and index.
    ISBN 1-55963-271-2 (cloth). — ISBN 1-55963-272-0 (paper)
    1. Endangered species—United States.   2. Wildlife conservation—
United States.   3. Wildlife conservation—United States—Case
studies.   I. Clark, Tim W.   II. Reading, Richard P.   III. Clarke,
Alice L.
QL84.2.E55   1994   94-18097
591.52'9'0973—dc20   CIP

Printed on recycled, acid-free paper 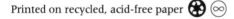 ∞

Manufactured in the United States of America
10   9   8   7   6   5   4   3   2   1

Contents

PART III

Theoretical Perspectives

PART IV

Lessons

Preface

This book is about improving endangered species conservation. And although our goal is shared by many, we think you will find our approach unique. We focus on two major targets for change that rarely come to mind when people speak of conservation: the professionals and the organizations who do the work of species restoration. The contributors to this volume have carefully considered how the accumulated knowledge and diverse perspectives of both practitioners and theorists can be used to improve these two critical elements of species recovery programs. This volume seeks to learn from lessons of the past in order to improve our chances for success in the future. The multiple perspectives presented here, if applied, promise to improve endangered species conservation significantly.

The book was written for the wide audience of people interested in endangered species conservation. Professional conservationists, managers and administrators, nongovernmental activists, analysts, researchers, professors and students, and the public at large will find this book useful. Several chapters are written by field biologists with many years' experience studying endangered species and laboring in the decision-making processes designed to save them. Other chapters are written by noted scholars of natural resource management, including population ecology, sociology, organization theory, and policy. Taken together, the book contains a new perspective on endangered species restoration and offers a host of practical suggestions, or lessons, for improving our individual and collective performance.

The book is the outgrowth of a small conference held at the University of Michigan in Ann Arbor on 8–9 January 1993, during which the contributors met and exchanged perspectives. This was a lively and productive interaction as participants approached endangered species conservation from diverse backgrounds. The conference and book were supported by the Frank H. and Eva B. Buck Foundation.

The encouragement and commitment that Robert Buck provided was essential and is very much appreciated. Dean Garry Brewer of the School of Natural Resources and Environment, University of Michigan, provided additional funds and support for the conference and book. Lynn Chase of the Chase Wildlife Foundation also provided essential funding. Catherine Patrick supported this book from the beginning in several important ways. Barbara Dean of Island Press professionally guided us through the publication process. Together these five people made the project possible and deserve special recognition.

Each chapter was reviewed by the three coeditors and at least one other authority. We wish to thank these reviewers: Ken Alvarez (Florida Park Service), Michael Bean (Environmental Defense Fund), Wayne Brewster (National Park Service), Denise Casey (Northern Rockies Conservation Cooperative), E. Frances Cassirer (Idaho Department of Fish and Game), Richard Conner (USDA Forest Service), Peyton Curlee (Northern Rockies Conservation Cooperative), Scott Derickson (National Zoological Park), Dan Doak (University of California, Santa Cruz), John Eisenberg (University of Florida), Steven Forrest (Sierra Club Legal Defense Fund), Jay Gore (USDA Forest Service), Lou Hanebury (U.S. Fish and Wildlife Service), Robert Hole (Mississippi State University), Larry Irwin (NCASI Inc.), Rachel Kaplan (University of Michigan), David W. Laist (Marine Mammal Commission), James Layne (Archbold Biological Station), Mary Maj (USDA Forest Service), Vicky Meretsky (University of Arizona), Jim Pissot (National Audubon Society), Steve Primm (University of Colorado), Herb Raffaele (U.S. Fish and Wildlife Service), John E. Reynolds III (Marine Mammal Commission), Victoria Saab (USDA Forest Service), William Settle (U.S. AID), Michael Soulé (University of California, Santa Cruz), Peter Stroud (Werribee Zoological Park, Australia), John Wargo (Yale University), Ron Westrum (Eastern Michigan University), Jim Wiley (U.S. Fish and Wildlife Service), and Gerald Wright (University of Idaho). The external reviewers provided a valuable contribution to the quality of our book. We apologize to any we may have omitted.

A number of other people, many of them students at the School of Natural Resources and Environment, University of Michigan, helped us with a myriad of organizational tasks and kept detailed notes of the conference proceedings that proved valuable in writing the book. Our thanks to Sara Barth, Geoff Brown, Denise Casey, Lynn Gooch, Jim Havard, Joel Heinen, Cathy Patrick, Margo Smit, and John Stoddard.

We especially want to thank the authors of this volume for contributing their experience and perspectives and for their time, effort, and cooperation. We greatly value their insights and their willingness to share them with a larger audience. Without exception, their chapters are valuable. To all these people, to those previously mentioned, and to the many who helped indirectly, our sincerest gratitude.

Improvements in natural resource conservation will happen when people have a genuine interest in the subject, a willingness to share ideas and experience, and a cooperative commitment to the outcome. The many people who contributed to this book in various ways exemplify that constructive approach. We thank you all.

PART I

Overview

A compelling rationale and an effective strategy for protecting endangered species will require recognition that contemporary [endangered species] problems are the result of socioeconomic and political forces.
STEPHEN R. KELLERT

The management of endangered species is typically a complex matter, involving constituencies with differing priorities and different courses of action.
MICHAEL E. SOULÉ

I

Introduction

**Tim W. Clark, Richard P. Reading,
and Alice L. Clarke**

Twenty years' experience with the U.S. Endangered Species Act (ESA) has shown that having a law on the books is not, in itself, a solution to the major problem of the erosion of the earth's biological diversity. A few statistics about the endangered species problem, nationally and globally, are noteworthy—for the more we learn about the biotic enterprise, the more "we come to realize that we are in the midst of an extinction episode of historic proportion" (Kohm 1991:4). A conservative projection of species loss is more than 20 percent of the planet's biodiversity within the next decade or two (Wilson 1989). Extinctions are estimated to be taking place at the rate of about 100 or more species per day. Losses in rain forests are about 10,000 times greater than natural "background" extinction rates (Wilson 1992). In the United States, an estimated 675 plant species may become extinct by the year 2000 (BioScience 1989). About one-third of all freshwater fish species are being seriously harmed by environmental degradation (Allan and Flecker 1993). About 3,000 plant and animal species are considered official candidates for protection under the ESA, but some 300 may already be extinct because the U.S. Fish and Wildlife Service (FWS) has limited ability to process the paperwork and implement protection programs (Meese 1989). Finally, fewer than half a dozen endangered species have been fully recovered under the ESA in the last twenty years (GAO 1992).

Are there plausible explanations for these dismal facts? Perhaps the ESA is an unworkable law; perhaps it addresses elements and processes that have no bearing on the disappearance of species. Another possibility is that extinction is inevitable given the scale and pace of human alteration of landscapes and no herculean human efforts can stop its course. Or perhaps it is simply that implementation of the existing law has been weak—that the "implementation gap" (Ascher and Healy 1990:3–4) belies the significance, power, and potential of this precedent-setting policy. This seems the most conservative, least fatalistic, and most probable of the explanations. But having narrowed down the possibilities, we have still to explicate the substance of the implementation process. Where is its locus? Who controls it? Who is responsible for it? Who pays for it? What are its functions? What does it actually accomplish? What limits it? What is its relationship to the law and to other aspects of society at large? Which elements play major roles and which are relatively insignificant? Which can be manipulated and which are immutable? What is the "logic," if any, of the implementation process?

Understanding policy implementation is a complicated task that many have undertaken. (See, for example, Pressman and Wildavsky 1978 and Nakamura and Smallwood 1980.) From these studies, from our own firsthand experience in endangered species work, and even from popular understanding of the issue, it is clear that two elements stand out from the rest in terms of their magnitude and centrality, their control over other elements, the leverage available to improve their performance, their ultimate responsibility for the outcomes, and, indeed, their very role in society as embodiments of so much of humanity's problem-solving knowledge and skills (Westrum and Samaha 1984; Osborne and Gaebler 1993). These two elements are the professionals and the organizations that participate directly in endangered species restoration programs, and it is these that we scrutinize in this book. Our basic argument, then, is this: poor implementation of the ESA is itself a major cause of the continuing decline of species, and professionals and organizations are significantly responsible for the quality of implementation.

Shrinking numbers of species and other taxonomic units are not (and should not be) the only measure of these enterprises. Indeed, there have been a number of major reviews and analyses of the ESA implementation process itself (Yaffee 1982; GAO 1988; Tobin 1990; Keiter and Holscher 1990; Grumbine 1992; Houck 1993). These studies have taken a variety of perspectives and focused on various aspects of ESA implementation. But time and again, implementation is cited as a primary locus of weaknesses, and repeatedly the organizations involved, specifically government agencies, are blamed for the failures. The professional component of implementation is cited much less often. Perhaps individuals are seen simply as agents of their employing organizations; maybe there is a desire to avoid personalizing the issues; perhaps some analysts do not see the professions as institutions with a coherent (and influential) set of perspectives, techniques, and knowledge. We see, however, that professionals do play a major role in ESA implementation and cannot be separated from their organizations. Each enables the other to perform at its best, and each places strict limitations on the other's performance.

Collectively, the statistics of species attrition and these studies suggest that the endangered species problem is increasing faster than we as a society—through our professions and organizations—are able to cope with it. The current approach to implementing species recovery in the United States is not working as well as many people expected. What can turn this situation around? We believe that improvements can begin by learning how endangered species programs actually work. An abundance of experience has been gained by a variety of people involved in endangered species recovery efforts over the last twenty years: field biologists, government managers of wildlife, lands, forests, or other natural resources, agency administrators, academics, industries and other commercial interests, nonprofit conservation groups, and the public. Endangered species policies and programs have such far-reaching implications for society that a great number of people are directly or indirectly involved in the process, are interested in its progress, or are affected for better or worse by it. This immense body of experience is worth drawing out—from the most minor and specific details to the broadest generalizations. Organizing and comparing this information, articulating its insights, and feeding that insight back into the public policy process, through both prescription and implementation of the ESA, is essential. At the same time, pertinent theory has been developed in a variety of disciplines, including professional practice and organizational management, two areas that bear directly on how well endangered species policy is implemented. This knowledge should inform our search for improvements.

The Response to Extinction

Biologists define extinction, which is basically a population-level process, as the irrevocable loss of a biological element (Gilpin and Soulé 1986). Because extinction is viewed largely as a biological phenomenon, the dominant professional and organizational response has been to focus on biology, obscuring critical nonbiological dimensions. Thus most endangered species restoration programs are staffed and led by professional biologists, usually working for government agencies. And the professional responses to endangered species recovery have become well established and institutionalized in the structure and operating procedures of government bureaucracies (see Kerr 1984).

Professional Performance

One major weakness in efforts to recover endangered species lies in the kind of knowledge, problem-solving approaches, and outlook that professionals bring to this difficult and important work. Unfortunately, there are blind spots in traditional biological responses to endangered species conservation (see Westrum 1986). A major aspect of the problem arises because, as Schaller (1992:47) has noted, "conservation problems are social and economic, not scientific, yet biologists have traditionally been expected to solve them."

All professionals—from physicians and lawyers to biologists and organizational designers—share certain characteristics even though they differ in their knowledge, problem-solving approaches, and outlook. Professionals are experts with considerable experience and skill in solving particular kinds of problems. Professional knowledge comes packaged in disciplines and organized in universities around departments such as biology, sociology, and political science. Each discipline trains its professionals in specialized knowledge via specialized language. Professional associations are maintained on the job through professional societies and departments and subunits in employing organizations. Each discipline is held together by central concepts around which its knowledge is organized. A central concept of biology, for example, is neo-Darwinian evolution. Their training and the kinds of knowledge they possess very much affect how professionals see and carry out their work.

Professionals also bring to their tasks certain styles of problem solving that usually are deeply entrenched in the philosophy and techniques of their discipline. Most professionals strive for rational, systematic inquiry in problem solving as well as effectiveness and efficiency. Nevertheless, the fundamental differences in philosophy and methods have profound implications for how work is done. Schön (1983) describes two models of professional problem solving. "Technical rationality" employs the philosophy of experimental science and argues that reliable knowledge comes only from rigid experimentation. This model tends to be highly quantitative; it is noncontextual; it seeks universal laws. "Reflective practice," on the other hand, understands that reliable knowledge comes from a variety of sources, some less "hard" than others (retroductive approaches, for example, or logical inference). This model is both quantitative and qualitative; it seeks contextual understanding; it encourages continual self-reflection in practice. These two models require very different modes of professional behavior—which, in turn, have very different consequences for ESA implementation.

Professionals also come equipped with a certain outlook or sense of identity that determines how they see their role, what kinds of information are relevant, and what methods are appropriate. As the saying goes, "To the person with a hammer, every problem is a nail." Differences in disciplinary training and problem-solving approaches account for this difference in outlook, as do different psychological, sociological, and political dimensions of professions and careers. Each profession socializes and intellectualizes its adherents into more or less distinctive outlooks and identities. Traditional wildlife biologists, for example, tend to be inculcated with the "heritage, structure, code, and norms . . . to maintain their traditional identities against the tempers of social whim" (Romm 1984:15). Because of individual and disciplinary differences in their knowledge, problem-solving approaches, and outlooks, then, professionals in endangered species conservation tend to have markedly different responses.

A variety of problems show up in professional practice in endangered species recovery programs. The knowledge may be outdated; the wrong kind, wrong mix, or wrong number of professionals may be used; professional knowledge may not be matched to the task; problems may be defined in terms of one discipline's knowledge without the benefit of other relevant knowledge; methods may be inappropriate; the problem may not respond to traditional approaches; the program may lack qualified leadership. Or professionals may be averse to social conflict; their sense of duty to employers may conflict with their sense of professional responsibility; they may view others' values as irrational or misinformed. Certainly these professional problems emerge in the chapters that follow; they can occur in any endangered species restoration effort. Often they are attributed mistakenly to personality quirks, lack of commitment, lack of knowledge or understanding, or other personal issues. Sometimes such problems are lumped under terms—such as "biopolitics"—that offer little by way of useful description or insight for understanding the problem.

Several writers have suggested educational programs to rectify these problems. Romm (1984), for example, offers four suggestions for improving university professional education in natural resource management: students should receive professional courses in resource policy that stimulate their curiosity and increase their understanding of public policy processes, develop their capacity to analyze causes and sources of leverage, impart conceptual and communication skills, and develop the integrity they will need to maintain a professional stance even in stressful situations. These ideas reinforce Schön's (1983) model of reflective practice. Imparting this perspective to professionals already in the agencies, however, is much more problematic.

Working to improve professional practice would certainly benefit endangered species conservation efforts, but it is not sufficient. Because much of professional practice is shaped by the organizations with which and in which professionals work, we must also focus on improving organizational performance.

Organizational Performance

A second major weakness in endangered recovery efforts, as we see it, lies in the organizational systems used to accomplish the task. Many organizations of different kinds are involved in endangered species conservation, and they vary in their abilities and performance. As at the professional level, problems at the organizational level too have been recognized: "Endangered species bureaucracies are really no different from any other bureaucracies. They follow the same immutable law of short-term self-interest, and this goal tends to dominate all others. The tragedy of endangered species bureaucracies is that in pursuing this goal they can force the species they are charged with protecting to serve bureaucracy rather than the reverse." (Snyder and Snyder 1989:259). Although such comments locate the problem, they do not offer further insight into its specific aspects or possible solutions. It is here that the social sciences can make significant contributions.

Many people find it hard to view organizations as a variable that is itself subject to management in endangered species conservation. The office buildings, management plans, and employees are easy to see, but the organizations themselves are abstract and vague. Yet the professional world is dominated by organizations of diverse kinds. Organizational experts have identified six major ways of looking at organizations (Argyris and Schön 1978). They may be seen as *groups* of people who interact and have a sense of collective identity, as *agents* or instruments for achieving social purposes, or as *structures* (as in "organizational charts"). Organizations may also be perceived as self-regulating *systems* that maintain certain consistent patterns of behavior or as *cultures*, or small societies, in which people create shared meanings, symbols, rituals, and cognitive schemes to regulate their interactions. They may also be viewed as *political systems* for governing and dealing with contending interests and powers.

Despite these different ways of viewing organizations, all organizations do have common properties. They are social entities, they are goal-directed, they are structured deliberately for certain activities, and they have definite boundaries (Daft 1983). Every organization has a structure (usually both formal and informal structures). It has certain tasks to do. It is run and staffed by people. It operates according to a set of goals and theories. It has critical issues of concern regarding its relationship with its surroundings and its internal operation. It serves particular groups of constituencies, clients, and beneficiaries, has various sources of legitimacy, and is accountable to various groups. Gordon (1983) groups these many characteristics at three levels: individual, group, and organizational issues. Individual issues include perceptions, interpersonal communication, motivations, stress, and personal development. At the group level, which includes varying lateral, hierarchical, or functional subgroups within an organization, the issues include leadership and management patterns, power relationships, and group performance. At the highest level, organizational issues include structural and cultural aspects, the organization's ability to adapt in response to changing tasks or conditions, and design and contingency considerations. Different combinations of all these factors create different kinds of organizations—mechanistic, authoritarian, and bureaucratic at one extreme and organic, participatory, and flexible at the other. Obviously these two organizational extremes, which differ vastly in their structure and functioning, perform very differently in endangered species conservation efforts.

The ability of organizations to solve endangered species problems is a product of their structure, culture, and managerial systems as well as the setting in which they operate. (See also Warwick 1975; Perrow 1986; Hall 1987.) In government bureaucracies—the most prevalent organizations involved in endangered species recovery—information may be distorted as it moves up and down long hierarchies, professional roles may be narrowly defined, intelligence failures may cripple efforts, science may be weak or delayed in its use, and detrimental delays may occur in taking needed actions—to mention just a few of the common problems. Although difficulties in endangered species programs are often couched in the language of technical biological problems, many of them in fact can be traced to how the organization is structured and how it functions.

It is possible, however, to design and manage organizations in ways—specifically nonbureaucratic ways—that match the complex, urgent, multifaceted task of restoring species. Alternative designs might include flexible, creative, and highly responsive systems, such as high-performance teams. Although few endangered species programs explicitly consider the range of options, organizational performance may represent one of the best targets for achieving effective, long-lasting improvements. This book, therefore, focuses on organizations, as well as on professionals, as opportunities for improving the way we go about conserving and restoring species threatened with extinction.

About This Book

This book differs from all preceding books on endangered species conservation of which we are aware. It is the first to engage not only biologists but social scientists in the analysis and discussion of endangered species conservation. In doing so, it seeks an interdisciplinary explanation for the weak performance of endangered species recovery efforts and an exploration of alternatives. On 8–9 January 1993, a small conference at the University of Michigan in Ann Arbor brought together a range of practitioners and theoreticians to share their knowledge and seek lessons for ESA improvement. This book, featuring contributions from nearly all those participants, is an outgrowth of that conference. It focuses on ESA implementation, in practice and in theory, on the structures in place (both professional and organizational) to implement the policy, and on the constraints and opportunities for improvement. We call for other groups—government agencies, nonprofits, academics—to carry out similar analyses and discussions in an honest, systematic manner and on a regular, case by case basis. The results should be made widely available as a basis for learning and improvement throughout the conservation community.

The cases and perspectives presented in this book represent some of the most highly visible endangered species in the United States and the best range of knowledge about ESA implementation. Case studies include the northern spotted owl (*Strix occidentalis caurina*), black-footed ferret (*Mustela nigripes*), grizzly bear (*Ursus arctos horribilis*), Florida manatee (*Trichechus manatus latirostris*), red-cockaded woodpecker (*Picoides borealis*), California condor (*Gymnogyps californianus*), and Florida panther (*Felis concolor coryi*). In addition, an Australian endangered species, the eastern barred bandicoot (*Perameles gunnii*), and three U.S. candidate species, the Coeur d'Alene salamander (*Plethodon idahoensis*), harlequin duck (*Histrionicus histrionicus*), and wolverine (*Gulo gulo*), are discussed for comparison.

These cases, of course, are not representative of all endangered species conservation programs. Most of the species discussed here have garnered a high public profile, have obtained a large share of the money available to save endangered species, have benefited from considerable scientific and management attention, have occupied center stage for years (in some cases decades), and have been perceived largely as "charismatic" animals. In many ways, they have received the very best talents, time, and resources our country has to offer. There are no plants, invertebrates, or "obscure" species here—species that are little known to the public, attract few resources, and rate a low priority with agency managers and politicians. In fact, these obscure species fall victim to the same professional and organizational weaknesses that appear throughout this book—perhaps to an even greater extent. If we are unable to conserve or restore well-funded, well-researched, and high-profile species, the prospects for poorly funded, poorly understood, little known, and "uninspiring" species are bleak indeed.

In practice, none of the recovery programs described in this book nor any that we know of have taken an integrated, interdisciplinary approach to endangered species conservation—an approach incorporating biologists, policy experts, sociologists, psychologists, organizational consultants, conflict managers, and others. Collective efforts like this remain only ideal models untried in practice. This book encourages the application of such models, experimentation, learning, communication, and improvements from the realm of theory to the world of practice.

Part I of the book presents an overview of the ESA in addition to this introduction. Part II has nine chapters that examine recovery programs for eleven species and impart lessons learned from firsthand experience. Part III consists of six chapters presenting theoretical perspectives from several social and biological sciences and offering recommendations for program improvement. Part IV is a synthesis of the observations and lessons from the preceding chapters.

In Chapter 2, J. Alan Clark introduces the idea of federal protection of endangered species, the ESA's major provisions, regulations, internal agency politics, and judicial interpretations. In assessing the ESA's effectiveness, Clark reviews eight major criticisms of the ESA and concludes with five recommendations for improving the act's implementation.

In Chapter 3, Steven L. Yaffee examines the northern spotted owl recovery program, focusing on the case as an indicator of the sociopolitical context of endangered species management. He reviews the current situation and describes how we got here. The owl issue is really many issues in one: the clash of values, the overall nature of decision-making processes, legal structure, and politics, all as an expression of human values and a set of strategic behavior. Yaffee offers insights, lessons, and recommendations throughout.

Chapter 4, by Richard P. Reading and Brian J. Miller, describes the black-footed ferret recovery program, a complex effort characterized by several professional and organizational weaknesses. The authors suggest that the overall program may well eventually restore the ferret despite its many problems. Reading and Miller discuss five problems—program structure, working groups, program control, the role and use of science, and measuring program success—as a basis for evaluating and making recommendations for improvement.

David J. Mattson and John J. Craighead focus on Greater Yellowstone's threatened grizzly bear population in Chapter 5. They summarize the recent history and current status of grizzly bear recovery efforts in and around Yellowstone. They explore competing definitions of the conservation problem and examine ESA implementation, focusing on information flows, information feedback loops, individual professional behavior, and program monitoring. Mattson and Craighead close with eleven lessons for improving future programs.

Richard L. Wallace examines the Florida manatee recovery program in Chapter 6. He provides extensive background for his analysis and discusses the difficulties in measuring success. Periodic crises have played an important role in moving the manatee program forward. Having an "outside" group formally intervene, evaluate, and provide assistance to the program during these crises proved invaluable. Wallace stresses the theme of organizational learning and offers a model for improving implementation.

Chapter 7 is by Jerome A. Jackson, who describes his experiences in the red-cockaded woodpecker recovery efforts. After giving a brief history of conservation efforts and describing his analytical perspective, he examines a host of variables that represent obstacles to professional participation. Among these are agency conservatism, implementation of specific features of the ESA, and levels of agency commitment. Jackson ends the chapter with several personal and professional lessons for improvement.

Noel F. R. Snyder focuses on the California condor in Chapter 8. Citing administrative difficulties that arose in the critical year 1985, he highlights key areas where weaknesses occurred in the condor program and explains what might have been done about them. Snyder focuses his examination on the strengths and weaknesses of recovery teams and recovery planning. These two variables, central to nearly all recovery programs, are the targets of his lessons for improving ESA implementation.

The Florida panther recovery program is the subject of Chapter 9 by Ken Alvarez. He examines the government's behavior in panther recovery efforts. After presenting background case material, he focuses on how well ESA implementation has worked and discusses several facets of the panther program, including refuge planning, captive population management, and public education. Alvarez closes with a lengthy list of professional and organizational ideas for improvement.

Chapter 10 is by Craig R. Groves, who describes programs for three little-known candidate species for ESA listing: the Coeur d'Alene salamander, harlequin duck, and wolverine. After briefly describing each program, Groves discusses four professional and three organizational lessons, focusing on problem definition, education and experience with nongame species, motivation and interest in candidate species conservation, program evaluation, agency structure and culture, and interagency cooperation and communication. He finishes with a number of recommended steps for designing and implementing successful conservation programs for candidate species.

Chapter 11, the last chapter in Part II, is by Gary N. Backhouse, Tim W. Clark, and Richard P. Reading. The recovery effort for an endangered Australian species, the eastern barred bandicoot, is a useful comparison to the U.S. cases. The authors describe the species' decline, research, and management as background to an examination of how initially weak professional and organizational responses to the declining population failed to conserve the species. They follow with lessons based on this experience and end with a description of new organizational arrangements, management operations, and the improved status of the species.

Part III begins with Chapter 12 by Steven C. Minta and Peter M.

Kareiva, who discuss how to improve basic conservation biology, a central aspect of ESA implementation. They suggest that conservation science may be struggling with a pivotal conceptual reorganization and that development of conservation theory is outpacing data acquisition. They recommend improvements in experimental design, methods, and conceptual organization so that information can be presented to leaders and policymakers without bias. The reliable knowledge that results would stimulate communication and consensus.

Chapter 13 is written by Julia M. Wondolleck, Steven L. Yaffee, and James E. Crowfoot. They examine the conceptual basis for designing and implementing alternative dispute resolution (ADR) processes and their relevance for endangered species conservation. After underlining the importance of how decisions are reached—a matter that has direct bearing on the viability of those decisions—they draw lessons from specific endangered species cases in which ADR processes have been utilized and conclude with specific recommendations for decision makers considering dispute resolution.

Ron Westrum takes an organizational perspective in Chapter 14. Recovery teams and other special groups are the basic unit of conservation work. He examines how intention, incompetence, ignorance, and ill fortune all lead to recovery team failure. Westrum goes on to describe the desirable qualities of teams, political control of species recovery, and key coordinating and motivating roles. Finally, he offers suggestions to agencies concerned with ESA implementation and recommends improvements to upgrade learning and performance.

Tim W. Clark and Richard P. Reading review a range of professional practice issues in Chapter 15. Two essential features of all ESA programs are the behavior of professionals and their ability to collaborate and share mental models across disciplines. Professionals must be able to communicate effectively. They must be aware of their own values, attitudes, and behavior as well as those of others involved in species recovery. At times all three may have to change in order to increase the chances for successful conservation. Finally, professionals must be good learners and encourage their organizations, too, to learn quickly and well.

In Chapter 16, Stephen R. Kellert examines a range of valuational, socioeconomic, and organizational factors in ESA implementation. These social science variables are often inadequately considered, viewed as only marginally important, or completely ignored because of the biases of wildlife professionals, the difficulty of understanding human behavior and organizations, and the political risks associated with efforts to manage these factors. Kellert analyzes three valuational, three socioeconomic, and four organizational factors of paramount importance and offers over a dozen lessons for improving ESA implementation.

In Chapter 17, Garry D. Brewer and Tim W. Clark present a policy sciences perspective on ESA implementation. Implementation is a key step in the public policy process, but it is often overlooked as a focus for careful attention and management. The implementation phase is examined in some detail here, as the authors discuss complex, multiparty implementation efforts, the foundations of the policy sciences, multidisciplinary viewpoints, how to orient to problems, and how to map a problem's context. They conclude that implementation is a social process and, as such, requires an explicit policy orientation by participants in endangered species conservation.

Part IV, Lessons, is written by the three coeditors. Here the book's purpose is summarized and its numerous practical insights are distilled into eight lessons: the nature of the endangered species restoration task, the professional and organizational knowledge and skills needed to meet it successfully, teamwork, government behavior, how ESA prescriptions could better facilitate implementation, learning, leadership, and more. These lessons can help at the local field level as well as the national level.

The need to improve ESA implementation is here and now. Indeed, we may well be at a critical juncture in the struggle to conserve the biological diversity of the United States and the world. The big question that remains now is whether we will capitalize on this opportunity.

Acknowledgment

Denise Casey of the Northern Rockies Conservation Cooperative critically reviewed this chapter.

References

Allan, J. D., and A. S. Flecker. 1993. Biodiversity conservation in running waters. *BioScience* 43:32–43.

Argyris, C., and D. A. Schön. 1978. *Organizational Learning: A Theory of Action Perspective*. Reading, Mass.: Addison-Wesley.

Ascher, W., and R. Healy. 1990. *Natural Resource Policymaking in Developing Countries: Environment, Economic Growth, and Income Distribution*. Durham, N.C.: Duke University Press.

BioScience. 1989. Extinction countdown for U.S. plants. *BioScience* 39:276.

Daft, R. L. 1983. *Organization Theory and Design.* St. Paul, Minn.: West.

General Accounting Office (GAO). 1988. *Endangered Species: Management Improvements Could Enhance Recovery Program.* GAO/RCED-89-5. Washington, DC: Resources, Community, and Economic Development Division.

————. 1992. *Endangered Species Act: Types and Numbers of Implementation Actions.* GAO/RCED-92-131BR. Washington, DC: Resources, Community, and Economic Development Division.

Gilpin, M. E., and M. E. Soulé. 1986. Minimum viable populations: Processes of species extinction. In *Conservation Biology,* M. E. Soulé (ed.). Sunderland, Mass.: Sinauer Associates.

Gordon, J. R. 1983. *A Diagnostic Approach to Organizational Behavior.* Boston: Allyn & Bacon.

Grumbine, R. E. 1992. *Ghost Bears: Exploring the Biodiversity Crisis.* Washington, DC: Island Press.

Hall, R. H. 1987. *Organizations: Structures, Processes, and Outcomes.* Englewood Cliffs, N.J.: Prentice-Hall.

Houck, O. A. 1993. The Endangered Species Act and its implementation by the U.S. Departments of Interior and Commerce. *Colorado Law Review* 64:278–370.

Keiter, R. B., and P. T. Holscher. 1990. Wolf recovery under the Endangered Species Act: A study in contemporary federalism. *Public Land Law Review* 11:19–52.

Kerr, D. H. 1984. *Barriers to Integrity: Modern Models of Knowledge Utilization.* Boulder: Westview Press.

Kohm, K. A. 1991. Introduction. In *Balancing on the Brink of Extinction: The Endangered Species Act and Lessons for the Future,* K. A. Kohm (ed.). Washington, DC: Island Press.

Meese, G. M. 1989. Saving endangered species: Implementing the Endangered Species Act. In *Defense of Wildlife: Preserving Communities and Corridors,* G. Mackintosh (ed.). Washington, DC: Defenders of Wildlife.

Nakamura, R. T., and F. Smallwood. 1980. *The Politics of Policy Implementation.* New York: St. Martin's Press.

Osborne, D., and T. Gaebler. 1993. *Reinventing Government: How the Entrepreneurial Spirit Is Transforming the Public Sector.* New York: Plume/Penguin Books.

Perrow, C. 1986. *Complex Organizations: A Critical Essay.* 3rd ed. New York: McGraw-Hill.

Pressman, J. L., and A. Wildavsky. 1978. *Implementation.* Berkeley: University of California Press.

Romm, J. 1984. Policy education for professional resource managers. *Renewable Resources Journal* 2:15–17.

Schaller, G. B. 1992. Field of dreams. *Wildlife Conservation* (September/October):44–47.

Schön, D. A. 1983. *The Reflective Practitioner: How Professionals Think in Action.* New York: Basic Books.

Snyder, N.F.R., and H. A. Snyder. 1989. Biology and conservation of the California condor. *Current Ornithology Research* 6:175–267.

Tobin, R. J. 1990. *The Expendable Future: U.S. Politics and the Protection of Biological Diversity.* Durham, N.C.: Duke University Press.

Warwick, D. 1975. *A Theory of Public Bureaucracy: Politics, Personality, and Organization.* Cambridge: Harvard University Press.

Westrum, R. 1986. The blind eye of science: Every system of knowledge is also a system of ignorance. *Whole Earth Review* (Fall):36–41.

Westrum, R., and K. Samaha. 1984. *Complex Organizations: Growth, Struggle, and Change.* Englewood Cliffs, N.J.: Prentice-Hall.

Wilson, E. O. 1989. Threats to biodiversity. *Scientific American* 261:108–116.

———. 1992. *The Diversity of Life.* Cambridge: Belknap Press of Harvard University Press.

Yaffee, S. L. 1982. *Prohibitive Policy: Implementing the Federal Endangered Species Act.* Cambridge: MIT Press.

)

2

The Endangered Species Act
Its History, Provisions, and Effectiveness

J. Alan Clark

The primary instrument of federal efforts to protect endangered species is the Endangered Species Act of 1973 (ESA). This chapter begins with a brief history of federal efforts to protect endangered species and a summary of the ESA's major provisions. The chapter then looks beyond the actual language of the ESA and analyzes other factors, primarily legal, that affect the act's implementation. This analysis is followed by an assessment of the ESA's performance and a discussion of the primary criticisms of the act. A few suggestions for improving endangered species conservation through amending the ESA are then offered. Although endangered species conservation could be significantly improved through adequate funding and careful amendment of the ESA, the principal problems lie with implementation. This chapter recommends a multidirectional approach to these problems.

A Brief History of Federal Protection

The history of federal efforts to protect endangered species dates back to the beginning of the twentieth century. Responding to concern over the rapid decline of the passenger pigeon (*Ectopistes migratorius*) and other game birds, Congress passed the first federal wildlife law: the Lacey Act of 1900. The Lacey Act authorized federal enforcement of state wildlife laws, and Section 1 gave the secretary of agriculture authority to adopt all measures necessary for the "preservation, distribution, introduction, and restoration of game birds and other wild birds." (See Bean 1983:17–19 for further discussion of the Lacey Act and its history.) Congress took additional steps to protect endangered species by passing the Migratory Bird Treaty Act of 1918 (to protect birds migrating between Canada and the United States) and the

Migratory Bird Conservation Act of 1929 (to provide funding to acquire migratory bird habitat).

Slowly, political and public awareness of the growing problem of endangered species began to increase. The environmental activism of the early 1960s accelerated this growing awareness, and in 1964 the Department of Interior's Bureau of Sport Fisheries and Wildlife (now the U.S. Fish and Wildlife Service) formed the Committee on Rare and Endangered Wildlife Species. Based on the work of this committee, the Department of Interior published "Redbook—Rare and Endangered Fish and Wildlife of the United States—Preliminary Draft." This 1964 publication, more commonly known as the "Redbook," contained the first official listing of species the federal government considered to be in danger of extinction.

Two years later, Congress passed its first comprehensive endangered species legislation: the Endangered Species Protection Act of 1966. Though espousing the lofty goal of "conserving, protecting, restoring, and propagating selected species of native fish and wildlife" (Section 2(a)), the 1966 act did little except authorize efforts to acquire important habitat. Congress soon recognized the inherent weaknesses of this act and replaced it with the Endangered Species Conservation Act of 1969. This replacement act extended protection to certain invertebrates, increased prohibitions on illegal trade, and began the process that culminated in the Convention on International Trade in Endangered Species of Wild Fauna and Flora. As Bean (1983) has noted, however, improvements to protection of domestic species wrought by the 1969 act were relatively minor, and the 1969 act itself was soon superseded by the even stronger Endangered Species Act of 1973. (For further discussion of the 1966 and 1969 acts, see Bean 1983 and Rohlf 1989.) Though substantively amended several times (most significantly in 1978, 1982, and 1988), the ESA remains the primary instrument of federal efforts to protect endangered species.

The ESA is not the sole mechanism for protecting endangered species, however. In addition to the acts cited above, provisions to protect endangered species can be found in many other federal statutes (the Marine Mammal Protection Act of 1972, National Forest Management Act of 1976, Conservation Programs on Military Land (1988), Institute of Tropical Forestry Act (1990), and Driftnet Act Amendments of 1990 among others). Federal agencies are often quite active in their efforts to protect endangered species, as well, though most of these efforts are usually attempts to avoid contravening the provisions of the ESA. As Coggins (1991:67) has noted: "The four main commands of the Endangered Species Act—to conserve listed species, to avoid jeopardization, to avoid destruction of critical habitat, and to avoid taking—are now firmly entrenched in public natural resources law. The federal agencies realize that the presence of a listed species dramatically alters the management equation."

Protection for endangered species can be found in numerous international treaties (such as the Convention on International Trade in Endangered Species of Wild Fauna and Flora), many state laws and regulations (Griffin and French 1992), and even in local statutes (Dodd 1992). Many nonfederal programs provide greater protection for larger numbers of endangered species than does the ESA, though over a more limited geographic range (Griffin and French 1992). Moreover, many private organizations (e.g., The Nature Conservancy, Ducks Unlimited) make significant contributions to endangered species conservation through research, monitoring, and habitat acquisition (Griffin and French 1992). But as the ESA is both the symbolic and actual instrument through which the primary efforts to protect endangered species take place, only the provisions of the ESA are summarized here.

The ESA's Major Provisions

The major provisions of the ESA set forth eligibility and procedural requirements for listing species as endangered or threatened, provide various protections for listed species, prohibit federal agencies from engaging in actions that would jeopardize listed species or critical habitat, and create the framework for cooperative programs with the states.

The Listing Process

Section 4 of the ESA delineates the process and sets the standards for determining whether a particular species should be listed under the act. (The same criteria apply to delisting a species.) The U.S. Fish and Wildlife Service (FWS) and the National Marine Fisheries Service (NMFS) share jurisdiction over implementation of the ESA, and the listing process begins when an individual or group petitions either FWS or NMFS to list a species. These two services can also initiate the listing process themselves.

After a petition is received, the service has ninety days to decide whether the petition presents information sufficient to indicate that a listing may be warranted. If the petition does present sufficient information, the service has one year from the time it received the petition to make a final determination as to whether the listing is (1) warranted, (2) not warranted, or (3) warranted but service action on other listing proposals precludes immediate listing of the petitioned species. In making its listing determinations the service must consider threats to habitat, overutilization, disease, predation, inadequacy of current protection, and any other factors affecting the petitioned species' existence. The services are prohibited from considering the economic impact of their listing decisions. Instead they must make their determinations "solely on the basis of the best scientific and commercial data available" (Section 4(b)(1)(A)).

Although Section 3 stipulates that only species may be listed, the ESA's definition of "species" includes not only full species but subspecies and distinct vertebrate populations as well. In determining whether a petitioned species is a listable entity under the ESA, the services rely on "standard taxonomic distinctions and the biological expertise of [FWS] and the scientific community concerning the relevant taxonomic group" (50 C.F.R. sec. 424.11(a)). Section 3 also sets forth two separate categories under which species can be listed: "endangered species," which are species in danger of extinction throughout all or a significant portion of their range, and "threatened species," which are species likely to become endangered in the foreseeable future. Additions and deletions to the list of threatened and endangered species are made by agency rule making.

Protections for Listed Species

Once a species is listed under the ESA, it is entitled to numerous protections. Some of these protections are procedural in nature. But for a few exceptions, for example, Section 4 requires the services to develop and implement recovery plans and to designate critical habitat for each species listed. A more direct mechanism for protecting endangered species is Section 9's prohibition on the "taking" of listed species. Section 3(19) defines the term "take" to mean "to harass, harm, pursue, hunt, shoot, wound, kill, trap, capture, or collect, or to attempt to engage in any such conduct." Moreover, FWS has published regulations that further define "harm" (as found in the statutory definition of "take") to mean "an act which actually kills or injures wildlife. Such act may include significant habitat modification or degradation where it actually kills or injures wildlife by significantly impairing essential behavioral patterns, including breeding, feeding, or sheltering" (46 *Fed. Reg.* 54748, 54750). Section 8(A) implements the provisions of the Convention on International Trade in Endangered Species of Wild Fauna and Flora, which provides protection for designated species on the international level. (See Bean 1983:324–329 for a description and discussion of this convention.)

Unlike the protections provided under Section 7, which apply only to federal agencies, Section 9 applies to all persons, agencies, and organizations, and it prohibits the import, export, possession, delivery, receipt, transport, sale, and offer for sale of any endangered fish or wildlife, whether live or dead (including any part, product, egg, or offspring). Similar, though less protective, measures are also provided for listed plant species. Several exemptions to Section 9's prohibitions have been inserted into the ESA. For example, individuals of species of fish or wildlife held in captivity prior to their listing are excluded. And if certain delineated procedures are followed, Section 10 allows the services to grant a permit for the "incidental take" of listed species during otherwise lawful activities, such as scientific study and work with "experimental populations" (a population established by the service as part of a recovery plan). Incidental take permits can also be granted in conjunction with an approved Habitat Conservation Plan.

Although most of the ESA's protections explicitly address only endangered species, the act gives the services authority to extend these protections to any threatened species deemed in need of such protections as well (Section 4(d)). Acting on this authority, FWS has promulgated regulations (50 C.F.R. sec. 17.31(a)) that automatically extend the ESA's protection of endangered species to threatened species. Before this regulation, extension of ESA protection to threatened species was done on a case-by-case basis.

Federal Agency Actions

In addition to Section 9's prohibitions, which apply to all persons, Section 7 imposes additional procedural requirements on federal agencies. Under Section 7, federal agencies are prohibited from engaging in any action (including authorization, funding, and permit issuance) that is "likely to jeopardize the continued existence of any endangered species or threatened species or result in the destruction or adverse modification of [critical] habitat of such species" (Section 7(a)(2)). Though Section 7 does not apply to state, local, or private actions, its provisions do have profound implications, and Section 7 was the statutory basis for the Supreme Court's famous decision in *TVA* v. *Hill* (437 U.S. 153 (1978)). There the Court held that work on a nearly completed dam must cease because federal action was likely to jeopardize the continued existence of the snail darter (*Percina tanasi*), a small, minnowlike fish.

Under Section 7, federal agencies must consult with the service before engaging in action whenever there is a possibility that the action might jeopardize a listed or proposed species or adversely affect critical habitat. If the service determines that such a possibility exists, the agency proposing the action must prepare a "biological assessment," based on the best scientific and commercial data available, to make a conclusive determination of the proposed action's impact. If this assessment concludes that the proposed action is not likely to harm listed or proposed species or critical habitat, and the service agrees, the consultation ends and the agency may proceed with its proposed action. If the assessment concludes there may be an adverse affect, however, the agency must make suggestions for mitigation and engage in either formal or informal consultation with the service. Nearly all agency proposals are addressed through informal consultation.

When formal consultation is required or undertaken voluntarily, the service must prepare a biological opinion. If this opinion concludes that the proposed action will not jeopardize listed or proposed species or critical habitat, the consultation ends and the agency may proceed with the proposed action. But if the service reaches a contrary conclusion, it issues a jeopardy opinion, which usually includes reasonable and prudent alternatives. Statistics published by the U.S. Government Accounting Office (GAO 1992) and World Wildlife Fund (Barry et al. 1992) indicate that jeopardy opinions are uncommon and rarely stop a proposed action. The Section 7 consultation process is advisory in nature. Ultimately it is the agency proposing the action that determines whether the requirements of Section 7 have been met.

Cooperation with States

The ESA requires federal cooperation with the states, and Section 6 creates a process through which individual states can enter into management and cooperative agreements with FWS. Management agreements provide for administration and management of areas established for conservation of listed species, and cooperative agreements provide assistance and funding to states that establish and maintain an adequate and active endangered species conservation program. Once a state has an approved program and a cooperative agreement with the secretary of interior has been signed, the state program becomes eligible to receive up to 75 percent of the program's cost. If more than one state is involved in an approved program, funding for up to 90 percent of the program's cost may be available. Most states have taken advantage of Section 6's provisions and have entered into cooperative agreements for either plants or animals or both. Only Alabama and Louisiana have not signed any form of cooperative agreement (Ernst 1991). Funding for these cooperative programs has been inconsistent, however, and often insufficient (Ernst 1991).

Other Provisions

Several other ESA provisions are also worthy of mention. These include procedures for acquisition of habitat for listed species (Section 5), authority for citizens to sue to enforce the act (Section 11(g)), provisions for civil and criminal penalties for violations (Sections 11(a),(b)), and creation of the Endangered Species Committee (also known as the "God Committee" or "God Squad"), which, under certain circumstances, has authority to exempt agency action from the mandates of Section 7 (Section 7(e)). Familiarization with the ESA's provisions provides an essential foundation for understanding the federal process for protecting endangered species. Once this foundation is in place, other elements affecting implementation can then be addressed.

Beyond the Language of the ESA

Understanding a statute's provisions is the starting point for comprehending how programs created under that statute actually work. As Bean (1989:271) has observed: "Negotiating a treaty or passing a law is clearly not the end of the process; it is the beginning." For the real world of implementation is often astonishingly different from expectations based on an isolated reading of a statute. Yaffee (1982:69) notes that implementing the ESA might "not [be] much different from implementing other types of policy," and he meticulously documents how internal and external forces resulted in the ESA, a highly prohibitive policy, being implemented in a nonprohibitive manner.

Grasping the realities of implementation is critical not only to assessment of an act's success but also to any reasonable hope of improving the outcome. Consequently, the evaluation process must go beyond the language of the statute itself. "While policies are written in words on paper," Yaffee observes, "they exist only in the form of the individuals, organizations, and agencies that implement them and the nature of the information, resources, authority, and incentives that flow between these actors" (1982:9). Several additional factors, primarily legal, are closely tied to statutory language, and they too shape implementation—for example, agency regulations, internal agency policies, and judicial interpretation. These factors require analysis as well.

Regulations

Like any other statute enacted by Congress, regulations interpreting and applying the ESA's provisions are promulgated by designated federal agencies. And, as noted earlier, FWS and NMFS share jurisdiction over implementation of the ESA. Generally, NMFS has authority over marine and anadromous species (about 5 percent of listed species) while FWS has authority over all other species (about 95 percent of listed species; GAO 1992). Regulations promulgated by the two services are published in the Code of Federal Regulations and carry the force of law. Often a statute provides merely a shell of congressional policy, while it is left to agencies to flesh out this shell and provide the technical details required for implementation. For example, Section 4(d) of the ESA gives the services authority to enact regulations to protect listed species. Using this authority, NMFS crafted detailed regulations requiring the insertion of excluder devices in shrimp trawl nets to stop the inadvertent drowning of threatened and endangered sea turtles.

Internal Agency Policies

In addition to statutes and regulations, agency activity is often controlled by internal mechanisms such as policy statements, guidelines, working definitions, and legal opinions. Unlike statutes and regulations, these internal policies are generally not subject to judicial review. But these mechanisms do play a subtle, yet unequivocally significant, role in the implementation of endangered species policy. The standard that NMFS uses to determine whether an individual run of Pacific salmon is a "species" requiring ESA protection is not set forth in regulations, for example, but is delineated in a policy statement. And FWS's controversial decision to deny ESA protection for hybrids of listed and nonlisted species was based on an internal legal opinion (which has since been withdrawn).

Judicial Interpretation

The courts play a pivotal role in implementation as well—they are the final word on the interpretation of a statute or regulation. But courts defer to agency decision-making so long as the agency has not acted unconstitutionally or in an "arbitrary and capricious" manner—a difficult evidentiary standard to meet. But if Congress disapproves of how the agencies interpret, apply, or implement a particular statute or regulation or how the courts interpret it, Congress may amend the act to redirect agency implementation and provide clearer guidance for judicial interpretation.

To understand implementation of endangered species conservation programs, therefore, not only must the statute be analyzed and assessed but a spate of other elements, including regulations, internal agency policies, and the courts, must be considered. Later chapters reveal that the key parties, resources, motivations, information flow, professional and organizational norms, and many others facets of implementation must also be analyzed. By assembling all these pieces of the implementation puzzle, a clearer understanding will emerge. Only then can programs be assessed intelligently and effective suggestions be offered for their improvement.

Assessing the ESA

Assessments of the ESA often begin with a litany of statistics associated with the act's implementation. Here we consider several of these commonly discussed statistics.

ESA Listing Data

At present, just over 650 domestic species are listed under the ESA. For approximately 600 other domestic species, FWS has determined that adequate information exists to support listing under the ESA but other agency priorities are preventing their being listed ("Category 1" species). FWS has determined that more than 3000 other species may also be threatened or endangered but has decided that the current data are not conclusive ("Category 2" species). Combined the services presently list an average of forty-four species a year, and GAO (1992) calculates that, at this rate and disregarding any need to list non–Category 1 species, the services will need sixteen years just to list the Category 1 species.

Since 1973, only sixteen species have been removed from the endangered species list. Of these, seven species were removed because they became extinct, four species were removed because there were errors in the original listing data (population numbers were higher than originally thought, for example, or the species was technically disqualified from protection owing to hybridization with nonlisted species), and five species were removed because they recovered (GAO 1992). These figures do not take into account the numerous species whose ESA status was either upgraded or downgraded (changed from endangered to threatened or vice versa). Because these recovery figures appear to be low, criticisms of the ESA often focus on these data.

Success or Failure?

Intricately tied to criticisms based on these listing and recovery data are pronouncements of the ESA's success or failure. According to Rohlf (1991:274), for example, "the Act has had very limited success in achieving its stated goal of halting and reversing the trend toward species extinctions." Tobin (1990:257) concludes that "the [ESA] program can point to few successes, at least when measured against its statutory goal. Only a handful of species have been recovered despite more than two decades of expenditures. . . . Solutions to the problems of extinction are probably well beyond the marginal tinkering that now characterizes most policymaking." In mulling over the future of conservation legislation in the twenty-first century, Bean (1989:271) observes: "It is clear that most of the species . . . that should be protected under that act are not yet protected, many of the species that are formally protected have nevertheless continued to slip steadily toward extinction, and a relative handful appear to be making the recovery." Even the GAO (1988:18) has concluded that "measured against the logical, absolute standard, the small number of domestic species officially declared recovered would suggest that the program has been of limited success in recovering species."

But success or failure can be assessed from a different perspective. As Bean (1991:38) argues:

> The ultimate measure of success or failure of these efforts [under the ESA], however, is whether the species that are the objects of the act's concern face a more or less secure future. . . . If the number of recovered species seems few, it must be remembered that fifteen years is a very short time in which to expect dramatic results. During that period, however, the foundations for future recoveries have been laid. For many species, the likelihood of eventual recovery has increased because research done under the [ESA] has made it possible to understand better the causes that threaten their

survival and to identify the actions needed to remedy these threats. For others, we may only have bought additional time. Additional time is no small matter, however, for it may prove to be vital time in which to design more long-lasting solutions.

Irvin (1992:644) agrees with Bean and suggests that "the more appropriate measure is the number of species whose condition has stabilized or improved as a result of ESA protection." Citing 1990 FWS data, Irvin (1992) notes that the status of 41 percent of listed species was either stable or improving, as compared with 38 percent whose status has continued to decline. Irvin (1992:644) goes on to state: "Given that the ESA is analogous to an emergency room which handles only the most dire cases, supporters of the [ESA] argue that it has been a success by getting hundreds of species off the operating table, though not yet out of the hospital."

Appropriateness of Assessment Criteria

In assessing the ESA, the outcome is directly related to how success or failure is initially measured. If assessment of the ESA is based on the number of species removed from the endangered species list because they have recovered, the ESA appears to be a dismal failure. But if success is measured by slowing extinction rates or by improving species status, the ESA might just as well be considered a roaring success. What *is* success? Is it increasing the population of a species from fourteen to a hundred? Is it recovering a species sufficiently so that it has a significant probability of surviving for fifty years? A hundred years? Five thousand years?

Because assessments and pronouncements of success or failure are so amorphous, the ESA's relative success or failure will not be discussed further in this chapter. Nevertheless, only the naive assume the ESA has been smoothly and effectively implemented. In fact, the ESA's effectiveness has been subject to criticism from such diverse groups as agricultural interests, resource extractors, developers, environmentalists, biologists, academics, politicians, and the implementing agencies themselves. Criticism often provides a reasonable starting point for effective improvements to legislation and its implementation. Consequently, we turn now to several of the most common criticisms of the ESA.

Criticisms of the ESA

Authorization for the ESA most recently expired in September 1992, and ensuing public debate over the ESA's reauthorization produced numerous criticisms of the act. Rohlf (1991, 1992), Grumbine (1992), Irvin (1992), and the Endangered Species Roundtable (1992) have presented helpful compilations of the most common of these criticisms. Grumbine (1992:95) has listed what he considers to be the ESA's ten "most glaring defects," for example, and Irvin (1992) describes three primary criticisms presented by opponents to the ESA and suggests five ways to improve the act. Rohlf (1991:275) offers six "biological reasons" to support his assertions that the ESA "doesn't work" and that "flaws in the [ESA] itself significantly contribute to its ineffectiveness in conserving biodiversity." And the Endangered Species Roundtable (1992), a coalition of ranchers, resource extractors, and others, has presented its own lengthy list of the ESA's faults. As the observations of these and other commentators are representative, they are summarized in the following paragraphs.

Imprecise Standards for Delineating Species Status

The ESA sets up three different levels of species status: endangered, threatened, and recovered. Endangered species are those in danger of extinction. Distinctions between these categories and their varying degrees of protection have been the focus of a number of criticisms. Rohlf (1991:276) criticizes the ESA because it "makes no reference to quantitative or even qualitative parameters of what constitutes a 'danger' of extinction." Because there is no explicit statutory standard for determining when a species is in danger of extinction, the services make this determination on a case-by-case basis. Rohlf is troubled by these ad hoc determinations. The Endangered Species Roundtable (1992) is critical of FWS's regulatory decision to automatically extend to threatened species the protections mandated for endangered species. And since the ESA does not delineate an absolute standard for determining when a species has "recovered," the Endangered Species Roundtable (1992) charges that delisting under the ESA is a "myth."

Inappropriate Units of Protection

Much current debate over the ESA has focused on the units of protection afforded under the act. Some critics claim the ESA's protective scope is not large enough; others claim the scale is too large or otherwise inappropriate. Noss (1990:243) has suggested that "perhaps the final answer will be that there is no best scale, . . . and that conservation strategy must address multiple levels of organization." The following overview of the debate over the scale of ESA protection, therefore, should prove illuminating.

One of the ESA's primary purposes is "to provide a means whereby the ecosystems upon which endangered species and threatened species depend may be conserved" (Section 2(b)). Many critics have concluded that the ESA is ineffective in conserving these ecosystems (see Hutto et al. 1987; Noss 1991; Grumbine 1992) and suggest that the act's scale of protection needs to be broader than the present single-species approach. Winckler (1992:74) argues that, in light of serious funding deficiencies, a single-species approach is inefficient: "If our goal is to preserve as many kinds of plants and animals as possible, it makes little sense to spend limited funds on heroic steps to rescue a handful of near-extinct species. A more effective strategy would focus on protecting ecosystems that support maximum biological diversity." But Grumbine (1992:95) concludes that the ESA "is not capable of solving species-level biodiversity dilemmas, let alone habitat- or ecosystem-scale problems." Recent legislative proposals that would address these large-scale units, such as ancient forest, ecosystem, and biodiversity protection bills, have failed to attract sufficient political support for passage. Others have criticized the ESA for not providing sufficient protection on a small enough scale. For example, Rohlf (1991) has predicted that metapopulation and patch dynamics are likely to become increasingly important in an era of accelerating habitat fragmentation and concludes that the ESA provides insufficient protection of metapopulation habitats.

The three taxonomic categories that can be listed under the ESA are species, subspecies, and distinct vertebrate populations (Section 3(16)). These categories have been the source of much recent debate. Many critics and politicians assert that the ESA should protect neither subspecies nor populations. For example, Interior Secretary Manuel Lujan asked, in reference to the controversy over the Mount Graham red squirrel: "Do we have to save every subspecies? Do we have to save [an endangered species] in every locality where it exists? . . . The red squirrel is the best example. Nobody's told me the difference between a red squirrel, a black one, or a brown one" (Lancaster 1990:A1). Bean (1992:4) cites other federal officials echoing similar sentiments: "The Inspector General of the Interior Department, in his 1990 report, off-handedly suggested that perhaps the Act should be limited to full species, so as to make its costs more commensurate with the levels of funding Congress has been providing. Former

[FWS] Director Frank Dunkle has recently proposed that the Act not protect subspecies . . . that are common elsewhere."

Under the ESA, distinct populations of plants and invertebrates are given no protection. A number of environmentalists (among them Murphy 1991 and Bean 1983, 1992) are critical of this inconsistency, which has led Grumbine (1992:95) to accuse the ESA of playing "taxonomic favorites." The question of what should qualify as a listable population is drawing increasing attention. But neither service has published a comprehensive set of regulations or guidelines for determining what should qualify as a distinct vertebrate population (though NMFS recently published a policy statement delineating criteria for categorizing populations of Pacific salmon; 56 *Fed. Reg.* 58612, 1991).

Additional debate centers on whether the ESA should protect hybrids between listed and nonlisted species. FWS originally asserted that hybrids were protected, but it has subsequently reversed its position. At present, FWS has withdrawn all former opinions and has refused to take any position on the hybrid debate (Bean 1993: pers. comm.). There have been several consequences of this debate and uncertainty over ESA protection for hybrids: the delisting of the Mexican duck (*Anas diazi*); insufficient attention to the plight of the now extinct dusky seaside sparrow (*Ammodramus maritimus nigrescens*); and attacks on the legality of ESA listings for a number of species, including the Louisiana black bear (*Ursus americanus luteolus*), Florida panther (*Felis concolor coryi*), red wolf (*Canis rufus*), and gray wolf (*Canis lupus*).

Undue Attention to High-Profile Species

Aside from the issue of which taxonomic unit should be protected under the ESA, there has been considerable debate over the attention given to "high-profile" species in terms of listing, recovery, and funding priorities. Generally, high-profile species are those that enjoy a high level of public support, receive the attention of politicians, or are embroiled in controversy. In 1988, GAO reported that, despite contrary instructions from Congress, FWS continued to focus its recovery expenditures on high-profile species. Responding to GAO's report, Congress amended the ESA to prohibit the services from considering taxonomic classification in prioritizing recovery plans, and FWS revised its priority guidelines. Even so, high-profile species continue to receive a disproportionately large share of service resources. A 1990 Department of Interior report found that 50 percent of available recovery funds was spent on just ten species (Griffin and French 1992). As Grumbine (1992:95) concludes: "Favoritism has been translated by the agencies into funding decisions that benefit only a few dozen high-profile species."

Insufficient Protection of Habitat

In 1982, Congress amended the ESA to require that critical habitat be designated at the time a species is listed only to the "maximum extent prudent and determinable" (Section 4(a)(3)). Prior to this amendment, the ESA mandated that critical habitat be designated every time a species is initially listed. This amendment has been the source of great controversy. Some critics (such as the Endangered Species Roundtable 1992) claim that designation of critical habitat is destructive to the economy, while others (such as Irvin 1992) criticize the ESA for allowing consideration of economic consequences in determining critical habitat or for removing the requirement that designation be mandatory. Rohlf (1992) contends that because designations of critical habitat are so controversial, the services hide behind statutory and regulatory loopholes and avoid designating critical habitat. And according to GAO (1992), critical habitat has been designated for only 16 percent of all listed species.

Rohlf (1991) is also highly critical of Section 9's provisions for granting incidental take permits given in conjunction with an approved Habitat Conservation Plan (HCP). He believes that under the HCP process, species recovery is compromised and important habitat is negotiated away. Critics of the ESA also make note of the high cost and low number of successful HCPs (Endangered Species Roundtable 1992; Rohlf 1992). Most HCPs cost millions of dollars (Rohlf 1992), and since the ESA was amended in 1982 to allow for the creation of HCPs, only eleven HCPs have been completed and enacted (GAO 1992). Nevertheless, a number of environmentalists (see Bean et al. 1991 and O'Connell 1992) espouse the virtues of the HCP process, carefully documenting the "success" of such plans and promoting HCPs as an important mechanism for protecting species on private lands and balancing economic concerns.

Inadequacy of the Recovery Plan Process

Though the ESA was amended in 1978 to require development and implementation of recovery plans for all listed species, today only 61 percent of listed species have approved recovery plans (GAO

1992). Aside from this low rate of recovery plan development, a number of environmentalists (see Grumbine 1992 and Snyder 1992) have criticized recovery plans per se, finding that they do not utilize state-of-the-art information and models, that they inappropriately reflect political pressure and agency bias, that they are disconcertingly inflexible, and that they become outdated long before their expiration or even implementation. Nonenvironmental critics (such as the Endangered Species Roundtable 1992) also find fault with recovery plans, but tend to criticize other aspects, such as high costs and alleged lack of opportunity for public participation.

Inadequacy of Interagency Consultation

Most Section 7 consultations, as noted earlier, are resolved informally. In fact, from 1987 through 1991 some 89 percent of all Section 7 consultations were resolved informally (GAO 1992). Rohlf (1991) is critical of this high percentage of informal consultations: where there is no formal consultation, no information on the decision is required to be made public. Consequently, Rohlf argues, the public is effectively shut out of the decision-making process. Grumbine (1992:98) criticizes the consultation process for what he views as its indefensible distinction between major and minor actions, as only major actions require a biological assessment.

Inappropriate Discounting of Uncertainty

The tendency of agencies and courts to refuse to consider future risks—such as environmental stochasticity and cumulative effects—to species' long-term survival greatly concerns Rohlf (1991). Under existing regulations, only currently proposed federal actions are included in cumulative-effects analysis (Grumbine 1992). Actions that are part of a larger plan but are not slated for immediate implementation are not included. Consequently, a slow erosion of species stability may ensue. Stochastic events could be devastating to a species' survival—particularly where recovery efforts are limited by survival probability levels that have been set dangerously low. For example, NMFS has suggested that recovery efforts be made only until the target species reaches a threshold where it has a 95 percent chance of persistence for a mere 100 years (Thompson 1991). Such criteria for species viability in recovery efforts have come under significant criticism for their seeming indifference to the uncertainty and limitations of current scientific data, for inadequate consideration of the potential impact of cumulative effects and stochasticity, and for the short timeframe of their persistence parameters in assessing viability.

Insufficient Consideration of Economic Factors

The ESA permits consideration of economic factors at many stages of ESA procedures, including the development of HCPs and the designation of critical habitat. In extremely controversial confrontations involving the ESA, the Endangered Species Committee, which has the power to exempt a project from certain ESA requirements, can be invoked. The language of the act does not, however, permit consideration of economic factors at the listing stage (though some would argue that such factors surreptitiously come into play during implementation). This refusal to consider economic factors in the listing process has been the source of much controversy, and some groups have called on Congress to amend the ESA to permit consideration of economic concerns during listing decisions (Irvin 1992). Others contend that deciding whether a species is endangered is strictly a biological question and insist that the ESA should continue to require the services to exclude economic factors in deciding whether to list a species (Irvin 1992).

Improving Implementation Through Improving the ESA

Most of the foregoing criticisms are not, at their core, primarily biological in nature but concern implementation. Indeed, a number of commentators have recognized the importance of implementation with regard to criticisms of the ESA. (See, for example, Clark and Harvey 1990; Karr 1990; Yaffee 1991; O'Connell 1992.) Karr (1990: 245) maintains that "implementation of endangered species legislation has created a bureaucratic quagmire," and O'Connell (1992:140) blames the ESA's perceived failures not on the act itself but on its implementation: "The Endangered Species Act is in fact a remarkably prescient statute that has been plagued since its adoption by ineffective implementation." As the possibility of confronting implementation problems through the ESA itself should not be overlooked, a few suggestions for amending the ESA are offered here as examples of the kind of tinkering that might be beneficial. First, however, the issue of funding must be addressed, as criticism of the ESA is often inextricably linked to criticism of ESA funding levels.

ESA Funding Inadequacies

One of the most common responses to criticisms of the ESA is the argument that implementation problems do exist but are the result of insufficient funding. O'Connell (1992:142) puts the primary blame for ineffective implementation on "inadequate funding" and concludes that this weakness is largely responsible for the problems Rohlf (1991) cites in his criticisms of the ESA. The Endangered Species Roundtable (1992) says that attempts to implement the ESA demonstrate that our resources are inadequate to protect all the listed species. (The Roundtable, however, argues for less protection, not more funding.) Bean (1991:39) comments that "if any lesson is clear after seventeen years of experience under the Endangered Species Act, it is that the threat of extinction is far greater than it was appreciated to be in 1973 and that the resources needed to address the problem are far greater than those that have been made available thus far." Bean (1992:3) attributes much the ESA's perceived failings to lack of adequate funding: "It is abundantly clear that more resources are needed than are currently being made available for the Act's administration . . . and this is the inescapable conclusion of those who decry the 'failure' of the Act to accomplish more than it has to date."

O'Connell and Bean have convincingly concluded that a significant infusion of funds into endangered species programs would dramatically increase the effectiveness of ESA implementation. Clearly, the ESA cannot be implemented effectively without sufficient funding. Programs to recover many listed species, particularly those for plants and invertebrates, have been allocated little or no funding to date. Though some funding shortages are due to a basic lack of congressionally allocated funds, other shortages are the product of internal agency budget distributions, with high-profile species receiving a highly disproportionate share of available funding. Regardless of the internal and external causes of funding shortages, the need for more funding is a stark reality. Six hundred Category 1 species await processing. Three thousand Category 2 species await research and assessment. Allocation of more funds would allow the services to hire additional employees to engage in research and assessment and to accelerate the administrative processes required under the ESA. But adequate funding is only a starting point for effective implementation.

As will become all too painfully apparent from many of the case studies described in this book, adequate funding is no guarantee that implementation will be effective. Programs such as those for the California condor (*Gymnogyps californianus*), Florida panther (*Felis con-*

color coryi), northern spotted owl (*Strix occidentalis caurina*), Yellowstone grizzly bear (*Ursus arctos horribilis*), and black-footed ferret (*Mustela nigripes*) have received substantial, if not more than adequate, funding. Yet those involved in these programs will delineate in detail how these well-funded and high-profile programs nevertheless suffered severe implementation problems. As we will see, the solutions to the implementation problems of these—perhaps all—endangered species conservation programs lie outside the issue of funding.

Amending the ESA Itself

So the question remains: aside from providing additional funding, what actions can be taken to improve the ESA and its implementation so that our endangered species conservation programs are more effective? One answer is to amend the ESA itself. A few suggestions are offered here as examples of ways the ESA might be amended to improve implementation.

Amend the ESA to Address Biological Concerns. Congress has the power to amend the ESA to address the act's "biological" and nonbiological shortcomings as delineated by many of the act's critics. For example, the ESA could be amended to provide greater protection for ecosystems, to protect populations of invertebrates and plants, to prohibit the taking of plants, to provide greater protection of listed species on private lands, to require immediate designation of critical habitat for all listed species, to require timely completion and implementation of recovery plans, to allow designation of nonoccupied habitat to protect species dependent on metapopulation dynamics, to strengthen enforcement provisions, to accelerate listing procedures for candidate species, and to mandate utilization of specified state-of-the-art formulas and models.

Amend the ESA to Limit Agency Discretion. Statutes, implementing regulations, internal policies, and judicial interpretation provide the primary policy and legal foundations upon which agency implementation is based. In addition, agencies are prohibited by law from exceeding the bounds of the authority implicit or explicit in the statutes they are charged with implementing. But agency implementation often veers far from congressional intent, and the ESA could be amended specifically to address the implementation problems resulting from inappropriate agency regulations, policies, and actions.

One standard legislative response to such intentional or unintentional derailment is to attempt to eliminate bureaucratic discretion by amending the statute to make it as detailed as possible. For example, congressional distrust of the Environmental Protection Agency's ability to resist political pressure under the Bush administration was one of the primary reasons the Clean Air Act Amendments of 1990 ran over six hundred pages. The hope of many in Congress was that by making the statute hyper-specific, agency regulations could not deviate significantly from congressional intent. Moreover, the courts would be more likely to find inappropriate agency action to be "arbitrary and capricious," which is difficult to prove under more general statutes. When the intent of the statute is sufficiently clear, courts have a stronger legal basis for determining whether an agency has exceeded its authority.

Many environmentalists support varying degrees of this hyper-specification to achieve both biological and nonbiological goals (Murphy and Noon 1991; Rohlf 1992). Rohlf (1992:144) says: "It is the ESA's generality that permits politically driven officials to interpret and implement the Act almost as they wish." He also observes that "if in the ESA itself Congress established a biologically explicit listing/recovery threshold . . . this standard would remain constant regardless of the political philosophy of the White House occupant. Moreover, a statutory expression of precise congressional intent would provide courts a measuring stick, enabling them to strike down agency interpretations inconsistent with this intent" (Rohlf 1992:145). As Karr (1990:244) notes: "Effective implementation depends upon clear and concise definition of goals" and a proper choice of indices for measuring success. Similarly, Murphy and Noon (1991) argue that amending the ESA by changing the definition of critical habitat to incorporate more detailed, scientific concepts would result in improved status for many species.

Though some criticisms of the ESA might be addressed effectively through hyper-specification, the solution to many problems associated with the act may not be amenable to such an approach. In discussing the ESA, Greenwalt (1991:34) comments: "Laws which provide little room for flexibility . . . are doomed in the long run." Amending a statute can take many years, and it may take several additional years to promulgate regulations implementing these amendments. Because of the rapidly evolving discipline of conservation biology, as well as the highly volatile and often precarious status of many endangered species, it may not be wise to institutionalize models, formulas, prioritization schemes, and the like that might become outdated months after the act is amended to incorporate them. In any case, a recalcitrant agency or administration is still likely to find ways around even hyper-specific statutory language.

Alter the Standard of Judicial Review and Basis for Standing to Sue. Judicial deference to an agency's discretion could be significantly affected by an act of Congress. Under the current standard of judicial review, courts defer to agency decision making so long as it is not "arbitrary and capricious" or unconstitutional. Such a standard may be reasonable with regard to many statutes, as Congress is free to "correct" agency decisions with which it disagrees. But Congress is often loath to engage in controversial debate or to spend time making small or technical amendments. The legislative correction procedure may take years, and judicial challenges to agency decision making are increasingly time-consuming, expensive, and unsuccessful. When dealing with the urgency of endangered species conservation, such a time frame is often ill-advised and could jeopardize species or even lead to their extinction.

Consequently, Congress might consider amending the ESA or enacting new legislation to change the arbitrary and capricious standard, at least as it applies to the ESA. A replacement standard could reflect existing legislative and judicial standards, such as "substantial evidence." Alternatively, courts could be given the power to engage in *de novo* review of agency action. Legal standing to sue agencies could also be expanded so that, regardless of the judicial review standard, more opportunities for challenging agency action would be available. By holding agencies to a higher standard and increasing the basis for standing to sue, the hope is that legal challenges would be more effective. But the courts themselves may not provide the desired outcome. As Bean (1993:pers. comm.) notes: "The effect of imposing a more rigorous standard of review is to shift more of the responsibility for deciding often technical scientific issues away from an agency with at least some substantial biological expertise . . . to judges, nearly all of whom have none."

Pursue a Multidirectional Approach. Apart from adequately funding and amending the ESA, there are many other avenues to improving endangered species conservation programs and these options should be pursued. As Bean (1989:270) notes: "One must recognize also that one cannot expect too much from legislation." The ESA is not the entire problem; it is but one component. Social and governmental problems as a whole affect the implementation of all regulatory programs, including the ESA. Though all these elements must be confronted for substantive improvement of endangered species conservation, this chapter has focused strictly on assessing the role of the ESA itself.

Lessons

The ESA may not be the most appropriate or effective mechanism for protecting biodiversity, ecosystems, or even endangered species. Nonetheless, the ESA is the primary instrument through which such protection has been made. Many commentators consider implementation of the ESA to be a success. From several perspectives, this assessment seems merited. But other assessments of the relative success of the ESA conclude that the act's implementation does not reflect legislative aspirations and objectives or the realities of current scientific knowledge. And, based on listing and recovery data, these critics conclude that the ESA, as implemented to date, is a failure. Some commentators blame implementation disappointments on funding inadequacies; others fault the ESA itself. While monetary infusions would greatly facilitate implementation of seriously underfunded ESA programs, the case studies discussed throughout this book demonstrate that money alone will not solve the ESA's implementation problems. And though careful adjustments to the ESA's wording may indeed benefit implementation, legislative amendments are unlikely to provide satisfactory improvement.

The foundations for improving implementation of endangered species conservation programs can be built from many sources—including increased funding, statutory amendments, and application of a number of theoretical concepts discussed in other chapters, such as enlisting a policy orientation (Chapter 17) and improving organizational and professional theory (Chapters 14 to 16) and practice. These sources should not be limited to those set forth in this chapter or elsewhere in the book. For example, heightening public and political awareness of the importance of conserving biodiversity through educational efforts could also have a positive impact on endangered species conservation programs. And though supporters of the ESA may wish to boast of its "success," attention should nonetheless be paid to the act's critics, for they can provide a starting point for constructive improvement of the ESA and its outcomes. Through a rationally prioritized, multidirectional approach, implementation of endangered species conservation programs can be significantly improved.

Acknowledgment

Michael Bean critically reviewed a draft of this chapter.

References

Barry, D. L., L. Harroun, and C. Halverson. 1992. *For Conserving Listed Species, Talk Is Cheaper Than We Think: The Consultation Process Under the Endangered Species Act.* Washington, DC: World Wildlife Fund.

Bean, M. J. 1983. *The Evolution of National Wildlife Law.* Rev. ed. New York: Praeger.

―――. 1989. Conservation legislation in the century ahead. In *Conservation for the Twenty-First Century,* D. Western and M. C. Pearl (eds.). New York: Oxford University Press.

―――. 1991. Looking back over the first fifteen years. In *Balancing on the Brink of Extinction: The Endangered Species Act and Lessons for the Future,* K. Kohm (ed.). Washington, DC: Island Press.

―――. 1992. Issues and controversies in the forthcoming reauthorization battle. *Endangered Species Update* 9:1–4.

Bean, M. J., S. G. Fitzgerald, and M. A. O'Connell. 1991. *Reconciling Conflicts Under the Endangered Species Act: The Habitat Conservation Planning Experience.* Washington, DC: World Wildlife Fund.

Clark, T. W., and A. H. Harvey. 1990. Implementing endangered species recovery policy: Learning as we go? *Endangered Species Update* 5:35–42.

Coggins, G. C. 1991. Snail darters and pork barrels revisited: Reflections on endangered species and land use in America. In *Balancing on the Brink of Extinction: The Endangered Species Act and Lessons for the Future,* K. Kohm (ed.). Washington, DC: Island Press.

Dodd, L. B. 1992. Endangered species protection through local land-use regulations. In *Transactions of the Fifty-Seventh North American Wildlife and Natural Resources Conference,* R. E. McCabe (ed.). Washington, DC: Wildlife Management Institute.

Endangered Species Roundtable. 1992. Endangered Species Act: Time for a change. Unpublished white paper.

Ernst, J. P. 1991. Federalism and the act. In *Balancing on the Brink of Extinction: The Endangered Species Act and Lessons for the Future,* K. Kohm (ed.). Washington, DC: Island Press.

General Accounting Office (GAO). 1988. *Endangered Species: Management Improvements Could Enhance Recovery Program.* GAO/RCED-89-5. Washington, DC: Resources, Community, and Economic Development Division.

————. 1992. *Endangered Species Act: Types and Numbers of Implementation Actions.* GAO/RCED-92-131BR. Washington, DC: Resources, Community, and Economic Development Division.

Greenwalt, L. A. 1991. The power and potential of the act. In *Balancing on the Brink of Extinction: The Endangered Species Act and Lessons for the Future,* K. Kohm (ed.). Washington, DC: Island Press.

Griffin, C. R., and T. W. French. 1992. Protection of threatened and endangered species and their habitats by state regulations: The Massachusetts initiative. In *Transactions of the Fifty-Seventh North American Wildlife and Natural Resources Conference,* R. E. McCabe (ed.). Washington, DC: Wildlife Management Institute.

Grumbine, R. E. 1992. *Ghost Bears: Exploring the Biodiversity Crisis.* Washington, DC: Island Press.

Hutto, R. L., S. Reel, and P. B. Landres. 1987. A critical evaluation of the species approach to biological conservation. *Endangered Species Update* 4:1–4.

Irvin, W. R. 1992. The Endangered Species Act: Prospects for reauthorization. In *Transactions of the Fifty-Seventh North American Wildlife and Natural Resources Conference,* R. E. McCabe (ed.). Washington, DC: Wildlife Management Institute.

Karr, J. R. 1990. Biological integrity and the goal of environmental legislation: Lessons for conservation biology. *Conservation Biology* 4:244–250.

Lancaster, J. 1990. Lujan: Endangered Species Act "too tough," needs changes. *Washington Post,* 12 May 1990, p. A1.

Murphy, D. D. 1991. Invertebrate conservation. In *Balancing on the Brink of Extinction: The Endangered Species Act and Lessons for the Future,* K. Kohm (ed.). Washington, DC: Island Press.

Murphy, D. D., and B. R. Noon. 1991. Exorcising ambiguity from the Endangered Species Act: Critical habitat as an example. *Endangered Species Update* 8:6.

Noss, R. F. 1990. Can we maintain biological and ecological integrity? *Conservation Biology* 4:241–243.

————. 1991. From endangered species to biodiversity. In *Balancing on the Brink of Extinction: The Endangered Species Act and Lessons for the Future* K. Kohm (ed.). Washington, DC: Island Press.

O'Connell, M. 1992. Response to "Six biological reasons why the Endangered Species Act doesn't work—and what to do about it." *Conservation Biology* 6:140–143.

Rohlf, D. J. 1989. *The Endangered Species Act: A Guide to Its Protections and Implementation.* Stanford: Stanford Environmental Law Society.

————. 1991. Six biological reasons why the Endangered Species Act doesn't work—and what to do about it. *Conservation Biology* 5:273–282.

————. 1992. Response to O'Connell. *Conservation Biology.* 6:144–145.

Snyder, N. 1992. The California condor as an example of problems in organization of endangered species efforts. Unpublished MS on file with the author.

Thompson, G. G. 1991. *Determining Minimum Viable Populations Under the Endangered Species Act.* NOAA Technical Memorandum. NMFS F/NWC-198.

Tobin, R. J. 1990. *The Expendable Future: U.S. Politics and the Protection of Biological Diversity.* Durham, N.C.: Duke University Press.

Winckler, S. 1992. Stopgap measures. *Atlantic Monthly* 269:74–81.

Yaffee, S. L. 1982. *Prohibitive Policy: Implementing the Federal Endangered Species Act.* Cambridge: MIT Press.

———. 1991. Avoiding endangered species development conflicts through interagency consultation. In *Balancing on the Brink of Extinction: The Endangered Species Act and Lessons for the Future,* K. Kohm (ed.). Washington, DC: Island Press.

PART II

Case Studies: Experiences and Lessons

As vital as high-quality empirical work and hypothesis testing are in conservation biology, we get the job done only by translating scientific findings into policy and management prescriptions, and then by getting those prescriptions implemented on the ground.

REED NOSS

Wildlife biologists and their colleagues in forestry, range sciences, and conservation biology have been swept up, in just a few short years, into public debates which have taken them from the status of sequestered experts to that of key players.

BARRY NOON

No matter how strong a recovery plan is, the ultimate extent of implementation of the plan is primary. If provisions of the plan are ignored or continual exceptions are made . . . there is little hope [for species recovery].

JEROME A. JACKSON

3

The Northern Spotted Owl
An Indicator of the Importance
of Sociopolitical Context

Steven L. Yaffee

The northern spotted owl (*Strix occidentalis caurina*, Figure 3-1) became a research and management concern in the mid-1970s. What is amazing is how much was known then about the nature of the problem facing the owl's survival and its likely solution, yet managers were unable to deal with it effectively over the ensuing twenty years. In the mid-1970s, a set of researchers and managers in the Pacific Northwest knew that the owl was threatened, that old growth was the issue, and that changing federal forest management was the only way to ensure the owl's survival. Indeed, at the first meeting of the Oregon Endangered Species Task Force, an interagency group of scientists and managers created in response to budding public interest in nongame wildlife, the group decided to take on the spotted owl as its first priority—in part because they viewed it as a problem easily solved (Oregon Endangered Species Task Force 1973). At the same time, economists warned that trends in timber harvest would result in a timber supply gap, a roughly twenty-year period between the time that old growth would be logged out and second growth stands would come on-line as a viable supply for mills (Beuter et al. 1976).

Although we knew enough about owl biology and timber economics in the early to mid-1970s to forestall the inevitable crisis, little was done in the 1970s, and not enough in the 1980s, to avoid a major controversy. By the early 1990s, the spotted owl conflict had ensnared thousands of people ranging from the families of loggers in small towns in Oregon to the president of the United States. And by early 1993, a solution had neither been put into place nor even clearly articulated. While the spotted owl case provides insights into the effectiveness of federal endangered species policy as established

FIGURE 3-1.   The northern spotted owl. (Photo by T. Fleming.)

by the Endangered Species Act (ESA), it also points to the importance
of the broader context in which endangered species policy is made
and implemented. Much of the spotted owl story took place while the
owl was not a federally listed endangered species, and many of the
key decisions were made under the auspices of other laws. And the
fact that the owl case affected economies and institutions to such an
extent reflected changes in regional and national politics more than
anything else.

This chapter focuses on describing the broad context of endangered
species management, as illustrated by the spotted owl case, and its implications for organizational management and professional practice.
We can talk about organizational strategies and professional roles
until we are blue (or green) in the face, but without a supportive institutional environment, the approaches will not succeed. This
chapter describes the current situation (as of early 1993), explains
how we got here, and then details several observations about the con-

text of endangered species management. Throughout the chapter, lessons are drawn to show how managers and decision makers can do better in future endeavors.

The Current Situation and How We Got Here

The spotted owl itself is a remarkably nonthreatening bird. One of three subspecies of spotted owl that range from Canada to Mexico, the northern spotted owl is a medium-sized, brown and white bird with a 43-inch wingspan. It has a lifespan of about fifteen years and eats small mammals such as squirrels and wood rats (*Neotoma* spp.). The owl does not appear to be particularly frightened of humans and indeed often appears curious about humans and human activity in its environment.

It seems relatively clear that the owl thrives in the old-growth Douglas fir (*Pseudotsuga menziesii*), western hemlock (*Tsuga heterophylla*), and mixed conifer forests of the Coast Range and the west side of the Cascades in Oregon, Washington, and northern California. The owl requires some of the characteristics of old growth, including a closed canopy, considerable downed woody material, and standing dead trees (snags). While a small portion of its population may survive in second-growth redwood (*Sequoia* spp.) stands in northern California that exhibit these same characteristics, most of the population resides where large blocks of relatively unfragmented old growth exist.

The owls use a lot of this habitat. As summarized by the U.S. Fish and Wildlife Service (FWS) in its 1992 draft recovery plan, the median home range of observed owl pairs ranges from 1411 to 14,271 acres, including some 615 to 4579 acres of old-growth and mature forest (U.S. Department of the Interior 1992). The agency estimates that roughly 7.6 million acres of suitable habitat remains, with a 1 to 2 percent annual decline (U.S. Department of the Interior 1992). While the extent of owl habitat that existed before settlement is unknown, it seems likely that no more than 10 percent of presettlement habitat remains. As of 1991, a total of about 3500 pairs of northern spotted owls had been located; only 250 pairs were found on lands previously reserved from timber harvest (U.S. Department of the Interior 1992). Since much of the area's private and state-owned lands have been logged, most of the remaining habitat lies on federally owned land, primarily that managed by the Forest Service (74 percent of habitat rangewide) and the Bureau of Land Management (12 percent). In Oregon and Washington, it is estimated that more than 95 percent of remaining habitat is federally owned (U.S. Department of the Interior 1992).

As of early 1993, almost all old-growth logging on federal lands had been stopped by court action. The Forest Service was embarking on a new process to devise a defensible owl management plan after failing to craft one in at least four previous attempts. In the early 1970s, the agency refused to deal with the issue, claiming that land set-asides to protect the owl would violate multiple use principles (Nietro n.d.:2–3). In the early 1980s, the agency tried to deal minimally with the owl in its regional guide for the planning process in the national forests in the Pacific Northwest, saying that final decisions would be made in the planning process for the thirteen owl forests in the Northwest mandated by the National Forest Management Act.

Environmental groups appealed the agency's nondecision and forced it to treat the owl issue seriously. Over the next three years, the Forest Service put considerable effort into preparing a supplemental environmental impact statement on the spotted owl in the Pacific Northwest, using an elaborate analysis including population viability analysis and extensive public participation. The result was a decision that neither the agency's biologists nor the public supported. Indeed, the agency was taken to court by both timber and environmental groups, and the court agreed with the nongovernmental groups.

Hoping to regain the moral high ground of technical legitimacy, the agency put Jack Ward Thomas, one of its most credible scientists, on the line as a member of an Interagency Scientific Committee in 1990. Yet when the Forest Service tried to adopt the committee's recommendations as its owl management plan, both the interest groups and the Bush administration said "no way" and the agency was back in court. Federal District Court Judge William Dwyer called for another analysis, placing the blame squarely on the administrative agencies: "More is involved here than a simple failure by an agency to comply with its governing statute. The most recent violation of NFMA exemplifies a deliberate and systematic refusal by the Forest Service and the FWS to comply with the laws protecting wildlife. This is not the doing of the scientists, foresters, rangers, and others at the working levels of these agencies. It reflects decisions made by higher authorities in the executive branch of government" (Dwyer 1991:33–34). Another environmental impact statement was prepared in January 1992, and once again environmental groups appealed. In May 1992, Judge Dwyer deemed it inadequate. He gave the Forest Service until September 1993 to come up with a legally defensible plan.

In early 1993, the timber sale program of the BLM, which manages a considerable amount of the former Oregon & California Railroad (O&C) lands in western Oregon, was also stopped by court injunctions. Operating with a dominant-use mandate favoring timber production on the O&C lands, the BLM never took the issue seriously. The agency prepared a couple of plans and all were appealed. Agency leaders finally asked the God Squad—the Endangered Species Committee, a set of seven cabinet-level administrative officials with the power to exempt projects from the provisions of the ESA—to exempt a handful of timber sales. While many observers of the process felt that the sales did not meet the criteria for exemption specified in the ESA, five of seven committee members voted to grant an exemption in May 1992 after officials at the highest levels of the White House staff told at least one member of the committee that his job was on the line. When environmental groups appealed the decision, the courts agreed that the exemption was improperly given.

For its part, the FWS had to finalize a recovery plan for the spotted owl after it was forced to list the owl and its critical habitat by court action. In 1987, Green World, a small Massachusetts environmental group, petitioned the FWS to list the owl as an endangered species. At first the agency refused to review the owl's status. Then, in 1988, agency staff examined the status of the subspecies and declined to list it under the ESA after FWS Director Frank Dunkle intervened. The environmental groups appealed the decision, and the court agreed with them. Calling the FWS decision "arbitrary and capricious," the court asked agency officials to explain how they arrived at their conclusion. Not surprisingly, FWS leaders found that new information warranted reopening the status review, and from that review they decided that the owl did indeed warrant listing as a threatened species under the ESA.

When the 1990 listing declined to identify critical habitat because it was "not determinable" at the time, environmental groups cried foul and the courts agreed. In May 1991, the FWS proposed 11.6 million acres of public and private land as critical habitat, revising that figure downward to 6.9 million acres in its final designation published on 15 January 1992 (U.S. Department of the Interior 1992). The magnitude of these numbers, and the fact that they involved private lands as well as public, got everyone's attention, generating extraordinary political pressure from all sides of the issue.

At the same time, the issue of owl and old-growth conservation is not just about bureaucratic intransigence and political machinations. It has real costs and benefits for many. Timber-dependent communities in the Pacific Northwest faced a precipitous loss of jobs causing social disruption that was tangible and compelling. Mills faced uncertain supplies of raw materials, creating a difficult decision-making situation for mill owners and managers who had millions of dollars of investment on the line. The resultant fears, anxiety, and hostility boiled in the cauldron of public opinion in the Northwest during the late 1980s and early 1990s. In 1989, for example, the Forest Service stopped sending employees dressed as Smokey the Bear and Woodsey the Owl to local parades after receiving death threats against them. In 1990, several school boards attempted to ban *The Lorax*, by Dr. Seuss, from their libraries.

For their part, environmental groups advocated a real and technically valid concern about the continued fragmentation of the old-growth forests of the Pacific Northwest region, as well as the loss of interior-dwelling species dependent on those forests. Some thirty-two species were considered to be old-growth obligates in the early 1990s, and subsequent research suggests that there may be many more. Groups that had been successful at stopping the timber programs of the Forest Service and BLM in the Northwest were also concerned about the price of success: would they win the owl battle, yet lose the endangered species war, by triggering a backlash against the ESA? They worried about the mobilization of the so-called wise use movement, an ostensibly grass-roots movement that favored continued commodity production from public lands. And they worried about environmental policymaking at a time of a weak national economy.

Overall, as a matter of public policy and resource management, the twenty years of effort to deal with the owl is an amazing situation. It is not that little was done. Indeed, huge amounts of energy and administrative resources were spent dealing with the owl issue over this period, including the entire careers of a significant number of scientists, managers, and lobbyists. High level agency officials and political leaders spent an immense amount of time coping with the escalating owl situation, distracting them from other important agency and public business.

Even more remarkable, perhaps, is that the owl case in many ways represents a best-case situation in sensitive species management. Compared to the situations facing many other rare and endangered species, the owl was in pretty good shape. For most of the case history, a great deal of information about the biology, demographics, and ecology of the owl was available. The subspecies did not belong in intensive care, as is the case with many listed species where only a handful of individuals remain. Moreover, the problems facing the owl were recognized at a time when there were slack resources to deal with them. The early 1980s recession in particular provided great opportunities to change the management of certain segments of national forest land. The agencies with management responsibility were capable. The Forest Service in particular had access to considerable expertise and humanpower, and all parties faced considerable incentives to get the conflict resolved in a timely fashion.

Why, then, in what was in many ways a best-case situation, is the history of owl management so discouraging? Why did the conflict fester for so long? While there are many answers to these questions (Yaffee 1994), one set of responses deals with the nature of the issue itself and the setting in which organizations, experts, and decision makers formulated and implemented courses of action. The owl issue, like many endangered species management controversies, was a surrogate for a much broader set of concerns. And the problematic way in which it was handled reflects the nature of the broader sociopolitical context of public policy decision making—including multiple and conflicting values, public choice processes that are ineffective at dealing with issues like the owl controversy, a legal situation that is complex and ambiguous, and a wide array of legitimate and illegitimate political behavior. Just as children and owls strongly reflect the environment in which they live, policy and management decisions are shaped by their sociopolitical context. To understand why things are the way they are, professionals and organizations need to understand this context. To ensure that good technical ideas are implemented effectively, they need to be able to deal with, and influence, this sociopolitical environment.

Many Issues in One

The owl controversy was not just about owls and sensitive species management. In fact, this is generally the case with endangered species management issues. What often appear to be simple conflicts pitting one interest against another, such as owls versus jobs, are often multidimensional conflicts between a variety of substantive, political, personal, and bureaucratic interests. As a result, such conflicts draw the attention of hundreds of individuals, groups, and agencies, all with a variety of motives. As these issues become intertwined, each subissue becomes harder to separate and deal with. In aggregate, seemingly small issues often generate large stakes, making resolution difficult.

The range of issues lurking below the surface of the owl controversy was remarkably broad. While the case clearly affected the long-term survival of the northern spotted owl, how it was resolved also had meaning for other environmental objectives. For some groups, the controversy was partly a battle to preserve additional wildlands, regardless of their value as endangered species habitat. Additional wilderness designation was unlikely, following the passage of Oregon and Washington wilderness bills in the 1980s, and the owl controversy was a way for environmental groups to leverage additional set-asides. It was also a way to challenge the direction and implementation of land management policy by the Forest Service, the BLM, and state and private landowners in the Pacific Northwest. Owl management also became a surrogate for a broad discussion about the appropriateness and effectiveness of federal endangered species policy. Should balancing between ecological and economic objectives be allowed? Should equal protection be provided to all levels of taxonomic significance? How should ecosystem-level conservation concerns be integrated into a law that was quite clearly focused on protecting fragile species?

The owl issue also had considerable impact on the future of the federal resource management agencies, their efficacy, values, and methods. Was multiple use appropriate and possible? Was commodity production an important objective for national forest management? Were technically derived solutions to multiple-objective problems possible? For the agencies themselves, how these questions played out had considerable meaning for the size of their future budgets, the scope of their mission, and the degree to which they would control their own futures. The owl case also became a testing ground for evolving methods of wildlife management, including such concepts as population viability analysis, management indicator species, and adaptive management. The solutions to technical questions—is it *possible* to grow owls or old growth, for example—got intertwined with more value-laden questions like whether it is *appropriate* to grow owls via active management such as captive breeding.

Clearly the owl management controversy encompassed a set of socioeconomic questions. What is the long-term viability of the timber industry in the Pacific Northwest? How dependent is the region on timber as a component of its economic base, and how healthy is this level of dependency? The case also clearly raised questions about the long-term economic and social health of a set of timber-dependent communities. Could the impact of regional economic transformations caused either by productivity improvements in mills or losses due to changing federal land management be mitigated? And since roads and schools in many communities were heavily subsidized by timber receipts, the controversy even raised questions about the financing of essential public services.

How the spotted owl controversy was resolved had international ramifications as well. Since wood products and raw logs are a vital component of international trade, both with Canada and the Orient, how federal lands were managed would have an impact on global markets. Owl management approaches also sent important messages to other countries about biodiversity protection and wise land management. To environmentalists, the question became: if we cannot protect our own biological diversity contained within the old-growth forests, how can we expect Brazil to listen to our pleas to protect tropical rain forests? Timber supporters responded that if we do not produce timber domestically, the demand will increase on sources of supply in the tropics, and so we would be better off getting wood fiber from forest areas that can be managed sustainably and for which government has the means of regulating forest practices to ensure environmental protection.

Basic policy uncertainties underlay the owl controversy too. How extensive should be the rights of nongovernmental interest groups to challenge public policy choices? What is the obligation of government to dependent communities and losers in regional economic transformations? Who should prevail when local, state, regional, and national interests conflict? What is the role of science and scientists in public policy decision making and land management?

The spotted owl case was also very much about the articulation and validation of a set of dominant social values: how much is an endangered species worth, and who gives way when economic and environmental quality objectives seem to conflict? We do not get together, all 250 million of us, once a year to decide what our values are. Instead, the choices we make as a society do this implicitly. This mode of value legitimation makes seemingly small management controversies much more significant than they might otherwise be.

Many endangered species management situations reflect this underlying reality. Issues that pit the preservation of the Pacific salmon (*Oncorhynchus* spp.) against multiple water users of the Pacific Northwest, the Delta smelt (*Hypomesus pretiosus*) against water users of the Sacramento River basin, the red-cockaded woodpecker (*Picoides borealis*) against timber interests in the Southeast, the California gnatcatcher (*Polioptila californica*) against real estate concerns in southern California, and numerous endangered fish populations against development in the upper Colorado River and many other places reflect the use of endangered species management processes to force more fundamental value choices.

Most of these endangered species versus development situations are indicators of a much more fundamental set of problems and choices. Most represent fully appropriated environmental systems that are sorely in need of serious planning, and the Endangered Species Act becomes the policy lever of last resort. We have gotten into land use planning via the back door, and the ESA-created decision-making processes were not intended or designed to accomplish these broader ends. The burden they carry is unfair; the mismatch of cramming a large set of very big square pegs into a fairly small round hole is doomed to create continuing controversies that are neither efficient nor likely to be resolved in a lasting manner.

Since many endangered species management situations become catchalls for a variety of regional issues, for those of us interested in effective management of sensitive species, it becomes vitally important to:

- Create and contribute to alternative processes, such as regional planning mechanisms, to work out some of these issues before we get to blaming them all on a fish or a bird.

- Rejuvenate endangered species management processes to enable them to work more effectively on a broader set of issues. These improvements would include facilitating multispecies recovery plans and using habitat conservation planning effectively. They might also involve training biologists in such nontraditional topics as negotiation, public finance, and economic development so that they can contribute to the resolution of development questions and the mitigation of the negative sociopolitical effects of endangered species management.

- Educate a broader set of publics about the values inherent in protecting biological diversity by making the case that economic and environmental sustainability are linked. The historic boom-and-bust cycles of many communities dependent on resource extraction suggest this interdependency, as does the environmental condition of many areas of Eastern Europe.

The Impact of Sociopolitical Context

Beyond suggesting the complexity of environmental and resource issues, the owl case also indicates the importance of the context in which organizations and professionals function. While sometimes incredibly motivated professionals and highly effective organizations can overcome problems in their external setting, often their best efforts are frustrated by limited support in the world around them. The context for endangered species management is set by:

- Societal values, including the knowledge base that contributes to value formation

- The nature of overall decision-making processes

- Legal structure

- Politics as an expression of human values and a set of strategic behaviors

This context forms the petri dish in which organizations and individuals either thrive or die. It also has implications for what organizations and professionals involved in endangered species management should do in order to be effective. The balance of this chapter describes these four elements of the sociopolitical context of endangered species management and what they imply about organizational and professional roles and behavior.

A Clash of Values

The owl case was clearly a clash between a post–World War II Forest Service and an evolving set of public environmental values. The Forest Service recreated itself after World War II as a multiple-use agency with timber production as its primary organizational objective in western Oregon and Washington. It did this for good ideological and bureaucratic reasons. Timber production promoted the American dream of a single-family wood-frame house in the suburbs. It allowed the agency to maximize its budget, power, and influence by generating considerable political capital. Timber as primary objective also meant that agency leaders could maximize their control over the workforce by having narrow measures of organizational success that could be monitored easily.

In the 1950s and 1960s, agency leaders built an organizational image and style that made them highly effective and yet resistant to change when change became necessary. Their "can do" self-image made the service very successful as an action-oriented agency that got things done with an esprit de corps that was envied by many other federal agencies. Forestry's emphasis on maximizing yield through deliberate, technical analysis gave the agency a way to slip the bonds of the seeming irrationality of politics. And the development of a leadership elite of World War II veterans from a conservative, rural, middle-class background allowed the agency to build a homogeneous culture that simplified many of the tasks of organizational management.

At the same time, postwar affluence—and access to public forests provided by the development of a nationwide road network—created the medium in which a new appreciation of the noncommodity values of public lands could grow. Urbanization generated a population that grew increasingly divorced from the raw material supplies used to make the goods they consumed with such relish. The Vietnam War and the Watergate incident fed a long-standing but nascent cynicism about the effectiveness of government, and the war on poverty and the late 1960s and early 1970s environmental movement caused some in the expanding middle class to question the appropriateness of endlessly striving for material satisfaction.

A changing knowledge base, particularly in the 1970s and 1980s, reinforced a sense of a broader set of values in natural resources. In the 1960s, for example, old growth was considered to be a biological desert. In the 1970s, species-level diversity was valued and encouraged by creating more edge for wildlife through openings in the forest—an objective that coincided with timber harvest via clearcuts. Ironically, the forest fragmentation that resulted from this approach to management created problems for interior-dwelling species like the spotted owl, and advances in conservation biology and landscape-level ecology led scientists and environmentalists to push for ecosystem-level conservation and protection.

The spotted owl controversy in great part reflected the clash between, on the one hand, the values of the federal forest management agencies and their economic and political constituents and, on the other, a broader set of public values concerning the ecological, spiritual, aesthetic, and recreational dimensions of public wildlands. The old-growth controversy was the logical place for this clash to occur, since the Pacific Northwest was in many ways the last frontier of uncut forestlands in the continental United States. The logging of original forests that started with European settlement on the East Coast had made its way westward to the Pacific Northwest. The incredible value of the clear, straight timber of the Douglas fir forests in the Northwest ensured a battle of monumental proportions. While earlier skirmishes conducted under the wilderness banner could be dealt with by giving preservationists higher-elevation "rock and ice," while loggers worked the lower lands, the two met head-on in the mid-elevation forests that were habitat for the owl and other old-growth-associated species.

While the owl situation represents an extreme case of value conflict, many endangered species management controversies involve exactly such conflicts, making them much more important and symbolic, and more difficult to resolve, than they would be otherwise. The value content of management choices has several meanings for organizational and professional behavior:

- Agencies should do more in the way of value scanning. Such scanning would include learning about the values and interests of those outside the agency and thinking about the character of the future and what it implies for effective resource management.

- Agencies need to diversify their own value base and view their potential constituents much more broadly than either Service has done in recent years. Moreover, diversification means expanding the characteristics of agency workforces and not trying to socialize staff members into narrow norms of behavior. Just as a forest monoculture is unlikely to be able to adapt to such random events as disease and fire, an organizational monoculture is unlikely to adapt to changing public desires.

- Agencies and professionals must create decision-making processes that clarify and expose value differences, acknowledge the legitimacy of a diverse set of values in public land management, and find ways to meet the interests of different groups without forcing unnecessary trades between those holding different values.

• Education is needed that builds understanding in two directions: those outside the agencies must be informed and involved in order to build their understanding of what agency staff value and why; in turn, agency staff must become more aware and supportive of the diverse cultures and values inherent in a pluralistic society.

• Professionals must be careful not to hide their own value choices, or those of agency leaders, in scientific and technical analyses. The obligation of endangered species managers and researchers is to do good science, be watchful for value-laden assumptions, and explain their results to decision makers and the public. Moreover, they must recognize that resource professionals function within a larger society whose members have real stakes and legitimate interests in the outcome of endangered species management.

The Characteristics of American Decision-Making Processes

Not all the blame for the spotted owl controversy should be laid at the door of the Forest Service and the FWS. To some extent the seemingly endless debate was fueled by the underlying characteristics of American decision-making processes. Fragmented responsibilities and knowledge, incentives that promote short-term decision making at the expense of long-term efficacy, a tendency to foster competitive behavior at the expense of cooperative activities that may yield creative solutions—all set up a difficult situation for management controversies like those in the owl case. For example, the mechanisms by which we decide public policy, and the authority granted to land managers, are fragmented. Fragmented authority and responsibility are in part historical artifact. The federal public land systems are systems only in that they share a common label: national park, national forest, national wildlife refuge. In some places, they represent an after-the-fact attempt by managers to organize a scheme of management and rationalize a series of objectives where none was present at the time lands were set aside.

No one would argue, for example, that the BLM lands in western Oregon were ever intended to serve as a key component of a biodiversity protection scheme. It was only through the good graces of God and the bankruptcy court that the Oregon & California Railroad lands reverted back to the federal government. Wildlife refuges are by and large isolated chunks of wetlands with considerable value to grow ducks. National forests are a serendipitous gift from Gifford Pinchot and Teddy Roosevelt, who took a small window of opportunity to set aside much of what became the national forest system. All of these lands have potential value as core areas for the protection of biological diversity, but virtually none were set aside with that purpose in mind. They exist as a fragmented, somewhat chaotic landscape, with multiple owners and multiple management objectives, and the ESA is one of the few forces that might conceivably unify their intended direction.

Other dimensions of fragmented authority and responsibility exist by design. The federalist system distributes power and control quite deliberately across some eighty thousand units of government and between executive, legislative, and judicial branches. One of the major reasons for creating such a system was to check the power of any one component and provide an effective means of representation for all kinds of interests. But the result is a checkerboard of intents, tendencies, and decisions that can be mutually inconsistent and result in a state of public sector impotence where decisions rarely hold. The owl case demonstrates the fluidity of endangered species management decisions, as decisions made in one arena are appealed to the next and so on.

While fragmentation often yields enduring conflict, it also produces other unfortunate effects. At its best, it can provide only piecemeal solutions to what are often crosscutting problems. At its worst, it diminishes the accountability of each decision-making or implementing element, so that thorny problems often fall through the interstitial cracks of a fragmented set of managers and decision makers. The Forest Service and the BLM could avoid taking significant action while the owl was not a federally listed species. The FWS could avoid listing the owl because agency officials could claim that the land management agencies would do the right thing. Congress could avoid acting because it was the executive branch's problem, while the Forest Service could claim that congressional appropriations mandated a high cut level that they had no discretion to overturn administratively. Fragmented authority and responsibility not only diminish accountability but allow inaction without serious penalty.

Fragmentation also means that information is divided between locations and organizations, each gathering and interpreting it differently. Fragmentation of information often leads to controversies over data, as strategic battles over purposes are played out as technical debates over the state of knowledge. Even worse, fragmented knowledge leads to inferior solutions to important problems, as decision makers are unaware of key information. Decision makers in the owl case, for example, lacked a definitive map of ecologically significant old growth for more than fifteen years because of the state of the Forest Service's and BLM's inventory information, and neither agency was particularly motivated to clarify the old-growth picture.

Other aspects of American decision-making processes created a difficult situation for those interested in endangered species conservation. Many of the incentives facing key actors in the policy process promote behavior that is rational in the short term yet often inappropriate over the long run. The protection of sensitive species and the conservation of biological diversity make compelling sense, really, only in the long term. When faced with a proposed development action or a specific timber sale, for example, it is hard to argue that the project's marginal impact on a species or subspecies should be grounds for stopping the action. Yet the cumulative effects of multiple decisions that appear rational in the short term can lead to catastrophic effects in the long run. Similarly, it is hard to argue that the protection of one subspecies or isolated population is absolutely critical but relatively easy to argue that the protection of diversity makes a great deal of sense in the long run.

The incentives that function in many American decision-making processes are explicitly short term in nature. Presidents and agencies function on a four-year time horizon, for example, and members of Congress function on a two-year (or six-year) cycle. Biting the bullet and adopting tough deficit reduction measures is a politically irrational decision for many both inside and outside the Beltway, yet everyone agrees that if serious measures are not taken, our children and grandchildren will face an even tougher situation.

In the spotted owl case, one can cite many examples of short-term rationality overtaking long-term rationality. For timber interests, harvesting as much old growth as possible in the near term helps maintain jobs and profits—regardless of the long-term impacts on timber supply. For politicians, maintaining the flow of political capital that comes from continued federal activities that benefit local economies makes short-term sense—even if it leads to precipitous socioeconomic impacts in the long term. For regional environmental groups, continuing to press their advantage in the courts and stop all old-growth harvest appeared to make sense—even if it resulted in a backlash against the ESA in the long run. And for agencies like the Forest

Service, finding ways to minimize the impact of changing public policies on their workforces led to a policy of maintaining an ambitious
timber program nationwide—even if it cost them political and technical credibility and eroded their land base.

Finally, the nature of American decision-making processes promotes an appearance of zero-sum decision making, where one side
wins and the other side loses. A compelling set of literature on negotiation indicates that zero-sum decision making promotes competitive
and strategic behavior, since all the disputants can do is divide up the
fixed set of values on the table (Lax and Sebenius 1986). So-called
win/win solutions are neither sought nor thought possible. Casting
the owl controversy as a situation pitting owls against jobs was not
only inaccurate but tended to limit the ability of decision makers to
craft effective solutions to the real underlying issues.

These fundamental problems of American decision-making processes suggest what individuals and organizations interested in endangered species management should be doing:

- Dealing with the problem of fragmentation points to the importance of boundary-spanning institutions—that is, groups or
  forces that help people integrate across fragmented jurisdictions,
  responsibilities, and disciplines. It means creating interagency
  working groups, both ad hoc and formal. It also means setting up
  multiparty working groups—where public and private-sector organizations come together to work on common problems—and
  assisting such groups to work effectively.

- Dealing with the short-term/long-term problem means that
  agencies must work harder at imagining alternative futures, and
  learning about the future, in ways that might be threatening to
  traditional norms and constituents. It also means that agencies
  should experiment with different ways of doing things so that a
  diverse set of approaches can weather changes in knowledge and
  values. An important role for public land managers is to create
  and preserve options available for future publics and decision
  makers. And given the nature of the multiparty bargaining
  processes that create direction for society, public agencies must
  act as advocates for the interests of future generations of humans
  and other lifeforms.

- Dealing with the creativity problems resulting from a zero-sum,
  competitive context means changing the incentives operating in
  agencies and in the broader set of policymaking processes to

promote innovation and learning. Creativity and risk taking need to be fostered, not penalized, as is often the case in public-sector agencies. Alternative dispute resolution processes can be used to structure decision making so that they encourage collaborative problem solving and yield enduring solutions to public problems where they can be found.

The Statutory Basis

The third component of the sociopolitical context is a collection of statutory requirements and mandates that are vast and conflicting. It is important to remember, and somewhat surprising to realize, that the northern spotted owl did not become a federally listed species until 1990. Most of the action in the owl controversy occurred under the auspices of the National Forest Management Act, the National Environmental Policy Act, the Federal Land Policy and Management Act, and numerous congressional appropriations bills. Much of the judicial record resulted from challenges based on NEPA, NFMA, and the Administrative Procedures Act, not lawsuits brought under the provisions of federal endangered species law. Since these laws often conflict with each other, agencies can justify whatever direction they want to set. The Forest Service and the BLM were at least partly justified in implementing aggressive timber sale programs in the Pacific Northwest, since they were called for through allowable sale levels established in the congressional appropriations process.

It is a mistake to view endangered species management solely through the lens of the ESA. Other public and private laws and policy instruments will have as much or even more impact on the endangered species problem in the future. How national forests, national parks, and BLM lands are managed, and how water is allocated, will have a major effect on the state of biological diversity in the western United States. How federal and state agricultural policy, farming practices, and wetlands conversion guidelines are carried out will be a major determinant of the status of sensitive species in the rest of the country. Moreover, state wildlife statutes and local land use ordinances have as much impact on wildlife habitat as federal endangered species or public lands law. The ESA is an important law, providing a necessary, last-resort safety net, but other elements of policy and on-the-ground behavior will have a more significant effect on the broader biodiversity problem.

It is also a mistake to view the collection of laws, and particularly the ESA, as a set of commands whose primary impact is to mandate certain behavior. Laws function by structuring packages of incentives so that individuals and organizations are led to act in certain ways. They rarely succeed by absolutely prescribing correct behavior—as the uneven response to Prohibition, the variable response to speed limits and seat belt laws, and the problematic history of the ESA's implementation indicate (Yaffee 1982). Rather, they work by altering the strategic position of different forces operating in the implementation context.

While land managers had no clear obligation to protect the owl under the ESA prior to the late 1980s, the existence of the ESA, and the possibility that the owl might become a listed species, did have a major impact on administrative behavior. Indeed, much of the Forest Service's activity came in response to the fear of listing. To have the owl listed as an endangered or threatened species meant that Forest Service staff would have the FWS looking over their shoulders and influencing on-the-ground management. Since agencies value control more than almost anything else, opening up their decision-making process to others was the source of many a nightmare for agency officials. For some proponents of owl protection, the recognition that listing the owl might result in a political backlash that could harm the ESA kept their demands reasonable and ensured their participation in the inevitable political negotiations. And for many land managers who believed in multiple use, listing the owl would mean that they had failed, since sustainability is a prerequisite to wise multiple-use management.

Recognizing the complex statutory basis for endangered species management has important implications. First: It means that we should find ways to integrate endangered species and biodiversity objectives in other legal instruments and agency objectives. These instruments would include those that influence the management of public resources, such as NFMA, as well as those that influence activities on private lands, including state forest practices acts, the swampbuster provisions of the federal farm bill, and dredge and fill policies created under Section 404 of the Clean Water Act. It is only to the extent that endangered species objectives are integrated into the patchwork of laws and behaviors influencing resource use that species and the ecosystems in which they survive will receive adequate protection under the law.

Second: It is time to seek some changes in federal resource management law. I would leave the ESA relatively untouched, implementing it as a program to protect species who belong in intensive care.

Biodiversity and ecosystem conservation legislation should move beyond this approach—not as a replacement for the ESA, but as a complement to it. This legislation could take many forms: creating better pathways for information exchange between agencies, mandating modes of analysis such as gap analysis, ensuring that biodiversity values are adequately examined through the NEPA process, raising the priority of diversity objectives in managing federal public lands, and providing incentives for state or regional programs. It might also include a recasting of the public land systems—overlaying a dominant-use mandate on a subset of federal lands while not necessarily adjusting ownership. Hence national biodiversity conservation lands could be established as core areas to protect future options around which management might take place.

Third: Resource professionals must understand that laws do not function by dictating behavior. Rather, they influence a much more subtle set of dynamics that encourage or allow behavior to move in certain ways. Protective behavior has been encouraged by the existence of reporting requirements established by NEPA and NFMA, for example, and the opportunity for outside groups to act in the courts has been promoted by the standing to sue granted by the ESA. Both of these activities should be maintained: honest information should be provided to the public and other agencies about the state of ecological affairs, and interest groups should have the power to challenge administrative actions. Professionals should also understand why educating and involving the public is critical to effective endangered species management. Indeed, many of the cases described in this book point to the value of outside interest groups, since little agency action was forthcoming without them.

The Impact and Role of Politics

Politics and political behavior are very real components of the arena in which organizations and professionals struggle to protect endangered species of plants and animals. Politics will always be at play in endangered species decision making, and to some extent this is appropriate. Political behavior is one way that human values are expressed, and political institutions are one of our society's standard mechanisms for making choices that allocate resources to different elements of present and future society. The classic political statement—who gets what when—is very much evident in many endangered species management choices. Because many issues are often tied up in endangered species decision making, most of these decisions are allocation decisions that express a set of human values and belong at least partly in the realm of political decision making.

At the same time, political decisions should be informed by reliable information and guided by values that are deemed ethical and effective for the long-term health of the society. Administrative processes that have a basis in science need to generate honest information about the benefits, costs, and impacts of alternative direction and, as well, express the level of uncertainty in a situation. To the extent that society's values are clearly defined and ethical guidelines exist, decision making should relate to these values and norms. And to the extent that these values are in flux and fragmented across the political landscape, administrative decision makers should seek to build support for specific management directions or adjust them to fit the values evident in society at large.

In the owl case, political behavior was seen in several forms. Lobbying by interest groups and posturing by members of Congress to appeal to (and reflect the interests of) their constituencies were evident from the mid-1980s onward. Such behavior can be seen as a legitimate and ultimately effective way to influence the major value choices at stake in the outcome of the owl controversy. At the same time, the politicization of scientific judgment that took place in the Forest Service, BLM, and the FWS in the Reagan and Bush administrations was a much less legitimate exercise of political power. By having political appointees dictate what scientific studies could conclude—as was true with the first decision by the FWS not to list the owl and somewhat true in the Forest Service Supplemental EIS work—politicians at the highest levels of government denied us an honest assessment of the impact of various choices.

High-level political officials also delayed the ultimate outcome of the owl controversy and squandered some of the slack available to assist in making a final decision. By the mid to late 1980s, it was clear that the owl case would result in an adjustment of forest and wildlife policy to bring it in line with shifting public values and scientific understanding. By not crafting an effective solution or creating the political will to support it, the Bush administration used up valuable time and energy that could have been employed on many other public policy problems.

Creating the political will to make a decision on the owl case would not have been an easy task, for the owl case also provides a very good window into the problematic state of political organization in present-day America. Political power in the United States is dispersed across more people and organizations today than perhaps at any time in our country's history. As some of the historic political organizing forces have weakened in recent decades (political parties, broad social institutions such as religious groups, and leadership in legislatures such as the U.S. Congress), other forces such as interest groups and agencies have increased in significance and led to the fragmentation of political power. Interest groups are explicitly organized around specific issues, and the multiplicity of access points into decision making and changes in campaign finance have afforded them greater power to influence decisions.

Increased fragmentation of political power means that it has become increasingly difficult to create the political will to act on just about anything. Fragmented interests and dispersed power represent one of the most serious dilemmas facing our society, and its implications touch most aspects of social policy, from deficit reduction and health care reform to environmental policy. In the owl case, fragmentation of political power led to the state of impasse evident over the past few years. The result has been delay and unstable decisions—leading to endless rounds of agencies taking action; having the action appealed and stopped by the courts, political executives, or legislatures; leading to a new action that gets appealed; and on and on. While preservation interests did benefit by the impasse in forest management decisions in the late 1980s and early 1990s—in that it allowed them to build national political support and wait out the Bush administration—it is hard to argue that this pattern of behavior is an efficient or effective approach to deciding public policy. Opportunities are lost, real problems get worse, and scarce resources are spent to no good end. Indeed, in many endangered species cases, inaction results in a deteriorating situation, as candidate species die waiting to get into the emergency room.

The fundamental political realities of endangered species management suggest several key lessons for scientists and decision makers. First: Agency officials must build political concurrence for the actions they seek to take. Their actions must fit the interests of a set of supporters, and a diverse set of interests must support the proposed actions if they are to endure. Laws are only as good as their underlying political support, since they can be undermined in implementation or by the lawmaking body that created them. As one senator mentioned several times about the ESA in the owl case: "We made the law and we can change it."

Second: To build political concurrence, resource agencies must shift from their traditional narrow constituencies to build broad coalitions of support that remain a reflection of present-day values. The bad news for the Forest Service and the FWS in recent years is that traditional groups, including timber and hunting interests, have declined in power. The good news is that there is a remarkably broad set of coalition members waiting to be tapped: commodity groups, environmentalists, recreation groups, scientists and researchers, retirees, and educators. The agencies must remember to keep these coalitions in flux, however, so that they remain current and are not dominated by a single interest. One of the big lessons of the owl case is that the Forest Service and the FWS got trapped by their traditions, but the case also reminds us that there are great opportunities in a situation of dispersed political power.

Third: To build political concurrence, agency officials must work with individuals and groups outside the agency to build understanding and support for the actions they seek to take. Snail darters and spotted owls will receive continued protection only if a number of key decision makers and their political constituents are willing to support the action (or at least not actively oppose it). Because many of the values of endangered species protection are not obvious to congressional representatives, resource management agencies and professionals and conservation interest groups must work hard to explain why action is needed. FWS botanists may find it obvious why it is important to protect a colony of a rare plant, but a former businessman from Texas and a lawyer from Oregon, both members of Congress, may not understand what the problem is.

Fourth: The political side of endangered species management suggests the need for professionals to recognize the strategic dimensions of agency and technical decision making. The decisions they make, the information they acquire, the analyses they produce—all represent valuable information to political forces that seek strategic advantage. Information is one source of power in this realm, as is the credibility provided by apparent technical legitimacy. The upshot is that professionals should be careful about what they do and how it is done. The current debate about ecosystem management provides a good illustration. It should give us pause that both ardent preservationists and commodity interests are advocating an ecosystem approach to biodiversity protection. To the preservationists, an ecosystem approach means more protection. To the commodity interests, it means that certain species, including many that are most critically endangered, can be written off in pursuit of a broader, vaguer goal. The media is portraying the situation as one in which the species approach does not work and hence needs to be replaced. To scientists who recognize the technical validity of an ecosystem approach, I urge caution as they influence this debate. Their words have strategic significance in ways they may not mean.

Fifth: The primary source of legitimacy for resource agencies and professionals in the broad political process is the quality of their technical expertise and information—and this has important ramifications. They must be at the forefront of their science, do good science, and be honest about it. They also have a responsibility to use their technical knowledge to educate and inform decision makers and the public at large and to advocate in an honest way for what they view as good solutions. These same agencies and professionals have an increasingly legitimate role, as well, as the crafters of decision-making processes that can create technically valid, win/win solutions. Some of these roles may feel awkward indeed to the current crop of agencies and professionals who have been taught that good science is done in isolation from politics, management objectives, and people. While the roles of educator, advocate, and facilitator may not be traditional for resource management professionals, today they are critical.

Lessons

While the spotted owl case illustrates many of the intraorganizational dynamics and problems reflected in the other case studies in this book, it has a broader message: The sociopolitical context is vitally important. While management for the owl has an impact exceeding many of the other cases reported in this book, its overall message about the importance of values, politics, and the character of decision making applies as well to most of these other situations. And many of the endangered species issues in public view today remind us of what happened in the development of the owl case. Management controversies over the California gnatcatcher, Pacific salmon, red-cockaded woodpecker, and Delta smelt focus public attention on the habitat needs of far-ranging species juxtaposed against regional economic activities. It should not be surprising, therefore, that endangered species management is as much about organizing and dealing with humans and human institutions as it is about dealing with plants and animals. The success in future efforts to protect biological diversity will depend in large part on how well agencies and professionals understand and act within this sociopolitical context.

References

Beuter, J. H., K. N. Johnson, and H. L. Scheurman. 1976. *Timber for Oregon's Tomorrow: An Analysis of Reasonably Possible Occurrences.* Research Bulletin 19. Corvallis: Oregon State University School of Forestry, Forest Research Laboratory.

Dwyer, J. 1991. *Memorandum Decision and Injunction: Seattle Audubon Society et al. v. John L. Evans et al.* No. C89-160WD. Seattle: U.S. District Court, Western District of Washington.

Lax, D. A., and J. K. Sebenius. 1986. *The Manager as Negotiator: Bargaining for Cooperation and Competitive Gain.* New York: Free Press.

Nietro, B. n.d. *Chronology of Events Related to the Spotted Owl Issue.* Portland: U.S. Bureau of Land Management, Oregon State Office.

Oregon Endangered Species Task Force. 1973. Minutes of 29 June 1973.

U.S. Department of the Interior, Fish and Wildlife Service. 1992. *Recovery Plan for the Northern Spotted Owl—Draft.* Washington, DC: Government Printing Office.

Yaffee, S. L. 1982. *Prohibitive Policy: Implementing the Federal Endangered Species Act.* Cambridge: MIT Press.

———. 1994. *The Wisdom of the Spotted Owl: Policy Lessons for a New Century.* Washington, DC: Island Press.

4

The Black-Footed Ferret Recovery Program
Unmasking Professional and Organizational Weaknesses

Richard P. Reading and Brian J. Miller

Endangered species recovery programs could be greatly improved by addressing their professional and organizational weaknesses. Developing appropriate organizational arrangements early in the process can lead to efficient and effective programs characterized by a broad interchange of ideas, the absence of unproductive conflict, and successful restoration of the species. Such considerations, however, are not appreciated by most people working in endangered species conservation programs. As a result, professional and organizational performance is compromised and little learning takes place within and between programs. This has been the case, to varying degrees, in black-footed ferret (*Mustela nigripes*, Figure 4-1) recovery efforts.

Since black-footed ferrets were rediscovered in 1981, recovery efforts have often been hampered by destructive conflict. Although the ferret program's problems have been widely acknowledged (Carr 1986; May 1986; Weinberg 1986; DeBlieu 1991), the underlying causes have been interpreted in different ways. (See, for example, Clark and Harvey 1988 and Thorne and Oakleaf 1991.) Some have used theory and tools from policy and organizational sciences to analyze the problems and their causes. (See Clark and Harvey 1988; Clark 1989; Clark and Westrum 1989; Clark and Cragun 1991.) Their recommendations have seldom been used by government officials over the last decade, however, and recovery efforts continue to be plagued by unproductive conflict, inefficiency, and reduced effectiveness (Greater Yellowstone Coalition 1990; Reading 1991; Miller et al. 1993).

FIGURE 4-1.   The black-footed ferret. (Photo by T. W. Clark.)

In this chapter we describe professional and organizational performance in black-footed ferret restoration efforts, list the key lessons
from our involvement in these efforts, and conclude with some advice
for future improvements. We have been involved in ferret recovery efforts since 1983, working on a variety of ecological, social, and organizational studies in Wyoming, Montana, Colorado, Virginia (with captive animals), and Mexico. We have held several affiliations with
nongovernmental organizations (the Northern Rockies Conservation
Cooperative and the Smithsonian Institution), universities (the University of Wyoming and Yale University), and state and federal wildlife
and land management agencies (the Wyoming Department of Game
and Fish, U.S. Fish and Wildlife Service, and U.S. Bureau of Land
Management). Our experience suggests that professional and organizational issues lie at the core of most of the problems and controversies in the ferret program. Although the recovery effort still faces
several significant biological, ecological, sociological, and political
hurdles, we suggest that upgrading professional and organizational
performance provides the greatest opportunity for improvement.

We offer these analyses not to denigrate the many hardworking
people struggling mightily inside and outside the program but to point
out areas that require improvement, to offer recommendations, and

to promote a model program in which ferrets are recovered as efficiently and as quickly as possible. We hope that all will join us in realistically assessing problems in the management of the species and view our observations as we view them: as constructive criticism intended to benefit not only the black-footed ferret but the conservation of all species.

A Brief History

Black-footed ferrets are critically endangered, mink-sized (*M. vison*) mustelids. They are obligate associates of prairie dogs (*Cynomys* spp.), upon which they prey and in whose burrows they seek shelter and raise their young (Miller et al. 1988). Ferrets began disappearing following massive, government-sponsored prairie dog control programs initiated soon after the turn of the century (Miller et al. 1990). These programs, which continue at a smaller scale today, were successful in eliminating prairie dogs from an estimated 98 percent of their former range (Marsh 1984). As their prey base was destroyed, ferret populations became fragmented and began dying from various deterministic and random factors (Miller et al. 1988; Lacy and Clark 1989). Important among these was canine distemper—presumably introduced with European settlers and 100 percent fatal to black-footed ferrets (Carpenter et al. 1976; Thorne and Williams 1988).

By the time ferrets were first studied in 1964 in Mellette County, South Dakota, the species was already seriously endangered. Indeed, the species was on the first list of U.S. endangered species in 1964 and the U.S. Fish and Wildlife Service's endangered species priority list in 1976. Extensive surveying for additional populations proved unsuccessful, and when the Mellette County population disappeared in 1974, many people thought the species was extinct. Then a small population was discovered in a complex of white-tailed prairie dog (*C. leucurus*) colonies near Meeteetse, Wyoming, in the fall of 1981 (Carr 1986). Several agencies, universities, and nongovernmental organizations (NGOs) became involved in the program because of their legal mandates, land tenure, expertise, or interest in species conservation (Clark 1989). Although the species had formerly ranged across the Great Plains of North America, the U.S. Fish and Wildlife Service (FWS) gave the lead role in ferret recovery to the Wyoming Game and Fish Department (WGF).

The Meeteetse population was monitored and studied until 1985, when a combination of plague in the prairie dogs and canine distemper in the ferrets began driving it to extinction (Thorne and Williams 1988; Williams et al. 1988; Clark 1989). Eighteen individuals were captured from 1985 to 1987 and became the nucleus of a captive colony (Miller et al. 1988). For a review of the chronology of events in the history of ferrets and their management during the 1980s, see Casey et al. (1986), Clark (1989), and Reading and Clark (1990).

Captive propagation has succeeded in producing a number of animals, and today several hundred ferrets exist in eight facilities in the United States and Canada.[1] The challenge now is to locate or restore suitable habitat (that is, prairie dog colonies), successfully reintroduce ferrets into that habitat, and effectively manage these areas over the long term. Although research indicates that the extent of prairie dog versus livestock competition is exaggerated (O'Meilia et al. 1982; Uresk and Bjugstad 1983; Krueger 1988) and prairie dog control is not cost-effective (Collins et al. 1984), prairie dogs are still perceived as agricultural pests (Krueger 1988; Reading and Kellert 1993) and there is continued pressure to eradicate them—thus minimizing available ferret habitat. Conservation of the prairie dog ecosystem is a growing biodiversity issue throughout the West (Krueger 1988).

A revised ferret recovery plan completed in 1988 called for establishment of a minimum of fifteen hundred adult ferrets distributed in ten or more separate, self-sustaining populations throughout the species' former range (FWS 1988). Reintroduction began in the fall of 1991 with the release of forty-nine animals into south central Wyoming near Shirley Basin. Another ninety animals were released into the same area in the fall of 1992. Overwintering of at least two males and two females released in 1991 and successful reproduction by both females suggests that the technical aspects of ferret reintroduction are achievable. These technical aspects, however, may be much simpler and less problematic than the professional and organizational issues facing ferret recovery (see Reading 1993).

One of the next potential sites for reintroduction—and the site thought capable of supporting the largest number of ferrets in the United States—is in Phillips County, Montana (BFF Interstate Coordinating Committee 1989). Release of ferrets into Montana was originally scheduled for 1992, but the release was postponed until 1993 at the request of WGF and the governor of Montana, who sought to avoid a controversial wildlife issue in an election year (Petera 1992; Stephens 1992; Thorne and Russell 1992).

Professional and Organizational Performance

Recovery efforts since 1986 have dramatically increased ferret numbers in captivity and may eventually establish reintroduced animals into the wild. Indeed, many are already touting ferrets as a conservation success story (Cohn 1991; Ricciuti 1991; Thorne and Oakleaf 1991; FWS 1992; Obrecht 1993). Yet close analysis indicates that the program has been plagued since its inception by poor professional and organizational performance—problems likely to deter if not derail ferret recovery. (See, for example, May 1986; Clark and Harvey 1988; Clark 1989; Greater Yellowstone Coalition 1990; Reading 1991; Miller et al. 1993.) Although the ongoing problems are numerous, we focus here on five of the most important: program structure, working groups, excessive concern over program control, use of rigorous science, and measures of program success.

Program Structure

Structure—the pattern of interacting relationships, units, and practices in an organization, including hierarchy and underlying rules—is perhaps the most obvious organizational characteristic of a program. It is also a primary target for instituting enduring organizational change. Organizational structure may account for 50 to 75 percent of the variability in the way people behave in a group (Galbraith 1977). So understanding and managing organizational structure is critical for the efficient recovery of an endangered species.

A well-managed program is particularly important to ferret recovery because of the high degree of uncertainty and complexity associated with the restoration task, the absence of pertinent biological and ecological information, the presence of a suspicious, strongly antagonistic local public, fragmented authority resulting from the multiplicity of laws and interests involved, the need for high-quality information flows, and the high profile of the recovery program (Reading 1993). The resulting "task environment" places great strain on the program in the form of rapidly changing conditions, poor predictability, and intense external scrutiny and pressure. Organizational theory suggests that such complex, uncertain, and hostile task environments require an "organic" style of organization and management (Gordon 1983; Clark 1985; Westrum 1986). Such organizations should be "risk embracing" and flexible. They should have decentralized structures, rapid response times, and generative rationality—

that is, focus on generating ideas, on evaluation, and on future performance. Professionalism, critical review, rapid and open communication, limited hierarchy, and task orientation should all be stressed.

The black-footed ferret recovery program is organized around state and federal bureaucracies and uses semiformal working groups and advisory groups. The program's structure has varied slightly over the last decade, but the basic arrangements have remained largely unchanged. Soon after ferrets were discovered in Meeteetse, the FWS transferred lead agency status to WGF under authority of the Endangered Species Act. This decision was controversial because the species was originally found in twelve states and three nations and the recovery plan called for the establishment of ferret populations in several states (Carr 1986; FWS 1988). Upon attaining lead agency status, WGF rapidly moved the recovery program from an organic style of organization and management to a traditional bureaucratic style under its direct control (Clark and Harvey 1988; Clark 1989). Traditional bureaucracies are rigidly hierarchical, "calculatively rational" (development of future strategies is largely constrained by knowledge from past experiences), and heavily reliant on rules, regulations, laws, and standard operating procedures (Perrow 1986; Westrum 1986). The program has remained rigidly bureaucratic ever since.

The Wyoming program illustrates several problems common to traditional bureaucratic programs faced with highly complex, uncertain, and urgent task environments. WGF tended to be conservative and "risk aversive," reluctant to try innovative or intrusive biological methods of study (such as radio telemetry) or conservation (such as captive breeding or translocations from the original Meeteetse population). (See Bogan 1985; Carpenter 1985; Clark 1985, 1989; Wemmer 1985; Carr 1986; May 1986; Weinberg 1986.) Many problems resulted: unproductive conflict and distrust among parties, communication breakdowns and delays, diminished innovation, poor science, intelligence failures, and a shift from species recovery to control issues. For a review of these problems see Clark and Harvey (1988), Clark and Westrum (1989), Clark and Cragun (1991), Reading (1991), and Miller et al. (1993).

WGF now acknowledges the history of conflict but places the blame on a few personalities in other organizations (Thorne and Oakleaf 1991). WGF was largely successful in "eliminating" the key figures it considered problematic, including most nongovernmental field scientists, from recovery efforts. Still, Wyoming recovery efforts continue to be plagued by conflict. For a review of the ferret recovery program and its many problems over the first six years, see Clark (1989); for an opposing viewpoint, see Thorne and Oakleaf (1991).

In Montana, organizing and planning for ferret reintroduction began soon after ferrets were rediscovered in 1981. Although the Montana recovery program is organized around agency bureaucracies, it relies more heavily than Wyoming on a semiformal, interagency working group. Key parties in Montana include the FWS, the Charles M. Russell National Wildlife Refuge, the Montana Department of Fish, Wildlife, and Parks (MFWP), the U.S. Bureau of Land Management (BLM), and the Northern Rockies Conservation Cooperative. Until recently this cooperative structure appeared to help the Montana program proceed without all the conflict faced by recovery efforts in Wyoming (Reading 1993). No ferrets currently reside in Montana, however, and as the proposed reintroduction grows nearer, pressure on organizational arrangements has increased. Indeed, several of the problems faced by the Wyoming program have begun to emerge in Montana, differing only in scale and intensity. Thus it is crucial to identify the potential problems now and address them using sound organizational and management theory.

Recommendations for improving the management of recovery programs in the short term include:

- Develop flexible, decentralized program structures.

- Develop and maintain rapid, open, and reliable communication systems.

- Strive to accept risk.

- Be continually introspective, by means of periodic program reviews and evaluations.

Given the many structural, ideological, environmental, political, and behavioral forces acting to resist organizational change, however, manipulating the organizational structure is unlikely to succeed in the short term. Accomplishing these organizational goals will therefore often require the formation of special teams and groups. In addition, long-term recommendations include incorporating organizational considerations into university curricula for wildlife and conservation biologists and ensuring there are education programs and training workshops for people already working on species recovery.

Working Groups

Establishing special teams or task groups outside formal organizations permits "organic" structures to be embedded within larger bureaucratic arrangements. To maximize effectiveness and flexibility, these teams should be as independent as possible of the organizations their members represent. After all, teams formed internally are only as effective as organization leaders permit them to be. Outside, independent, and multiorganizational teams are often referred to as "parallel organizations." Gordon (1983) suggests that the most effective teams comprise collaborating professionals who view themselves as a cohesive unit with common goals and rewarding interactions. Such groups tend to be characterized by greater use of skills, knowledge, and innovative ideas, and the group atmosphere encourages rapid consensus, increased commitment, and sound decisions. The group strives to stay focused on the task and to keep communication rapid, comprehensive, frequent, and accurate. Conflict, if properly managed, can be productive. Strong leadership is a must. Finally, Gordon says that leaders and parent organizations should buffer groups and special programs from external pressure and disturbances.

Several working groups and advisory teams have participated in ferret recovery efforts in both Wyoming and Montana—Wyoming's Black-Footed Ferret Advisory Team (BFAT), FWS's Interstate Coordinating Committee, the Captive Breeding Specialist Group (CBSG) of the IUCN, the Montana Black-Footed Ferret Working Group (MWG), the Northcentral Montana Working Group, and more. During our involvement in the program, we found that these groups varied considerably in their effectiveness in clarifying tasks, using resources, learning from past experiences, reducing unproductive conflict, and ultimately moving toward ferret recovery. The successful groups were characterized by informal structures, strong leadership, generative rationality, and a diverse but relatively stable membership selected for their expertise and commitment to ferret recovery. Less successful groups were often composed of people included simply to represent certain organizations or interests, regardless of their expertise or commitment to recovery. These groups were characterized by calculative rationality and rigid bureaucratic structures dominated by one or two interests. Novel approaches were discouraged; the pressure for conformity was great. Some working groups, for example, displayed reluctance or even antagonism toward the idea of testing different reintroduction techniques—such as providing ferrets with experience of hunting or predator avoidance before their release or raising animals in semiwild arenas stocked with prairie dogs—and suppressed the proponents of such ideas.

Some groups provided a task-oriented, supportive atmosphere for ferret recovery that was sometimes quite different from the parent organizations. This approach fostered coordination, cooperation, continuity, and communication and helped people focus on the task of species restoration. Such groups were able to divide tasks among members, avoid duplication, and capitalize on each member's expertise and cognitive, financial, and other resources. Close working relationships in some groups helped alleviate many of the personality problems common in other programs (see Schön 1983). As members worked together, for example, they were better able to understand each other's biases and perspective, thus avoiding much unproductive conflict. The working groups facilitated good public relations by approaching the public as one cooperative and coherent group. Finally, some working groups enabled members to apply pressure on upper management by approaching issues as a united group. Members who had the support of upper-level managers in their own agencies often enlisted them to apply pressure on managers in other agencies to support the working group's initiatives, thus buffering the group's lower-level members. These factors helped some working groups maintain a task focus.

Although each of these groups emulated in some ways the high-performance teams recommended by Clark and Westrum (1989), they nonetheless could be improved by addressing five categories of problems. The first deals with the group's composition. One of the keys to high-performance teams is strong commitment to the task. Some groups in ferret recovery included members who were uncommitted or openly antagonistic toward ferret recovery, such as the Montana Department of State Lands within the MWG (see Reading 1993). Such members do not belong in a task group. This is not to say that their views should not be heard, understood, and possibly even incorporated into the program, but this should take place in a different forum. Moreover, some groups in ferret recovery underutilized outside specialists and professional advisory groups that might have contributed much to the program. Not only can outside expertise be used for specific projects on a temporary basis, but open and unbiased external reviews also help maintain task focus and bring credibility to the program. One of the key components of a group is strong leadership. In the ferret program, leadership was strong in some groups, such as the CBSG, but absent to muddled in others, such as the MWG (Reading 1993).

Second, several working groups were influenced by continual changes in membership. On the one hand, fluid membership provides an influx of new ideas and allows people with special expertise to enter and leave a program as their services are required (Quade 1975; Clark et al. 1989). Some individuals and groups, such as the CBSG, operated in this capacity, contributing enormously to the program's effectiveness. On the other hand, constantly changing membership can be disruptive, detract from the group's continuity, and adversely affect the group's ability to interact, communicate, cooperate, and understand each other. From early 1991 to mid-1993, for example, the coordinator of the FWS black-footed ferret recovery program changed twice. Each change was accompanied by altered responsibilities and program disruption. Similarly, disruption to the Montana recovery program followed assignment of new MFWP personnel to the program. Not surprisingly, gradual membership changes, such as the gradual change in FWS's Ecological Services representative to the MWG, tended to be more successful than rapid shifts. Slow transitions give members time to get to know newcomers and give new members an opportunity to learn how the group functions.

Third, role definition was problematic. Members of effective teams should be strongly task-oriented and encouraged to think creatively about problems and solutions, but in all ferret working groups some members identified more strongly with their organization than with the restoration task (Reading 1993). They viewed their role as a "political" representative of their employing agencies (see Wilson 1980). As one group member admitted, "I have to do what my boss says." This constraint often resulted in a reluctance to express personal views—especially if these views differed from the agency's goals, ideologies, or mandates. As participation by these members was limited to an expression of their agency's desires, they did not add much to the cognitive resources of the group. In extreme cases, agency goals began to take precedence over recovery goals, resulting in what Sills (1957) has called goal displacement—for example, maneuvering to gain control of the program. While it is important for groups to understand the constraints under which managing agencies must operate, it is beneficial for working groups to set aside affiliations while discussing strategies. This policy permits consideration of a much wider range of program options and results in greater effectiveness.

Agency concerns are important, but they can be incorporated after the biological imperatives are highlighted.

Fourth, some groups suffered from what Janis (1972) called group-think. Sometimes the pressure for group consensus is so great it stifles productive dissension, resulting in reduced creativity and effectiveness. Dissension and conflict do require careful management in order to maintain group cohesiveness, but they can be an invaluable means of improving performance. Some ferret groups, however, failed to utilize productive dissension for creativity and showed little expertise at conflict resolution. We witnessed several conflicts in both Wyoming and Montana that quickly spread to involve high-level decision makers. Dissension was usually viewed as negative and strongly discouraged.

Finally, there seemed to be little understanding of how to work effectively in a group atmosphere or contribute to a high performance recovery team (see Gordon 1983; Clark and Westrum 1989; Kaplan 1990). Participants also lacked understanding of several non-biological variables influencing successful recovery, especially socio-economic, organizational, and political considerations. Nor did they seem to understand how policy is made and altered during implementation. Each of these aspects of the ferret recovery program was underexamined and, at best, poorly addressed.

Establishing special teams and groups is pointless if they are simply used as extensions of a formal bureaucratic program structure. To be effective, special groups and teams should:

- Maintain informal structures that are flexible enough to enable specialists to enter the group, address specific, short-term problems, and then leave.

- Be led by a charismatic leader with strong personal, administrative, and technical skills.

- Maintain a goal orientation.

- Be buffered from outside pressure.

- Focus on generating ideas, communicating, evaluating, and cooperating, rather than on following standard operating procedures, rules, and regulations.

- Carefully manage conflict to keep it productive while avoiding the dangers of groupthink.

- Be staffed by a stable and diverse group of professionals included solely on the basis of their expertise, skills, and commitment to species recovery. Special teams should avoid including members simply to represent a particular organization or interest. Doing so only opens the door for the development of conflict over secondary goals such as control of the program.

Program Control

Whenever several parties are working toward a common goal, issues of control, based on power and authority, often come to dominate their interactions. Organizations can be used to control resources, influence policies, and ultimately change preferences and outcomes. As a result, organizations are a potential source of enormous social power. Power and authority relationships among key figures evolve as programs are carried out, although in many instances, traditional organizational relations and preexisting laws, regulations, and mandates strongly influence the development of relationships as new programs are initiated.

Gaining control of the program appeared to be important to several key figures in ferret recovery efforts. As a result, recovery goals frequently became secondary to, or were displaced by, control goals—often leading to unproductive conflict with other key figures. Although the intensity of the struggle may vary, conflicts over control of the ferret program have been constant and pervasive, influencing most aspects of the program. As a result, maintaining good relations between key figures has been a major challenge.

WGF controlled the national ferret recovery program for two reasons: FWS transferred lead agency status to WGF under authority of the ESA, and Wyoming controlled the ferrets. Having authority without accountability provided Wyoming with an avenue for greater control. Moreover, people in a position to change the situation often chose not to rock the boat. Even when they disagreed with policies enacted by Wyoming, several FWS representatives told us they believed that the perception of discord would produce bad publicity for the program and reduce fundraising opportunities. Thus most WGF decisions were unchallenged.

Maintaining program control appears to have become an important, often overriding, goal of WGF (May 1986; Weinberg 1986; Clark and Harvey 1988). Perhaps this battle was based at least partly on a conservative states' rights philosophy founded on the belief that state governments, as opposed to the federal government, should have primary authority within state boundaries. The Meeteetse ferrets were discovered during the Sagebrush Rebellion, in which western states—with the support of the Reagan administration—strongly opposed increasing federalism. Indeed, this philosophy may have been a primary factor in the transfer of lead status to Wyoming.

Several examples illustrate the importance of control in the ferret recovery program. First, BFAT was established in 1982. The group was chaired by a WGF employee, the field representative was a WGF employee, it was largely composed of political representatives without ferret expertise, and it did not use outside experts effectively. The purpose of the WGF-controlled advisory team was to make recommendations for ferret policy to WGF. By stacking the deck, BFAT consolidated control and bolstered legitimacy for the desired policies of the dominant organization.

Second, the events surrounding the inception of captive breeding further consolidated control. Although several biologists advocated captive breeding and reintroduction, or translocating wild juveniles, as a precaution against catastrophic loss of the Meeteetse population (Bogan 1985; Carpenter 1985; Clark 1985; Wemmer 1985), WGF initially resisted. In September 1984, under pressure from other participants, Wyoming decided to investigate captive breeding, but officials were reluctant to let ferrets leave the state because they feared loss of control (Carr 1986; May 1986; Weinberg 1986). In early 1985, after FWS representatives agreed to fund a new captive facility in Wyoming, WGF decided to capture six animals after the October 1985 census (Thorne 1987). Original plans to obtain genetically diverse animals were compromised after WGF failed to heed warnings by field researchers that the population was crashing during the summer of 1985 (Forrest et al. 1985; Clark 1989). Canine distemper was diagnosed in the ferrets that fall. As a result, the captive breeding stock originated from only a few animals collected largely from one small area. In 1988 plans were made to expand the breeding program to other facilities outside Wyoming. WGF was able to extend its control over those out-of-state facilities by refusing to sign the FWS recovery plan unless a WGF representative was named chair of the committee that coordinated the captive breeding program between institutions (Crowe 1988). Participating institutions were also forced to sign agreements giving WGF ultimate control over protocols and animals in their facilities and were excluded from the federal money that funded the Wyoming breeding center. The total captive population was controlled by Wyoming in two ways: one of their members chaired the captive breeding committee, and even though ferrets were permitted to be housed at other institutions, the subpopulation at each "satellite" center was too small to run effective comparative experiments. Thus all breeding and other population management depended on the main concentration of ferrets in Wyoming.

A third example of the importance of control in ferret recovery efforts concerns reintroductions. WGF moved to control reintroductions and associated activities by restricting the ability of participants to conduct research over which WGF had little control, such as radio telemetry. This move was accomplished by a series of restrictions and last-minute changes to previously agreed upon protocols, all of which prevented researchers from effectively collecting telemetry data (Reading 1991). Similarly, communication with local livestock operators was controlled by WGF. All interactions between researchers and the local people therefore required WGF as an intermediary—which often resulted in program delays, distrust between operators and researchers, and, ultimately, poor relations among agencies and poor monitoring of released animals.

Several researchers, especially nongovernmental participants, have been effectively forced out of the ferret recovery program by WGF— including many who collected the bulk of published ecological knowledge about the species. WGF is extremely sensitive to criticism. On several occasions researchers critical of WGF's positions or methods were asked (or forced) to leave the Wyoming recovery program. (Some were even asked to leave the state and not return.) This was true even of participants who informally approached WGF about their concerns over the program's science. Exclusion ranged from creating an unpleasant working atmosphere and placing restrictions on research that effectively compromised methods and experimental design to refusing to grant or renew research permits and even threats of arrest (Clark 1989, Reading 1991).

WGF was also able to delay reintroduction efforts in Montana, with assistance from Montana politicians, because WGF controlled the ferrets. WGF requested the delay ostensibly because more animals were needed for the captive breeding program, for the Wyoming reintroduction program, and for research (Petera 1992; Stephens 1992; Thorne and Russell 1992). Yet all the captive facilities approached or exceeded capacity at that time (Anonymous 1992), suggesting that WGF may have had other motives. Because of its control over the captive source population, WGF may be able to continue influencing reintroduction in Montana and other states.

Fourth, control issues have not been limited to the Wyoming program. As in Wyoming, the issue of states' rights versus federalism has played a role in Montana. MFWP has long sought state control of threatened and endangered species programs within Montana. FWS's reluctance to cede authority of the ferret program in Montana to MFWP has undoubtedly been influenced by its experiences in Wyoming.

Finally, by closing the planning and implementation process, WGF severely restricted its ability to incorporate or effectively solicit outside expertise and ideas. This constraint led to inevitable delays that were very costly. In 1985, for example, Wyoming refused to accept the field biologists' warning that the wild population was declining (Forrest et al. 1985). Further delays in initiating captive breeding probably reduced the number of founding individuals (and the genetic variability) (Miller 1988). Delays in incorporating breeding recommendations may have led to further genetic loss by failure to capture all potential matings. An estimated five founders have contributed to the current 350 adults (December 1992 IWG Meeting). The effects of inbreeding, already apparent, may develop into a major problem.

Struggles over control of recovery programs can severely reduce their efficiency (programs will be more costly than necessary) and their effectiveness (programs will be less successful at restoring species). Clearly, resolving or at least mitigating conflict over program control is desirable. To do so, however, requires early resolution of issues of authority and ultimate program control, strong leadership at all levels within the program, formation and utilization of special groups and teams as discussed earlier, occasional use of more formal processes (such as environmental dispute resolution), and formation of permanent multidisciplinary advisory teams with oversight authority. Such advisory teams could operate in cooperation with the agency mandated to implement the Endangered Species Act (the FWS for terrestrial species in the United States), acting as a source for recommendations and evaluation. Among other things, such a group could help maintain a recovery focus, address the many interdisciplinary problems facing recovery, and ensure the use of rigorous, reliable science.

The Role of Science

Of all the factors influencing endangered species recovery, the biological and technical aspects are the most obvious. (See almost any U.S. Fish and Wildlife Service recovery plan.) There is no doubt that these issues must be addressed or the result is almost certain to be failure. But often so little is known about a rare endangered species' biology and ecology, at least initially, that recovery efforts are infused with great uncertainty and complexity. The need for reliable knowledge is therefore critical, but it is often compromised and complicated by resource constraints, by the urgency of the recovery task, and by concerns over the effects of intrusive research on a rare species. Moreover, nontechnical concerns—such as program control struggles, organizational constraints, professional competency, and competing goals—often inhibit the collection or appropriate use of reliable knowledge in recovery programs.

A major issue among the key parties in the ferret recovery program has been the appropriate use of science and rigorous scientific methods. WGF has been criticized for its practice of excluding participants with opposing viewpoints, its unwillingness to incorporate more and better science into the program, and its inappropriate use of science and scientific methodologies, especially in the field of conservation biology. (See Clark and Harvey 1988; Greater Yellowstone Coalition 1990; Reading 1991; Miller et al. 1993.) Although recovery efforts in Montana have been more open, the importance of incorporating more rigorous science into the program appears to be little recognized by many participants there as well.

Telemetry research conducted during the 1991 reintroduction of ferrets into Shirley Basin, Wyoming, illustrates several of these points (see Reading 1991). With little notice or discussion and despite strong protests by researchers from FWS, universities, and NGOs, WGF altered reintroduction protocols that had been previously agreed upon. Although protocol called for radio tracking of all forty-nine ferrets reintroduced during 1991, for example, WGF prevented several animals from being radiocollared and forced the early removal of collars from several other ferrets. The reasons for the agency's decisions varied, but all were based on concerns over adverse impacts on ferret survival. Telemetry researchers strongly contested these decisions, arguing that the available data did not support WGF's contentions: there were simply no data to suggest that their telemetry unduly affected the ferret's survival. More important, they argued that without radio tracking there would be virtually no data on mortality factors or survival, making it almost impossible to judge the success of reintroduction. The issue was resolved by permitting only fifteen animals to keep their collars.

Ferret recovery efforts have also suffered because WGF has resisted incorporating adaptive scientific management. For example, although designated "experimental" under Section 10(j) of the ESA, the Wyoming reintroduction incorporated very little science or experimentation in the first two releases. Moreover, the research that did occur was weakened by restrictions placed on researchers by WGF (Reading 1991). Part of the rationale for WGF's unwillingness to include more science in the program became apparent at a 1990 meeting in which a high-level WGF official stated that he would not pay attention to ongoing ferret reintroduction research because WGF had already decided how ferrets would be released. At another meeting, the same official said that he did not want to compare two reintroduction techniques because an alternative technique might work better than the method proposed by Wyoming.

The lack of experimental design and rigorous science in reintroductions has been readily apparent. To date there has been no comparison of release techniques. For the first reintroduction of black-footed ferrets in 1991, WGF decided to release all forty-nine animals via elevated release cages on site and with supplementary feed (Oakleaf et al. 1991). Similarly, in 1992 all ninety ferrets were released from cages provided with supplementary feed (Oakleaf et al. 1992). Thus, after two years of reintroductions, we learned virtually nothing about methods of improving release techniques to increase survival. In addition, much of the "research" that has occurred in association with the reintroductions is ex post facto and so flawed in experimental design that the findings are meaningless (see Oakleaf et al. 1993).

Assessing the success of reintroduction has been similarly flawed. WGF has used spotlighting and snow tracking to examine survival. Both techniques have led to biologically insupportable conclusions. For example, snow tracking in March 1992 found a minimum of three ferrets still present in the core area. Solely on the basis of this evidence, WGF announced that at least seven animals had survived the winter. Later that year, four ferrets were discovered by spotlighting and again WGF concluded that "at least seven ferrets from the 1991 release survived their first winter, more than fulfilling the objective of 10 percent over-winter survival" (Obrecht 1993:2). The reasoning behind the agency's extrapolation from the observed three or four ferrets to seven has never been provided.

A more fundamental ideological difference underlies the conflict over the appropriate amount and intrusiveness of research in recovery efforts. WGF favors limited and less intrusive research in which ferrets are handled and monitored as little as possible (Reading

1991). This approach may have developed in response to earlier, negative, misinformed publicity proposing that the decline of ferrets in the wild resulted in part from excessive research. No evidence supports this assertion. In contrast, FWS, supported by many NGOs, has argued for more intrusive experimentation and science—testing different release protocols, examining survival among animals given different kinds and amounts of prerelease training and conditioning, and especially using radio telemetry to track the fate of released ferrets (Reading 1991). Such an approach would include continual testing, refinement, and retesting of different release techniques and should result in continuously improving reintroduction success rates. Apparently, adaptive scientific management will have to await testing on future reintroduction sites, for Wyoming has shown little interest in making this approach a priority.

Finally, ferret recovery efforts have suffered from a poor understanding of conservation biology, the limited use of experimentation, and the inappropriate use of science and scientific methods by several parties. These problems are compounded by severely limited external review by experts in biometry, small-population ecology, and conservation sciences (see the large number of unpublished reports and internal documents cited in Oakleaf et al. 1991, 1992, 1993), and they have seriously constrained the program's ability to assess data quality, to gain reliable knowledge, and to move toward more biologically and ecologically based program management.

Very few of the data collected since ferrets were removed to captivity in 1985 have been published or rigorously reviewed by outside professionals. Instead, much of the information appears in unpublished, internal agency reports or WGF completion reports (to comply with ESA Section 6 funding). Continual citation of unpublished and thus unreviewed reports lends a false air of credibility to manuscripts that often never proceed past a draft stage. Reports should be exposed to critical outside evaluation and review in a timely fashion. Otherwise the program is guided by inaccurate and misleading information with potentially inefficient, ineffective, or even disastrous consequences.

Because the risks are so high (the program's failure could lead to extinction), reliable information is essential to endangered species restoration efforts. The ability of recovery programs to acquire and use such information depends on a number of factors:

• Use of rigorous scientific methods

- Rigorous review by qualified professionals outside the program—especially biometricians—of all research proposals, data collection methods, analytic techniques, discussions of data, and proposed implications for future studies and management actions

- Encouragement of all parties constantly to assess the reliability of information used in the program

- Periodic formal program evaluations

- Development of expertise in state-of-the-art ecological theory and methods

- Incorporation of outside knowledge and expertise, including qualified researchers

- Creation of an incentive system that rewards both rigorous data collection and critical self-appraisal

One way to accomplish this last imperative is to incorporate acquisition of reliable knowledge explicitly into indices of program success.

Measuring Program Success

As Berg (1982:169) observes: "By different criteria, the same reintroduction may be judged as a failure or a success." Different participants often use different measures, standards, or time frames for measuring a program's success. Sometimes one method of evaluation can mask deficiencies in another.

Counting only ferret numbers in captivity and reintroductions has created an image of a black-footed ferret success story. These numbers are certainly a source of pride and represent the culmination of tremendous human effort on behalf of the species. Yet the numbers alone are deceiving because they do not account for inefficient utilization of resources, costs to recovery in terms of genetic diversity lost, opportunity costs lost, losses in public support for the program, future success, or many other facets of the recovery program that are essential if the program is to ultimately reach the objective of species recovery.

Some conservationists, including many in the ferret program, argue that only narrowly defined, "objective," biological criteria, such as number of animals surviving or population growth, should be used to assess a program's success or failure (see Reading 1993). Yet continuing to rely on narrow definitions of success ignores several important facts. First, a recovery program may succeed *despite* poor organizational arrangements. Aspects of the biology and ecology of the species, such as a high reproductive rate or improved habitat quality, could compensate for organizational shortcomings. Although it has been apparent to many working on ferret recovery that the program is less than ideal, ferret numbers are in fact increasing in captivity and reintroductions are showing signs of success. As newspaper and magazine articles around the nation celebrate the "black-footed ferret success story" and cite the ferret recovery program as a model to be emulated (Cohn 1991; Ricciuti 1991; Obrecht 1993), it seems that traditional measures of program success are insufficient. If organizational considerations are not better addressed, species whose biology and ecology are less amenable to recovery may go extinct.

Second, recovery programs may succeed in establishing a population without the use of rigorous science, experimental design, or adequate documentation. But the program would fail to learn how to replicate or improve future efforts. The current reintroduction of ferrets into Wyoming, for example, has produced wild-born litters and some survival despite poor organizational arrangements, inadequate science, and weak documentation (Reading 1991; Miller et al. 1993).

Third, since conservation dollars and managers' time are both limited, efficiency is important both to the program at hand and to other programs that must compete for similar time and funding. Recovery programs are generally expensive endeavors, and, relative to other programs, the ferret program has consumed enormous resources. Reintroduction programs should therefore strive to develop and refine organizational arrangements gradually to reduce the resources they consume (that is, increase efficiency) and increase the chances for successfully reestablishing the species in the shortest possible time (that is, increase efficacy). Throwing large amounts of money at a problem with no accountability attached to it creates the potential for empire building.

Fourth, programs should consider organizational costs and benefits. If relations among the key parties are damaged by conflict (even for a program deemed successful by the biological standards cited earlier), the strain could affect future programs. For example, stressed relations between FWS and MFWP that developed during the grizzly bear (*Ursus arctos*) and wolf (*Canis lupus*) recovery programs affected relationships between participants in the ferret program as well. In one instance, upper-level managers were forced to delay discussion of the ferret program when a conflict over grizzly bear management led to the expulsion of some federal employees from MFWP offices.

Fifth, just as strained organizational relations can have lasting negative effects, poor public relations can result in increased opposition to future recovery programs in the region and elsewhere. Conflict between the local public and past reintroduction and recovery programs has harmed ferret recovery efforts in Montana (Reading and Kellert 1993). Although Montanans expressed satisfaction with a local black-footed ferret management plan, for example, they remained opposed to ferret reintroduction because they feared future changes to the plan, as had occurred in an earlier elk (*Cervus elaphus*) reintroduction into the area. Moreover, careless public relations in the Wyoming ferret reintroduction resulted in a warning letter from the Wyoming Farm Bureau Federation to the Montana Farm Bureau Federation (Krause 1991) that heightened suspicion and weakened support among ranchers in Montana.

Sixth, controversial programs can result in undesirable changes in conservation laws or public attitudes. As a result of their experiences with endangered species recovery programs, including the ferret program, state agencies, ranchers, and resource extractors are arguing, searching, and lobbying for substantive changes to weaken the ESA. (See, for example, Endangered Species Roundtable 1992.)

And finally, reintroduction programs that fail to reestablish self-sustaining populations may nevertheless be successful in educating the local public about the importance of conservation, in developing values supportive of conservation efforts, in protecting or restoring habitat, in forming or strengthening local conservation programs, and in learning important information about the biology and ecology of the species and reintroduction techniques (Kleiman et al. 1991).

Because of the many implications of poorly performing endangered species restoration programs, measuring a program's success must move well beyond the narrow traditional indices of biology and ecology:

- Program effectiveness must be assessed at several levels using several different criteria. For example, a recovery program should be evaluated in terms of biological criteria (such as numbers of animals surviving or reproducing), valuational criteria (such as the change in public support or acceptability), and organizational criteria (such as the effectiveness of formulating and implementing recovery options).

- Participants in recovery programs must recognize the importance of nonbiological, nontechnical measures of program success. Broad, interdisciplinary education is therefore crucial.

- There must be a change in incentives and accountability so that the value of a supportive public, good working relations, and high efficiency are explicitly recognized.

Until we appreciate the full array of factors influencing program success, endangered species conservation and recovery efforts will continue to be costly, polemic, controversial, and ultimately unsuccessful at stemming the loss of biological diversity.

Lessons

Endangered species recovery is a highly complex undertaking. Only by cooperatively organizing expertise under a well-designed program structure will the effectiveness and efficiency of the process improve. Traditional bureaucratic arrangements, however, are poorly suited to address the complexity, uncertainty, and urgency of the recovery task. As a result, programs suffer from a variety of problems: inefficiency, communication breakdowns, inability to overcome the uncertainty and complexity of the recovery challenge, and more. Addressing these concerns requires the establishment of alternative informal structures, such as parallel organizations, better matched to the recovery challenge. Unfortunately, the effectiveness of parallel organizations is limited by the people involved. Often it is constrained by members chosen simply because they represent particular interests or are sympathetic to the organization in power. Instead, parallel organizations should be staffed by professionals with pertinent expertise and a commitment to recovery. The effectiveness of these groups should be periodically reviewed by qualified people outside the program.

Organizational effectiveness also requires that issues of program control be avoided or resolved early in a program's development using equitable and effective solutions. Otherwise, unproductive conflict and strained relations may develop and recovery goals may be displaced by program control goals. To avoid these problems may require formation of multidisciplinary advisory teams. Such a proposal is discussed in more detail in Miller et al. (in press b).

A professional advisory team can also provide rigorous external reviews of the scientific and technical portions of recovery programs. The importance of generating reliable knowledge cannot be overemphasized. Recovery programs should strive to obtain pertinent ecological data and to model both populations and habitat using state-of-the-art ecological methods (see Chapter 12). The ability to do so is often constrained, however, by organizational, sociopolitical, and economic considerations and by the competence of recovery program personnel. Addressing these constraints is a major focus of several chapters in this book. In addition, both continual education and use of outside expertise, including grants and contracts to qualified researchers, should be actively pursued.

Species can be recovered in spite of inefficiencies, an unsupportive or hostile public, poor relations among key parties, inadequate science, inappropriate organizational arrangements, and more—thereby concealing the program's inadequacies. Instead of simply tallying numbers of animals in captivity or in the wild, a program's success should be measured by a variety of standards, including program efficiency (measured in both time and resources), relations with the public, relations between key parties, biological and ecological knowledge gains, and genetic considerations.

Despite the glowing stories of success in newspapers, newsletters, and popular magazines, ferrets remain far from recovered and the ferret recovery program suffers from a host of professional and organizational inadequacies. As the program expands to new areas and incorporates new figures, there is an excellent opportunity to develop a national focus and avoid the deep-seated problems of past efforts by reorganizing the program's structure. Without reorganization, the same problems could easily recur as the ferret program moves into new states.

Note

1. Most are at the Wyoming Game and Fish Department's Sybille Wildlife Research and Conservation Unit near Wheatland. Smaller populations are kept at the National Zoo's captive breeding facility in Front Royal, Virginia; at the Henry Doorley Zoo in Omaha; at the Louisville Zoo in Louisville, Kentucky; at the Cheyenne Mountain Zoo in Colorado Springs; at the Toronto Zoo in Ontario; at the Phoenix Zoo; and in arenas stocked with prairie dogs near Pueblo, Colorado, by researchers from the U.S. Fish and Wildlife Service's National Ecology Research Center.

Acknowledgment

We thank Denise Casey, Steven Forrest, and Lou Hanebury for their critical review of the chapter.

References

Anonymous. 1992. Captive breeding report. *Drumming Post* 5(3):4–8.

Berg, W. E. 1982. Reintroduction of fisher, pine marten, and river otter. In *Midwest Furbearer Management*, G. C. Sanderson (ed.). Wichita: Kansas Chapter of The Wildlife Society.

BFF Interstate Coordinating Committee. 1989. Minutes of 5–6 December 1989, Golden, Colo.

Bogan, M. A. 1985. Needs and directions for future black-footed ferret research. In *Black-Footed Ferret Workshop Proceedings*, 18–19 September 1984, Laramie, S. H. Anderson and D. B. Inkley (eds.). Cheyenne: Wyoming Game and Fish Department.

Carpenter, J. W. 1985. Captive breeding and management of black-footed ferrets. In *Black-Footed Ferret Workshop Proceedings*, 18–19 September 1984, Laramie, S. H. Anderson and D. B. Inkley (eds.). Cheyenne: Wyoming Game and Fish Department.

Carpenter, J. W., M.J.G. Appel, R. C. Erickson, and M. N. Novilla. 1976. Fatal vaccine–induced canine distemper virus infection in black–footed ferrets. *Journal of the American Medical Association* 165:961–964.

Carr, A., III. 1986. Introduction. *Great Basin Naturalist Memoirs* 8:1–7.

Casey, D. E., J. DuWaldt, and T. W. Clark. 1986. Annotated bibliography of the black-footed ferret. *Great Basin Naturalist Memoirs* 8:185–208.

Clark, T. W. 1985. Organizing for endangered species recovery. Paper presented at Wildlife Management Directions in the Northwest Through 1990, 2–5 April 1985. Missoula: Northwest Section of The Wildlife Society.

———. 1988. The identity and images of wildlife professionals. *Renewable Resource Journal* (Summer):12–16.

———. 1989. *Conservation Biology of the Black-Footed Ferret (Mustela nigripes)*. Wildlife Preservation Trust International Special Scientific Report no. 3. Philadelphia: Wildlife Preservation Trust, International.

Clark, T. W., and J. R. Cragun. 1991. Organization and management of endangered species programs. *Endangered Species Update* 8(8):1–4.

Clark, T. W., and A. H. Harvey. 1988. Implementing endangered species recovery policy: Learning as we go? *Endangered Species Update* 5:35–42.

Clark, T. W., and R. Westrum. 1989. High–performance teams in wildlife conservation: A species reintroduction and recovery example. *Environmental Management* 13:663–670.

Clark, T. W., R. Crete, and J. Cada. 1989. Designing and managing successful endangered species recovery programs. *Environmental Management* 13:159–170.

Cohn, J. P. 1991. The new breeding ground: Zoo research programs are rescuing endangered species from extinction. *National Parks* 65(1/2):20–25.

Collins, A. R., J. P. Workman, and D. W. Uresk. 1984. An economic analysis of black–tailed prairie dog (*Cynomys ludovicianus*) control. *Journal of Range Management* 37:358–361.

Crowe, D. 1988. Memo from Wyoming Game and Fish Department to U.S. Fish and Wildlife Service, 1 March 1988.

DeBlieu, J. 1991. *Meant to Be Wild: The Struggle to Save Endangered Species Through Captive Breeding.* Golden, Colo.: Fulcrum Publishing.

Endangered Species Roundtable. 1992. Endangered Species Act: Time for a change. Unpublished white paper.

Forrest, S. C., T. W. Clark, L. Richardson, and T. M. Campbell. 1985. Black-footed ferret population status at Meeteetse, Wyoming. Unpublished report. Cheyenne: Wyoming Game and Fish Department.

Galbraith, J. R. 1977. *Organizational Design.* Reading, Mass.: Addison-Wesley.

Gordon, J. R. 1983. *A Diagnostic Approach to Organizational Behavior.* Boston: Allyn & Bacon.

Greater Yellowstone Coalition. 1990. *GYC's First 1990 Program Progress Report.* Bozeman, Mont.: Greater Yellowstone Coalition.

Janis, I. L. 1972. *Victims of Groupthink: A Psychological Study of Foreign-Policy Decisions and Fiascoes.* Boston: Houghton Mifflin.

Kaplan, R. 1990. Collaboration from a cognitive perspective: Sharing models across expertise. *Environmental Design and Research Association* 21:45–51.

Kellert, S. R., and T. W. Clark. 1991. The theory and application of a wildlife policy framework. In *Public Policy Issues in Wildlife Management,* W. R. Mangun and S. S. Nagel (eds.). New York: Greenwood Press.

Kleiman, D. G., B. B. Beck, J. M. Dietz, and L. A. Dietz. 1991. Costs of a reintroduction and criteria for success: Accounting and accountability in the golden lion tamarin conservation program. In *Beyond Captive Breeding: Reintroducing Endangered Mammals to the Wild,* J.H.W. Gipps (ed.). Oxford: Clarendon Press.

Krause, R. L. 1991. Letter from Wyoming Farm Bureau to L. Frank, Montana Farm Bureau Federation, Bozeman, 7 November 1991.

Krueger, K. 1988. Prairie dog overpopulation: Value judgment or ecological reality? *USDA Forest Service General Technical Report.* RM-154:39–45.

Lacy, R., and T. W. Clark. 1989. Genetic variability in black-footed ferret populations: Past, present, and future. In *Conservation Biology of the Black-Footed Ferret,* U. S. Seal et al. (eds.). New Haven: Yale University Press.

Marsh, R. E. 1984. Ground squirrels, prairie dogs and marmots as pests on rangelands. In *Proceedings of the Conference for the Organization and Practice of Vertebrate Pest Control,* 30 August–3 September 1982, Hampshire, England. Fernherst, England: ICI Plant Protection Division.

May, R. M. 1986. The black-footed ferret: A cautionary tale. *Nature* 320:13–14.

Miller, B. J. 1988. Conservation and behavior of the endangered black-footed ferret (*Mustela nigripes*) with a comparative analysis of reproductive behavior between the black-footed ferret and the congeneric domestic ferret (*Mustela putorius furo*). Unpublished Ph.D. dissertation, University of Wyoming.

Miller, B. J., S. H. Anderson, M. W. DonCarlos, and E. T. Thorne. 1988. Biology of the endangered black-footed ferret and the role of captive propagation in its conservation. *Canadian Journal of Zoology* 66:765–773.

Miller, B., C. Wemmer, D. Biggins, and R. Reading. 1990. A proposal to conserve black-footed ferrets and the prairie dog ecosystem. *Environmental Management* 14:763–769.

Miller, B., D. Biggins, L. Hanebury, and A. Vargas. 1993. Reintroduction of the black-footed ferret. In *Creative Conservation: Integrative Management of Wild and Captive Animals,* G. Mace and P. Olney (eds.). London: Chapman & Hall.

Miller, B., R. Reading, C. Conway, J. Jackson, M. Hutchins, N. Snyder, S. Forrest, J. Frazier and S. Derrickson. In press b. Improving endangered species programs: Avoiding organizational pitfalls, tapping the resources, and adding accountability. *Environmental Management.*

Oakleaf, B., B. Luce, and P. Hnilicka. 1993. Survival and movement of captive raised black-footed ferrets reintroduced in Shirley Basin, Wyoming: A comparison between black-footed ferrets released with and without telemetry collars. In *Black-Footed Ferret Annual Completion Report, 1992: Reintroductions.* Cheyenne: Wyoming Game and Fish Department.

Oakleaf, B., B. Luce, E. T. Thorne, and D. Biggins. 1991. Black-footed ferret reintroduction in Wyoming: Project description and 1991 protocol. Cheyenne: Wyoming Game and Fish Department, U.S. Bureau of Land Management, and U.S. Fish and Wildlife Service.

Oakleaf, B., B. Luce, P. Hnilicka, D. Biggins, and S. Torbit. 1992. Black-footed ferret reintroduction in Wyoming: Project description and 1992 protocol. Cheyenne: Wyoming Game and Fish Department.

Obrecht, J. 1993. Black-footed ferret. *Drumming Post* 6(1):2.

O'Meilia, M. E., F. L. Knopf, and J. C. Lewis. 1982. Some consequences of competition between prairie dogs and beef cattle. *Journal of Range Management* 35:580–585.

Perrow, C. 1986. *Complex Organizations: A Critical Essay.* 3rd ed. New York: McGraw-Hill.

Petera, F. E. 1992. Letter to G. Buterbaugh, U.S. Fish and Wildlife Service, Denver, 3 March 1992.

Quade, E. S. 1975. *Analysis for Public Decisions.* New York: American Elsevier.

Reading, R. P. 1991. Experimental black-footed ferret reintroduction in Wyoming and its implications for Montana. Unpublished report. Malta: Montana Bureau of Land Management.

———. 1993. Toward an endangered species reintroduction paradigm: A case study of the black-footed ferret. Unpublished Ph.D. dissertation, Yale University.

Reading, R. P., and T. W. Clark. 1990. Black-footed ferret annotated bibliography, 1986–1990. *Montana BLM Wildlife Technical Bulletin* 3:1–22.

Reading, R. P., and S. R. Kellert. 1993. Attitudes toward a proposed black-footed ferret (*Mustela nigripes*) reintroduction. *Conservation Biology* 7:569–580.

Reading, R. P., T. W. Clark, and S. R. Kellert. 1991. Toward an endangered species reintroduction paradigm. *Endangered Species Update* 8(11):1–4.

Ricciuti, E. R. 1991. The comeback kid. *Wildlife Conservation* 94(4):52–61.

Schön, D. A. 1983. *The Reflective Practitioner: How Professionals Think in Action.* New York: Basic Books.

Sills, D. 1957. *The Volunteers.* New York: Free Press.

Somit, A. 1986. Bureaucratic pathology, public administration, and the life sciences. In *Biology and Bureaucracy*, E. White and J. Loser (eds.). Lanham, Mass.: University Press of America.

Stephens, S. 1992. Letter to M. Sullivan, governor of Wyoming, Cheyenne, 5 March 1992.

Thorne, E. T. 1987. Captive propagation of the black-footed ferret in Wyoming. In *American Association of Zoological Parks and Aquariums, Regional Conference Proceedings,* 12–14 April 1987, Colorado Springs. Syracuse, N.Y.: AAZPA Publications.

Thorne, E. T., and B. Oakleaf. 1991. Species rescue for captive breeding: Black-footed ferret as an example. In *Beyond Captive Breeding: Reintroducing Endangered Mammals to the Wild,* J.H.W. Gipps (ed.). Oxford: Clarendon Press.

Thorne, E. T., and W. C. Russell. 1992. Black-footed ferret (*Mustela nigripes*): SSP report. In *Black-Footed Ferret Workshop Draft Report.* Apple Valley, Minn.: IUCN Captive Breeding Specialist Group.

Thorne, E. T., and E. Williams. 1988. Diseases and endangered species: The black-footed ferret as a recent example. *Conservation Biology* 2:66–73.

Uresk, D. W., and A. J. Bjugstad. 1983. Prairie dog ecosystem regulators on the northern high plains. In *Proceedings of the Seventh North American Prairie Conference,* 4–6 August 1980, Southwest, C. L. Kucera (ed.). Springfield: Missouri State University.

U.S. Fish and Wildlife Service (FWS). 1988. *Black-Footed Ferret Recovery Plan.* Denver: U.S. Fish and Wildlife Service.

———. 1992. *Endangered Species: A Summary of the Act and Service Activities.* Denver: U.S. Fish and Wildlife Service.

Warwick, D. 1975. *A Theory of Public Bureaucracy: Politics, Personality, and Organization.* Cambridge: Harvard University Press.

Weinberg, D. 1986. Decline and fall of the black-footed ferret. *Natural History* 95:62–69.

Wemmer, C. M. 1985. Black-footed ferret management and research: Views of a zoo biologist. In *Black-Footed Ferret Workshop Proceedings,* 18–19 September 1984, Laramie, S. H. Anderson and D. B. Inkley (eds.). Cheyenne: Wyoming Game and Fish Department.

Westrum, R. 1986. Management strategies and information failures. Paper presented at NATO Advanced Research Workshop on Failure Analysis of Information Systems, Bad Winsheim, Germany, August 1986.

Williams, E. S., E. T. Thorne, M.J.G. Appel, and D. W. Belitch. 1988. Canine distemper in black-footed ferrets (*Mustela nigripes*) from Wyoming. *Journal of Wildlife Diseases* 24:387–398.

Wilson, J. Q. 1980. *The Politics of Regulation*. New York: Harper.

Yaffee, S. L. 1982. *Prohibitive Policy*. Cambridge: MIT Press.

5

The Yellowstone Grizzly Bear Recovery Program
Uncertain Information, Uncertain Policy

David J. Mattson and John J. Craighead

Management of Yellowstone's threatened grizzly bear population (*Ursus arctos horribilis*, Figure 5-1) has a long history of often reported but poorly understood controversy. This history, though troubled, is a rich source of lessons about the management of a threatened species. Perhaps in no other arena is there so much need for adaptive organizational learning—given that the room for management error is so small and the consequences irreversible. By legal as well as biological definitions, endangered and threatened species are at great risk of extinction and require high-performance management to ensure their survival.

Here we examine individual and organizational behavior associated with grizzly bear conservation and offer some lessons that, if applied, could enhance the prospects for the bear's survival. Consequently we emphasize the performance of key figures and government agencies holding responsibility for research and management and will not dwell on the natural history of the population. Clearly, our analysis is bounded by our experiences and vantage point. Different experiences would highlight different factors. Accordingly, the views we express are not those of any organization or agency. Our experience with research and management of Yellowstone's grizzly bear population spans thirty-five years (1959–1993) and two major research projects that have involved us at the heart of the bear's recovery process. Thus we offer our informed perspectives with the hope that, through wise application of the Endangered Species Act, Yellowstone's grizzly bears will survive in perpetuity.

FIGURE 5-1.   The grizzly bear. (Photo by Bart O. Schleyer, Interagency Grizzly Bear Study Team.)

Some History and Context

To understand the present manifestations of grizzly bear research, management, and the recovery program in the Greater Yellowstone area, it is essential to understand some of the biology and history that has structured management programs and shaped risks and benefits for researchers and managers. In this section we review some relevant history and identify key players and processes.

Grizzly Bears and Humans

Grizzly bears are large-bodied, wide-ranging omnivores (Craighead 1976; Craighead and Mitchell 1982; Blanchard 1987; Blanchard and Knight 1991a), and, because of their size and feeding habits, they are direct and often physically threatening competitors for virtually all foods valued by humans (Mattson 1990; Craighead et al. 1994). Although the grizzly rarely causes human death or injury, the bear's size, claws, dentition, and demeanor make it at times, and under certain conditions, a threat to human safety (Herrero 1985).

Almost universally the cause of grizzly bear population declines and extinctions has been direct human-caused mortality and associated habitat usurpation by European settlers (Craighead and Mitchell

1982). Habitat modification is usually only a peripheral factor (Mattson 1990). Even though conflict between humans and grizzlies did intensify after European settlement, grizzly bears and aboriginal North Americans were dissociated even under primeval conditions (see Driver 1969), plausibly due to competition. The documented decline of grizzly bears started shortly after contact with European settlers (Storer and Tevis 1955; Brown 1985), and by the 1920s grizzly bears were extinct in over 95 percent of their former range in the conterminous United States. At that time they existed in only small isolated populations in remote areas (Servheen 1989).

Although habitat relationships of grizzly bears are complex, in general their survival is a function of a simple negative relationship with humans. This pattern is evident worldwide, regionally, and within occupied grizzly bear habitat (Craighead 1980; Mattson 1990). Today grizzly bears survive only in extensive wilderness areas—in the conterminous United States principally as four "populations" in the northern Rocky Mountains (Craighead and Mitchell 1982, Servheen 1989). Yellowstone's grizzly bears are one of the two largest remaining populations and comprise a significant portion of the estimated remaining seven to nine hundred bears (Servheen 1989).

Recent History

The first intensive field study of Yellowstone grizzly bears was conducted by Dr. John Craighead and Dr. Frank Craighead, Jr., from 1959 through 1970. (For an overview see Craighead et al. 1974, 1994; Craighead and Mitchell 1982.) During this study, grizzly bears concentrated at open-pit refuse dumps and used human-origin foods, both inside and outside the park. Late in the Craighead study (1967–1970), Yellowstone National Park (YNP) instituted a policy of "natural regulation" based on the agency's interpretation of a paper by Starker Leopold and his colleagues (Leopold et al. 1963) that called for national parks to preserve "the primitive scene." This policy included the termination of herd reduction programs for elk (*Cervus canadensis*) and bison (*Bison bison*) in the park and, more important for the bears, abrupt closure of the open-pit refuse dumps.

Later two conflicting hypotheses were posed—one by the Craighead research team based on analysis and interpretation of their extensive data base and one by Glen Cole, a research biologist for YNP, based on tenuous and typically anecdotal information. According to the Craigheads, abrupt dump closures would have substantial negative effects on the grizzly bear population by lowering the stability, quality, and abundance of food. This would cause wider-ranging movements into unprotected areas as well as increases in human/bear conflicts. As a consequence, they postulated catastrophic effects, one of which would be greatly increased bear mortality. Alternatively, Cole postulated that dump closures would not have catastrophic consequences to the grizzly bear population because additional bears, not accounted for in the Craigheads' research, did not use the dumps. These unaccounted bears, as well as declines in density-dependent mortality, would buffer negative population effects from dump closures.

Research over the last twenty years provides no conclusive basis for rejecting either hypothesis regarding the ultimate effects of the dump closures on population viability—due partly to the long life span of grizzly bears and the correspondingly long time required for reliable determination of demographic effects. In all other respects, however, the Craigheads' hypothesis was confirmed. Food supplies declined; population structure destabilized; and movements and conflicts with humans increased (Mattson et al. 1991, 1992; Craighead et al. 1994; Craighead and Craighead 1971). More important, of the 222 to 350 grizzlies estimated to have existed in 1967 (Cole 1973; Craighead et al. 1974), at least 127 were killed over a three-year period (1969–1972) during and immediately after dump closures, primarily by legal hunters and management agencies (Craighead et al. 1974, 1988). After 1970, YNP administrators prohibited further research by the Craighead team in the park and no other research program was in place until 1974 to monitor the effects of dump closures. This lapse in data collection (1971–1973) greatly complicated the interpretation of demographic data in the years that followed.

The Interagency Grizzly Bear Study Team (IGBST) was created in 1973 as a result of the controversy over dump closures. The team is administered by the National Park Service (NPS) and initially consisted of three members, one each from the NPS, U.S. Fish and Wildlife Service (FWS), and U.S. Forest Service, plus cooperating members from the states of Montana, Idaho, and Wyoming (Committee on the Yellowstone Grizzlies 1974). Since its inception, the IGBST has been the primary source of information on Yellowstone's grizzly bear population and has borne primary responsibility for research. The IGBST's formal membership now consists entirely of two NPS personnel, however, although research conducted by the Wyoming Game and Fish Department and Idaho Cooperative Park Studies Unit is informally included in the IGBST program. Not only does the team's current composition no longer reflect the diversity specified in its initial charter but, more important, it does not reflect the original recommendation of the National Academy of Sciences (NAS) committee that reviewed research needs in 1974 (Committee on the Yellowstone Grizzlies 1974)—that most research be conducted by nonagency scientists. Although the history of this attrition in personnel and inattention to NAS recommendations is obscure, it has resulted in a research team with less technical breadth and depth and subject to control by a single agency.

The Yellowstone grizzly bear population was listed as threatened under provisions of the Endangered Species Act (ESA) in 1975. Ecosystem-wide coordination of grizzly bear research and management was at first exercised by the Interagency Steering Committee, a group of research administrators and midlevel managers instituted in 1975 (Committee on the Yellowstone Grizzlies 1974). Among its other tasks, the Steering Committee provided general review and direction for the IGBST research program. The ESA also specified management oversight by the FWS that has been amplified since 1979 by the activities of a Grizzly Bear Recovery Coordinator, whose tasks include writing and revising a recovery plan for grizzly bears in the conterminous United States. Initially little was done after listing to modify existing management practices, although efforts to reduce availability of human-related foods, especially in YNP, were continued (Primm 1993).

In 1982 the IGBST and the Interagency Steering Committee recognized a crisis—based on two years of high bear mortality (1981 and 1982) and a draft demographic analysis suggesting that the population was still in decline. This high mortality was human-induced and attributed to continuing availability of human foods in communities surrounding YNP and on Forest Service lands. In response to this crisis, the Steering Committee dissolved itself and recommended the formation of a replacement committee comprised of high-level managers with decision-making authority to deal more effectively with the problems facing Yellowstone's grizzly bears. The Interagency Grizzly Bear Committee (IGBC), which embodied the recommendations of the Steering Committee, was subsequently formed in 1983 by an interagency memorandum of agreement (Primm 1993). More restrictive management of human-related foods and more consistent protocols for management were instituted under the auspices of the IGBC and elaborated in the most recent (1986) Interagency Grizzly Bear Guidelines (Mealey 1986). Subsequent to implementation of these guidelines, grizzly bear mortality is believed to have declined substantially.

In 1988 wildfires burned a large percentage of Yellowstone's occupied grizzly bear habitat—the most substantial restructuring of habitat and food resources in the Yellowstone area since closure of the open-pit dumps. Again, as with the dump closures, the ultimate effects of these fires on the population are likely to be revealed only after a long period of time, some twenty to thirty years. Although positive or neutral effects have been postulated (Greater Yellowstone Coordinating Committee 1989; Blanchard and Knight 1991b), there is a strong basis for hypothesizing long-term negative effects that derive from potential redistribution of bears to unburned areas on the ecosystem's less protected periphery and reduction in abundance of high-quality foods such as whitebark pine (*Pinus albicaulis*; Mattson and Reinhart 1994).

Current management of Yellowstone's grizzly bears is directly shaped by just a few key figures and documents. The 1993 recovery plan, individual National Forest Management Act forest plans, the Interagency Grizzly Bear Guidelines, and ESA Section 7 consultations with the FWS provide the primary guidance for grizzly bear management in the Yellowstone area. Because Section 7 consultation draws heavily on the recovery plan and was used in formulating forest plans and the guidelines, the recovery plan has served as the key document guiding grizzly bear management in the Yellowstone ecosystem (Primm 1993). The strategic direction of management, therefore, is largely dependent on the FWS recovery coordinator, who is the sole author of the recovery plan. There has been no formal recovery team or fixed program for formal consultation to guide the recovery effort or the development of the recovery plan.

The current political climate of the Yellowstone ecosystem is relatively conservative. Seven of the nine current congressional delegates from the states of Idaho, Montana, and Wyoming with districts that include Greater Yellowstone are conservatives with a history of favoring extractive, consumptive uses of public lands and limited government regulation on private land (Primm 1993). There are also currently a number of local industries—timber, livestock, mining, tourism—reliant on public lands for resources and income. In addition, the regional real estate industry has a stake in the transfer of private lands from commercial agriculture to smaller residential or recreational holdings. These political and economic interests have a direct stake in defining the problems associated with managing Yellowstone's grizzly bears and favor continuing land uses that have historically caused human/bear conflicts and grizzly bear mortality (Primm 1993).

When Are Grizzly Bears "Recovered?"

As we will see, there are two major definitions of the current grizzly bear situation. Most of the debate surrounding these competing definitions has been mired in technical rationalization and has not addressed the associated policy and organizational issues. We argue that uncertainty in scientific information and in ESA policy design, deference to current economic interests, different scientific conceptualizations, and discretion in agency implementation are at the heart of divergence between these two definitions. Consequently in this section we elucidate these normative issues and show how they have shaped perceptions and management of the Yellowstone grizzly bear population's status. Clearly, there is no single way of seeing the plight of Yellowstone's grizzly bears (see Clark 1993). Nevertheless, some perceptions are shaped more by biological information—that is, reliable information—than others.

From the outset, ESA policy design has introduced uncertainty into grizzly bear management by not specifying vital normative criteria. In particular, ESA policy prescription does not specify:

1. Timeframes for management (whether we are managing for persistence of species over 100, 200, or 1000 years)

2. Levels of confidence at which we want to manage (whether we want 70, 90, or 99 percent confidence of having a species persist for the specified time)

3. A protocol for managing uncertainty or allocating burden of proof

4. Explicit biological goals at the population level (whether we are managing demographically, genetically, or both)

Although biologists have something to contribute, all of these issues are fundamentally valuational and need to be clarified in future ESA legislative design. Furthermore, all of these issues are absolutely critical to providing a clear direction for management. Without legislative specificity, ESA implementation is left up to agency discretion and hence to political and economic influences.

The importance of uncertainty to management cannot be over-stated (Panem 1983; Committee on Applications of Ecological Theory to Environmental Problems 1986; Walters 1986). The available scien-tific information is virtually never so definitive and explicit that only one course of management action is obvious. Yet only one manage-ment choice and implicit hypothesis will be the most biologically de-fensible at any one time, recognizing that as the amount of reliable knowledge increases, the biologically most defensible management choices may change. Although the ESA calls for use of "the best avail-able data," it does not clearly require that threatened and endangered species management be based on the biologically most defensible data or hypotheses. Nor does it require that, in the face of scientific uncertainty, deference should be given to the postulated needs of species at risk.

Management of Yellowstone's grizzly bears over the last twenty years has been further complicated by considerable scientific uncer-tainty over the status, trend, and distribution of the population—pri-marily as a consequence of long-duration research projects that have yielded relatively small data sets. From the outset, sampling problems were severe because grizzly bears exist at low densities and are hard to observe in Yellowstone's primarily forested environment. Biases and incompatibilities among data attributable to different methods are not trivial if for no other reasons than the 35-year span of research in-volving two different studies and the study area's large size (about 23,000 km² according to Blanchard et al. 1992). But these inherent problems have been compounded by the behavior of agency man-agers and biologists—particularly during the transition between the Craigheads' and IGBST research programs. Most bears marked during the 1960s were either killed or had their individual markers removed, while research within the park was limited to indirect techniques based on visual observations. Not until 1974 was the IGBST allowed to radio-mark bears and collect the data necessary to develop indi-vidual life histories comparable to those obtained by the Craighead team. Thus, despite the duration and intensity of grizzly bear research in the Yellowstone ecosystem, it has been difficult to obtain reliable and precise estimates of population and habitat status and trends. De-spite three decades of intensive research and several million dollars' expense, we remain plagued by considerable scientific uncertainty. Ironically, a significant part of this uncertainty has resulted from the decisions of agency managers and biologists themselves—especially during the critical transition years following dump closures—and despite the recommendations of the NAS review committee.

As might be expected, this uncertainty in science and ESA's design sets the stage for ambiguous and contradictory definitions of problems facing Yellowstone's grizzly bears. It fosters disputation of management goals and appropriate methods for monitoring and managing the population and its habitat. It shifts much of the burden of making or reshaping ESA policy onto the shoulders of agencies and civil servants. As a result, the real ESA policy is made by those who implement it (Mazmanian and Sabatier 1983; Houck 1993; Chapter 17). This broad agency discretion in policy implementation, rather than bridging conflicting interests, more often contributes to polarization of constituencies and issues. More important, this combination of limited information and agency prerogative fosters management that is detrimental to the grizzly bear population's long-term survival—especially in cases where an agency has its own interests at stake or is captive to the interests of consumptive resource users (Yaffee 1982; Primm 1993).

Agencies often view uncertainty as license for preserving the status quo or promoting politically expedient ideologies (Pfeffer et al. 1984; Primm 1993). For many years the Forest Service underestimated the importance of information suggesting that road access associated with timber harvest harms bears; indeed, the service even used tenuous assessments to argue the benefits of clearcutting to Yellowstone grizzlies. More recently, because the FWS does not address the potential effects of scientific uncertainty in its 1992 recovery plan revision, criteria for recovery are, not surprisingly, compatible with existing consumptive land uses. Similarly, the definition and designation of prioritized management situation areas within the grizzly bear recovery zone (Mealey 1986) are also strikingly compatible with preexisting human activities and land uses. Not only have these designations failed to reflect the intent of Congress when it called for critical habitat designations (U.S. Congress 1977), but they are not based on any substantive biological evaluation of the habitat (Mattson and Reid 1991). This kind of rationalized agency management is a predictable consequence of the marked incompatibility between grizzly bears and humans and the opportunities to exploit uncertainty in order to provide services, jobs, and access for humans.

Thus it is not surprising that there are two major competing views of the status and future prospects of the Yellowstone grizzly bear population. One view is implicit to the most recent recovery plan (U.S. Fish and Wildlife Service 1993); the other is described in publications by Mattson and Reid (1991), Shaffer (1992), and Craighead et al. (1994). The FWS definition given in the 1993 recovery plan postulates that the Yellowstone grizzly bear population will be recovered within thirty to forty years. Furthermore, the plan assumes that population sizes derived from an analysis done in 1978 (Shaffer 1983)—assuming 95 percent probability of persistence for one hundred years—are appropriate standards for assessing long-term population viability. Furthermore, the draft FWS recovery plan assumes that potential catastrophic and long-term habitat changes are not explicitly relevant to assessing population viability or setting recovery goals. Finally, it assumes that indices with indeterminate or unexamined biases are adequate for monitoring population size, trend, and distribution in a way that produces reliable knowledge.

The alternative definition is that long-term viability should be assessed according to more recent advances in conservation biology that invoke longer time frames—especially for long-lived animals like grizzly bears—and that much larger populations and new management concepts are needed to ensure long-term persistence (Chapter 12). This definition asserts that potentially catastrophic or long-term habitat changes are important to assessing long-term population viability and setting recovery goals (see Pimm et al. 1988; Thomas 1990). Finally, proponents of this definition argue that status and trends of the grizzly bear population and its habitat should be monitored using direct estimates of critical parameters or by using indices with few (and controllable) biases.

These two definitions are in direct conflict. In practical terms, the FWS definition implies that the Yellowstone grizzly bear population has recovered—given that the population currently satisfies most if not all of the recovery criteria written into the most recent draft recovery plan (Craighead et al. 1994). Conversely, the alternative definition implies that the population is not recovered or that its status cannot be determined with sufficient reliability to allow any confident conclusions and, as a consequence, the Yellowstone population is still in need of full ESA protection.

These two definitions require substantially different management responses. The FWS definition clearly accommodates the status quo and seems to accommodate further human development. For this reason, the FWS definition is likely to be favored by local industries and hence garner the greatest local political support. In contrast, the alternative definition at most accommodates the status quo but implies reduction of current levels of human activity. This definition assumes that successful conservation of Yellowstone's grizzlies will require new economic concepts (Rees and Wackernagel 1992; Wackernagel and Rees 1992) and the adoption of less exploitative natural resource management (Craighead et al. 1994).

Differences in these two definitions are obviously a result of different values and perceptions. A bear population with the same characteristics could be considered "recovered" or "not recovered" depending on the time frame and level of confidence used for the evaluation and whether future catastrophes or uncertainties in habitat conditions were considered. In other words, the clash of definitions is rooted at least partly in divergent values focused on questions of how long we want grizzly bears in the Yellowstone ecosystem and how confident we want to be that a viable population will exist at the end of that time. From this perspective, it is likely that formal adoption of one or the other definition will ultimately be determined by the political clout of the opposing parties, including interests inside and outside the management agencies. If these political interests prevail, rather than the interests promoting conservation science, there is little reason to think that the Yellowstone grizzly bear population will survive.

Other differences in these two definitions relate to substantive scientific and biological issues that could well be addressed in the technical arena—such as the capacity of present analyses to provide reliable estimates of relevant population parameters and the adequacy of indices proposed for monitoring the population. But to date there has been very little scientific discussion of these issues. Not only has access to relevant data been limited by the government, but there has not been sufficient time to fully evaluate proposed uses of indices that were revealed in the recently revised recovery plan.

Thus, ESA policy design allows competing constituencies (such as the commodity extraction industries) to exploit scientific uncertainty in order to define the problems facing Yellowstone's grizzly bears in terms of their own perceptions of the situation. This reality has often been clouded by ad hoc technical rationalization. Because pro-consumption industries have historically held the greatest political clout and have consequently been able to shape agency behavior in favor of their agendas, grizzly bear management in the Yellowstone ecosystem has typically favored definitions that make the least imposition on resource extraction (Primm 1993). Under this arrangement, recovery will continue to be defined in the politically and economically most expedient ways—so long as ESA policy does not deal with critical normative issues (such as the parameters of population viability) and so long as the collection of critical technical information remains solely the domain of management agencies.

Implementing the Recovery Program

Formal endangered species policy is written in the ESA, but the real policy is made by the ESA's implementers. This section describes the reality of policy implementation as we have experienced and understood it. It obviously reflects our values, expectations, and perspectives, which are not shared by everyone involved in grizzly bear conservation. As mentioned earlier, several factors play themselves out in implementation of grizzly bear recovery: ESA policy prescriptions, science, agency discretion, and various political and economic interests. Lindblom (1980) calls this the "play of power." The interaction of these forces—not only through problem definition but in more specific aspects of the grizzly bear recovery program—are illustrated here by examining information flows, information feedback loops, monitoring of implementation, and professional behavior (see Morgan 1986 and Clark et al. 1989).

We have two reasons for taking this approach to examining implementation. First, there have been successes in managing Yellowstone's grizzly bears, such as the cleanup of human facilities and the adoption of a uniform standard for management. Yet it is also clear, as is usually the case in complex programs, that the program could have done significantly better. This, in part, speaks to our expectations. By examining both the strengths and weaknesses of past efforts, we may arrive at a better guide to future action. Second, it is also likely that a new era of more difficult management lies ahead (Mattson and Reid 1991). If this is true, then it is vital to review and discuss how information is collected, used, and legitimized in the grizzly bear recovery program as a basis for meeting foreseeable as well as unforeseeable management challenges.

Information Flows

The information used by key decision makers responsible for the recovery of Yellowstone's grizzly bears originates from a variety of sources including science, law, agency culture, and society at large

(Clark 1993). Information is typically distilled (reduced in detail and biased) as it is passed into agencies and transmitted through the hierarchy from technical specialists to top-level decision makers (Ingram 1973; Sabatier 1978). Some distillation is necessary, of course. Researchers and lower-echelon specialists need to reduce the information's volume and detail to accommodate the limited time and technical expertise of higher-level managers. Even so, most analysts agree that substantial distortion of information impedes effective management and can even put resources at risk (Ingram 1973; Sabatier 1978; O'Reilly 1980; Clark 1993). This is an especially apt concern in the case of Yellowstone's grizzly bears. Our experience suggests that biological information suffers great distortion—unlike the information received directly from elected politicians or that which is rooted in agency culture (see Primm 1993).

Ultimately, we found that most of the distortion was attributable to four factors: insufficient time and resources to locate and integrate available information, both theoretical and specific; insufficient training to provide the necessary theoretical and conceptual context; the agency's resistance to outside collaboration; and subtle but real coercion of lower echelons to meet the expectations of supervisors (Ingram 1973; Sabatier 1978; Pfeffer et al. 1984). When all four factors were at work, gross distortion of basic biological knowledge was virtually guaranteed.

Lack of adequate time and training clearly takes its toll on information collected by agency specialists. Many technical specialists have a short tenure but are required to become experts on numerous highly complex issues. An agency biologist in the Yellowstone ecosystem, for example, may have to deal with issues relating not only to grizzly bears but also to great gray owls (*Strix nebulosa*), boreal owls (*Aegolius funereus*), migratory birds, American marten (*Martes americana*), elk, black bears (*Ursus americanus*), moose (*Alces alces*), bison, and others. This predicament is sometimes exacerbated by both official and unofficial impediments to soliciting assistance or information from outside the agency, often asserted in terms of agency pride ("We can take care of the job ourselves"). Occasionally outside consultation is viewed as a constraint to established agendas. This is particularly true when these agendas emphasize consumptive resource uses potentially at odds with management actions favoring grizzly bears. Regardless, these factors hamper both the collection of reliable and timely information by agency specialists and the incorporation of new techniques and research findings into the management process.

Coercion is probably responsible for most distortion of biological information. Typically this coercion is a result of budget allocations, job performance standards, and agency cultures. Nonetheless it results in highly selective transmittal or reconfiguration of information (see Bacharach and Lawler 1980). Within the Forest Service, for example, there are often direct and quantifiable performance standards that pertain to timber production goals. These standards are often at odds with standards pertaining to grizzly bear conservation, which are much more ambiguous and more easily circumvented. Similarly, in most ranger districts far more budget dollars are tied to timber programs than to grizzly bear management. Thus endowment of a position and potential for career advancement are more closely tied to facilitating timber production than to conserving grizzly bears. Supporting these observations, Twight and Lyden (1989) found that the Forest Service was captive to timber interests, although Brown and Harris (1992) report encouraging evidence that this bias is slowly changing. In the same way, NPS career opportunities and budgets are more closely tied to providing visitor services, protection, and law enforcement than to enhancing conditions for grizzly bears. This allocation of emphasis by both the Forest Service and NPS can be fundamentally at odds with the ESA's directives for ranking the conservation of grizzly bears over other resource uses on public lands within the Yellowstone recovery area (Keiter 1991; Kuehl 1993).

On some occasions, coercion is more direct—particularly when the technical expertise and information compiled by a grizzly bear specialist conflicts with existing or proposed management programs. (See Bacharach and Lawler 1980; Twight and Lyden 1988; Brown and Harris 1992.) Direct coercion is often couched in terms of "being a team player" and typically involves slander, threats to career advancement, and curtailment of travel and professional responsibility. We personally observed at least eight instances of such coercion related to grizzly bear management. Coercion directed at people outside of agencies is not uncommon. It occurs through public news releases and interviews, and involves attacks on the person's character and motivations as well as misrepresentation of professional work. The cases of Adolf Murie after publication of *Ecology of the Coyote in the Yellowstone* (1940) and the Craigheads after submission of their grizzly bear management report (*Management of Bears in Yellowstone National Park*, 1967; Frome 1984) are two noteworthy examples.

Self-justification is a typical reflexive response for most organizations when threatened (Yaffee 1982; Mazmanian and Sabatier 1983).

In the case of Yellowstone's bears, for example, NPS newsletters and news releases in the 1970s frequently referred to a recovered back-country population of three hundred grizzlies, with little or no reference to the source or reliability of the information. Similarly, NPS news releases and information papers until recently stated that black bears were numerous in Yellowstone's backcountry when, in fact, no substantive data support this statement and approximately 340 black bears were known to have been killed during the 1960s. When these public-management tactics are employed, both the public and resource administrators are done a great disservice.

Sometimes grizzly bear researchers and technical specialists, perhaps in common with researchers in any politically heated situation (Westrum 1986), produced and passed on distorted or incomplete information to managers. Most notably, we observed inclusion of unreliable or misleading information in key management efforts or documents such as the 1993 FWS recovery plan and official cumulative-effects analysis (Mattson and Knight 1991). We could only partly attribute these deficiencies to coercion or inadequate time and resources.

Other causes of technical failure, especially among grizzly bear specialists and researchers, were internal in nature: an unwillingness to consult and collaborate with other (especially nonagency) experts and scientists and an inability or unwillingness to incorporate recent advances in population and habitat analyses and conservation biology into their work. For example, long-term viability of the Yellowstone grizzly bear population has been only cursorily addressed in formal management. Little of the technical expertise or conceptual advances in population viability modeling have been used; instead, point population estimates have been the main criterion for judging future population prospects. Consequently, management issues related to long-term population viability have not been adequately communicated to managers; instead, short-term considerations have been emphasized (Mattson and Reid 1991). Similarly, agency researchers and biologists have failed to directly address the importance of uncertainty to long-term management and some have even displayed outright hostility to the idea.

This unwillingness to incorporate the work of nonagency scientists has also been evident in approaches to evaluating grizzly bear habitat. Although a variety of techniques are applicable, depending on the specific management or study area conditions (Mattson and Knight 1989), supervised satellite mapping is perhaps the best approach to evaluating extensive nonroaded areas (Craighead and Craighead 1991; Craighead et al. 1982). Although the technology for this approach has been available for over a decade, agencies have either ignored it, or, when embracing it, have reinvented the wheel. The pioneering work by nonagency scientists using LandSat data to evaluate grizzly bear habitat in the northern Rocky Mountains of Montana (Craighead et al. 1982) has neither been used nor credited in subsequent Forest Service efforts. The most plausible explanations for this behavior are shortcomings in technical training, job insecurity, influence from high-level administrators, or simply an attempt to take undeserved credit.

Management is only as good as the information it gets—and uses. Our experience suggests that much of the information reaching grizzly bear managers in the Yellowstone area has not been reliable or complete. This problem is rooted in distortion arising from inadequate time, resources, and training, but also, more importantly, from coercion of agency technical specialists by upper echelons. Thus much of this problem can be ascribed to the actions of key decision makers and scientists, whether by design or not.

Information Feedback Loops

The speed at which feedback travels to key decision makers from politicians, from other organizations, and from the biological system itself in response to management actions varies according to a predictable pattern (Ingram 1973; Sabatier 1978). Virtually all managers and researchers involved with Yellowstone's grizzly bears seem to receive faster—and more forceful—feedback from their parent or government agencies, from various interest groups, and from elected political agents concerning the implications of their actions than they do from their own investigations of biological systems. As a result, managers appear to respond almost wholly to the social process rather than to biological knowledge when making decisions. This is inevitable and not necessarily harmful. But problems sometimes arise that thwart biological solutions to the problems. For example, most nongovernmental organizations and political agents have partisan agendas that may or may not coincide with ESA or reflect the most reliable biological knowledge. Consequently, the ability of managers to implement ESA with fidelity depends on how influenced they are by such feedback versus how well ESA's prescriptions are formally and informally supported by the agency's culture and structure. Our experience tells us that many grizzly bear managers and their administrators are highly responsive to local interests, because agency feedback emphasizes natural resource consumption, job continuity, and commodity interests over conservation.

Feedback concerning the impact of management decisions on Yellowstone's grizzly bears often carries little weight simply because of the long time lags involved. Until the recent completion of a thorough assessment by Craighead et al. (1994), for example, there had been insufficient information to assess conclusively the impacts of the late 1960s open-pit dump closures. Thus the clear consequences of an agency's management actions to Yellowstone's grizzly bears are rarely revealed to decision makers during their tenure, while the political consequences are immediate and forcefully manifest. This asymmetry suggests that managers and researchers responsible for threatened and endangered species should be shielded from local, short-term interests. Instead they should operate in an environment where reliable biological knowledge is given more weight, including an explicit mechanism for dealing with scientific uncertainty.

But even when management agencies are set up to use biological information, their response is still only as good as the information they receive. In the case of the Yellowstone grizzlies, feedback on population and habitat status has come largely from "unconfirmed" demographic indices. The current draft FWS recovery plan proposes to set targets and monitor population status based on annual counts of "unduplicated" females accompanied by cubs-of-the-year, the distribution of females with young within the recovery area, and known annual grizzly bear mortality—with the proviso that these indices are treated as running means (U.S. Fish and Wildlife Service 1993). As indicated earlier, all of these indices are severely biased and not yet subject to widespread critical review by the scientific community (Craighead et al. 1994). The FWS recovery plan also fails to set goals and provide a means for maintaining or monitoring habitat. This shortcoming could be catastrophic for the bears. Without some indication of habitat status and trend, untenable habitat conditions could go undetected by demographic indices until it is too late for remedial action. Furthermore, this shortcoming is at odds with the ESA prescription. Thus it appears that Yellowstone grizzly bear management will continue to be limited by unreliable and inadequate feedback from the biological system itself—the bear population and its habitat.

Monitoring Implementation

The FWS, which has monitored grizzly bear management since the Yellowstone population was listed under the ESA in 1975, has manifested all the shortcomings described above. This behavior, in turn, has led to highly variable performance (arguably a result of differences in state-level supervision) at higher administrative levels and resulted in the politicized context. Yaffee (1982) and Houck (1993) view the behavior of the FWS as the means by which a prohibitive law has been transformed into a discretionary permit system. As Houck (1993:358) states: "The risk is, of course, that the compromises arrived at through agency interpretation of the [ESA] will only prolong the process of species decline."

The inability of the FWS to provide consistent monitoring has been manifest in several ways. Virtually all research results to date suggest a profound incompatibility between grizzly bears and humans (Craighead 1980; Mattson 1990), for example, and timber harvesting, road building, human visitation, and development of private lands have all increased substantially over the last three decades. Yet the FWS issued only two Section 7 jeopardy opinions in twenty-one consultations on projects for one state's portion of the Yellowstone area between 1977 and 1992. Section 7 consultation apparently gave priority to reducing agency conflict and as a consequence primarily supported the status quo or nominal mitigation (Houck 1993). The FWS also failed to take the lead in defining issues, estimating problems, and providing or accepting reliable methods to monitor the population and its habitat. This shortcoming has been especially clear in formulation of the most recent revision of the FWS's grizzly bear recovery plan.

The 1974 NAS committee also provided review and nonbinding oversight as part of an investigation into the controversy surrounding the late 1960s closures of open-pit dumps. Such a committee is ostensibly a good idea—especially if it can provide a truly independent review focusing on scientific issues and using reliable information. Yet we observed that such a committee can also be subject to personal and partisan influences—especially when the agency that solicits the committee's input is the financial sponsor and the members are selected by an active participant in the controversy. This bias can influence the committee's schedule, its composition, its scope of investigation, and, ultimately, its final recommendations. Indeed, in this instance we observed that the NAS committee did not provide unbiased scientific input or professional scientific follow-up, showing that this approach to independent review is not always successful and that the NAS should establish better guidelines for nonbinding reviews in which a person is pitted against his government or employer.

Management of Yellowstone's grizzly bears under ESA clearly lacks adequate monitoring. This shortcoming can largely be attributed to the same basic mechanisms in the FWS that have made the Forest Service and NPS vulnerable to outside pressure, agency culture, and political bias. Again, remedial action will require restructuring of the agency's internal system of rewards and punishment and reconfiguration of feedback to emphasize biological information. Adequate monitoring may also require periodic intervention or review by some independent nonagency panel—given the resistance to conservation that inevitably arises when traditional economic activities are threatened (Heinen and Low 1992).

Professional Behavior

Although there is considerable debate on the appropriate roles of professionals in policy implementation (Clark 1986, 1988; Lindblom 1980; Mazmanian and Sabatier 1983; Harmon 1989), the requirements of professionalism are relatively clear—especially with regard to management of Yellowstone's grizzly bear population. Not everyone involved in grizzly bear research or management shares our views, but most people would concur that our society expresses its priorities through laws passed by the legislative branch, approved and executed by the executive branch, and upheld by the judiciary. We recognize that the process is not always clear. Nevertheless, we believe that it is incumbent upon government employees to execute these laws as fully and consistently as possible, according to the stated or apparent statutory intent, however this is interpreted in light of a sincere analysis. When this is prevented by higher authority, then it becomes the employee's duty to place professional ethics above loyalty to an agency. Admittedly, the employee then runs great risk of an official reprimand or even more severe reprisals should he or she persist. Faced with a highly bureaucratic agency backed by a closed culture and a substantial budget, the employee has little chance of winning.

Civil servants inevitably create the real policy while implementing laws (Lindblom 1980; Mazmanian and Sabatier 1983; Chapter 17), and real grizzly bear conservation policy in the Yellowstone area has been largely shaped by the nature of government bureaucracy and the influence of conservative, pro-consumption, local constituencies, all to the detriment of grizzly bears. Ironically, our society has given preservation of Yellowstone's grizzlies relatively clear priority through the ESA. Thus, unlike the ambiguous and sometimes conflicting laws governing other resources (such as multiple-use laws), managers have a message direct from Congress regarding their management of

Yellowstone's grizzly bears. The discrepancy between the ESA's prescription and the implemented policy suggests that some involved in management or research of Yellowstone's grizzly bears may not have exhibited the highest professional behavior.

It is also incumbent on public employees to use all available information to develop effective research and management programs as a means of fulfilling the ESA. This means that agency researchers and managers should solicit input and assistance from as many sources as possible. Any civil servant who uses data or position for self-aggrandizement or to secure a career breaches the public trust. In this light, we suggest that professionals are obligated to speed the flow of reliable knowledge into and out of agencies, to incorporate the best available information into management, and to update actions as new information becomes available. The agency's structure and culture must not only allow but promote this kind of professional behavior.

Our experiences suggest that too many employees in government agencies involved with Yellowstone grizzly bear research and management have been motivated by careerist goals rather than the primacy of law and information. Similarly, agencies have been resistant to adaptive problem solving: people who exposed problems were removed, silenced, or at best stonewalled; responsibility was ignored or compartmentalized; free, open communication within or among organizations was discouraged or, at most, grudgingly allowed; faulty technical performance was covered up; new ideas were crushed or, at best, perceived as problematic. Westrum (1988) calls such an organization "pathological" or "calculative," depending on the severity of the characteristics. Advocating the use of biological information in either organization is a decidedly risky professional proposition. It is not surprising, then, that a number of employees did not exhibit the professional ideal outlined above in grizzly bear research and management.

What are the prospects for changing the performance of pathological or calculative agencies by appealing to some professional ideal? The chances are poor to nil, for people's behavior has generally evolved to favor short-term self-interest over long-term benefits to society (Heinen and Low 1992). We observed that many people holding professional ideals who entered the agencies responsible for grizzly bear management were either forced out, retained permanently in a lower echelon, or obliged to abandon their idealism (see Westrum 1988). Conversely, many of those who "succeeded" seemed to be the careerists and pragmatists—the ones who were highly responsive to the expectations and agendas of superiors or influential partisans. The agency's selection for these opportunists created self-reinforcing feedback, which in turn created substantial conformity of thought (Twight and Lyden 1988, 1989; Brown and Harris 1992). This groupthink typically favored natural resource consumption over conservation and inculcated pathological or calculative behavior. We also observed professional "goal displacement"—that is, the pursuit of nontask goals, such as personal gain or agency control, over the goal of grizzly bear recovery.

Although one can profess high ideals of professional behavior, this is clearly not enough to change the way that agencies implement the ESA and manage grizzly bears in the Yellowstone area. People's natural risk aversion often combines with an agency's calculative and pathological structure and culture to select for individuals in key management and research positions who are responsive to agency values. Resolution of this problem lies in changing the agencies and creating systemic risks and benefits such that managers are naturally led to pursue fulfillment of the ESA. Resolution also entails transferring more research responsibility to outside researchers.

Lessons

Most of the lessons from our experiences researching and managing Yellowstone's grizzly bear population relate to the distortion and withholding of information, the system of rewards and punishment for action advocates and key decision makers, the allocation of research responsibilities, and the allocation of implementation or normative responsibilities. The eleven recommendations that follow address these four issues and are arranged more or less in order of importance. Some of these lessons are more likely to be the focus of action than others, given the political climate at local and national levels. Although this chapter is not the ideal forum for explicating the details of implementation improvements, we nonetheless offer these lessons in the hope that they can contribute to improved management of endangered species.

First, a thorough analysis of relevant statutes is necessary so that ESA policy prescription is made explicit—with subsequent hierarchical ranking of ecosystem-wide resource goals and then the development of spatially and temporally explicit management objectives. This step is critical because of the overlap and contradiction among the many laws governing natural resource management in the Yellowstone area and the need to put the ESA in a clear legal and social context. This analysis should be part of regional resource and forest planning and would involve ongoing assessment and updating by specialists. Although the strategic "vision" process carried out by the Greater Yellowstone Coordinating Committee in 1990 is a theoretical example of this type of effort, most analyses conclude that it did not address statutory obligations adequately or explicate an ongoing adaptive process (Clark and Minta 1994; Lichtman and Clark in press).

Second, a rigorous analysis of agency job performance standards and budget allocations is needed to ensure concordance with priorities and objectives derived from the analysis of governing statutes discussed above. Without this step, there will be continued differences in management goals and government agencies will continue to be vulnerable to capture by local partisans. Most likely, a semiautonomous board or panel is needed to perform this function.

Third, increased emphasis and new avenues are needed for soliciting information and expertise from outside the management or research agencies involved in grizzly bear recovery, as well as from nontraditional internal sources. This step could be facilitated by making pertinent changes in job descriptions and job performance standards and by making time and funds available. This end would also be served by diversification of grizzly bear research to include nonagency scientists as recommended by the Committee on the Yellowstone Grizzlies (1974). Research functions are currently concentrated in one federal agency and, consequently, are vulnerable. In this regard, the current research program should be subject to a periodic, critical, and wide-ranging review of its goals, methods, and performance.

Fourth, the flow of newsworthy scientific findings and resource information through agency channels must be upgraded. To accomplish this, public relations writers and staff journalists should adhere to the highest ethical code and undergo thorough review. Again, reports and books treating scientific subjects should also be submitted to outside peer review.

Fifth, better personnel training is needed—both at higher academic levels and on the job—with an emphasis on professionalism, a good knowledge of the law, and an understanding of public policy processes (Clark 1992). Discussions are especially needed in the areas of fulfilling laws, taking personal risks, and emphasizing the completion of tasks when pursuing career advancement. Accordingly, desired behavior should be rewarded by higher job performance ratings or other incentives. Although we recognize that substantive change in agency structure and culture will not be achieved simply by appealing to selfless idealism, our experience suggests that if given the opportunity most people are motivated by appeals to higher ideals. Other things being equal, they will act with greater integrity if such behavior is expected.

Sixth, an explicit protocol is needed that incorporates scientific uncertainty into the estimation of biological parameters (including future population prospects) and into management practices. Given the need for responsive and effective management of threatened and endangered species, this would entail an adaptive (Walters 1986) and an open learning approach (Clark et al. 1989) in which management actions are treated as hypotheses that require testing. Hypotheses selected as a basis for management action should also be the most defensible in light of the most current and reliable knowledge.

Seventh, improved methods of monitoring both the grizzly bear population and its habitat are required. Ideally the population should be monitored by direct estimates of relevant parameters when possible and acceptable indices when not. This would clarify the reliability of population monitoring data and their application to management.

Eighth, any revision of the ESA should include a clear statement of what the act is intended to achieve and why. Moreover, it should explicitly address the normative issues of time frame, level of confidence, and burden of proof. In other words, over what time frame and at what level of confidence are we managing for species or population survival? Should the burden of proof lie with demonstrating that proposed actions *will* harm grizzly bears or *will not* harm grizzly bears? The answers to these questions have paramount implications to management and ultimately species survival. Consequently, we think these issues should not be left solely to the discretion of civil servants but should be addressed by a forum of conservation biologists and high-level policymakers.

Ninth, a recovery team rather than a single recovery coordinator should guide recovery efforts. A single person in such a key position is likely to deal with partisan pressures less successfully than a team and could cause considerable damage to the recovery process if he or she falls prey to venal considerations. Moreover, a team offers a wider range of knowledge, insights, and ideas and would be correspondingly more successful at dealing with the overall complexities of the recovery process (Clark and Westrum 1989; Chapter 14).

Tenth, a permanent oversight committee should be established to address problems in all threatened or endangered species cases. This committee could examine the progress of recovery and the performance of key figures and organizations. Ideally, the committee would consist of highly qualified specialists, well versed in biological, organizational, and policy issues, who had a reputation for fairness.

Eleventh, interagency management groups like the Interagency Grizzly Bear Committee (IGBC) should be given a monetary stake in management of the Yellowstone grizzly bear population that is independent of individual agency budgets. While we recognize that this would necessitate some administrative restructuring and could lead to management that is related more to empowerment of competing agencies or individuals than to grizzly bear recovery, some incentive is needed to replace the IGBC's current approach.

Recovery of threatened and endangered species requires high-performance management. There is little room for error. With the irretrievable loss of biological diversity at stake, government agencies must strive to perform effectively in the face of uncertainty, complexity, and political pressure. The future of Yellowstone's grizzly bear population is contingent upon the ability of research organizations to collect reliable biological knowledge and the ability of management to incorporate this information expeditiously into decisions and actions. We contend that biological knowledge, rather than political knowledge, should exert the greatest influence on management. There have been many calls for greater sensitivity to the human dimension in threatened and endangered species management in this country. We agree with this need—but only insofar as this receptiveness to human issues supports the ESA mandate for species or population viability. In other words, we are deeply concerned about calls for further compromise and accommodation. Yellowstone's grizzly bears may have already been compromised to the brink of extinction.

The people and agencies responsible for the future of Yellowstone's grizzly bears require the support and the opportunity to make difficult decisions without fear of political or agency reprisal. If society does not wish to incur the costs implicit to these arrangements, it should reexamine its values and its laws. Meanwhile our experience suggests that permitting high-level government administrators and political appointees to reinterpret laws in implementation often creates a no-win situation and thwarts the intent of Congress and the ESA. So long as this state of affairs continues, grizzly bear populations in the United States will remain in jeopardy.

Acknowledgments

We would like to thank Michael Frome, Maurice Hornocker, Craig Pease, and Roland Wauer for the opportunities, insights, and help they have provided and Steve Primm and Gerald Wright for critically reviewing the chapter. We would also like to pay tribute to Frank C. Craighead, Jr., for the battles he has fought and to all the other dedicated and idealistic people involved in recovery of Yellowstone's grizzly bear population who have taken risks, lived by their ideals, and often incurred substantial personal costs. People as well as grizzly bears have paid a price.

References

Bacharach, S. B., and E. J. Lawler. 1980. *Power and Politics in Organizations.* San Francisco: Jossey-Bass.

Blanchard, B. M. 1987. Size and growth patterns of the Yellowstone grizzly bear. *International Conference on Bear Research and Management* 7:99–107.

Blanchard, B. M., and R. R. Knight. 1991a. Movements of Yellowstone grizzly bears, 1975–87. *Biological Conservation* 58:41–67.

———. 1991b. Reactions of grizzly bears to wildfire in Yellowstone National Park, Wyoming. *Canadian Field Naturalist* 104:592–594.

Blanchard, B. M., R. R. Knight, and D. J. Mattson. 1992. Distribution of Yellowstone grizzly bears during the 1980s. *American Midland Naturalist* 128:332–338.

Brown, D. E. 1985. *The Grizzly Bear in the Southwest.* Norman: University of Oklahoma Press.

Brown, G., and C. C. Harris. 1992. The United States Forest Service: Changing of the guard. *Natural Resources Journal* 32:449–466.

Clark, T. W. 1986. Professional excellence in wildlife and natural resource organizations. *Renewable Resources Journal* 4:8–13.

———. 1988. The identity and images of wildlife professionals. *Renewable Resources Journal* 6:12–16.

———. 1992. Practicing natural resource management with a policy orientation. *Environmental Management* 16:423–433.

———. 1993. Creating and using knowledge for species and ecosystem conservation: Science, organizations, and policy. *Perspectives in Biology and Medicine* 36:497–525.

Clark, T. W., and S. C. Minta. 1994. *Greater Yellowstone Ecosystem: Prospects for Ecosystem Science, Management and Policy.* Moose, Wyo.: Homestead Press.

Clark, T. W., and R. Westrum. 1989. High-performance teams in wildlife conservation: A species reintroduction and recovery example. *Environmental Management* 13:663–670.

Clark, T. W., R. Crete, and J. Cada. 1989. Designing and managing successful endangered species recovery programs. *Environmental Management* 13:159–170.

Cole, G. F. 1973. *Management Involving Grizzly Bears in Yellowstone National Park, 1970–72.* U.S. Department of Interior, National Park Service, Natural Resource Report no. 7.

Committee on Applications of Ecological Theory to Environmental Problems. 1986. *Ecological Knowledge and Environmental Problem Solving: Concepts and Case Studies.* Washington, DC: National Academy Press.

Committee on the Yellowstone Grizzlies. 1974. *Report of Committee on the Yellowstone Grizzlies.* Washington, DC: National Academy of Sciences.

Craighead, F. C., Jr. 1976. Grizzly bear ranges and movements as determined by radio-tracking. In *Bears—Their Biology and Management,* M. Pelton, J. Lentfer, and G. Folk (eds.). New Series 40. Morges, Switzerland: IUCN.

Craighead, J. J. 1980. A proposed delineation of critical grizzly bear habitat in the Yellowstone region. *International Conference on Bear Research and Management Monograph Series* 1:1–20.

Craighead, J. J., and D. J. Craighead. 1991. New system-techniques for ecosystem management and an application to the Yellowstone ecosystem. *Western Wildlands* 17:30–39.

Craighead, J. J., and F. C. Craighead, Jr. 1967. *Management of Bears in Yellowstone National Park.* Mimeo report. Missoula: University of Montana, Montana Cooperative Wildlife Research Unit.

———. 1971. Grizzly bear–man relationships in Yellowstone National Park. *BioScience* 21:845–857.

Craighead, J. J., and J. A. Mitchell. 1982. Grizzly bear. In *Wild Mammals of North America,* J. A. Chapman and G. A. Feldhamer (eds.). Baltimore: Johns Hopkins University Press.

Craighead, J. J., K. R. Greer, R. R. Knight, and H. I. Pac. 1988. *Grizzly Bear Mortalities in the Yellowstone Ecosystem 1959–1987.* Bozeman, MT: Montana Fish, Wildlife, and Parks, Interagency Grizzly Bear Study Team, Craighead Wildlife-Wildlands Institute, National Fish and Wildlife Foundation.

Craighead, J. J., J. S. Sumner, J. A. Mitchell, and J. T. Hogg. 1994. *The Grizzly Bears of Yellowstone: Their Ecology in the Yellowstone Ecosystem, 1959–1992.* Unpublished manuscript.

Craighead, J. J., J. S. Sumner, and G. B. Scaggs. 1982. *A Definitive System for Analysis of Grizzly Bear Habitat and Other Wilderness Resources.* Wildlife-Wildlands Institute Monograph 1. Missoula: University of Montana Foundation.

Craighead, J. J., J. R. Varney, and F. C. Craighead, Jr. 1974. A population analysis of the Yellowstone grizzly bear, Montana Forestry and Conservation Experiment Station, School of Forestry. *University of Montana Bulletin* 40:1–20.

Driver, H. E. 1969. *Indians of North America*. Chicago: University of Chicago Press.

Frome, M. 1984. Are biologists afraid to speak out? *Defenders* 59:40–41.

Greater Yellowstone Coordinating Committee. 1989. *The Greater Yellowstone Postfire Assessment*. Washington, DC: U.S. Department of Interior.

Harmon, M. M. 1989. The responsible actor as "tortured soul": The case of Horatio Hornblower. *Administration and Society* 21:283–312.

Heinen, J. T., and B. S. Low. 1992. Human behavioral ecology and environmental conservation. *Environmental Conservation* 19:105–116.

Herrero, S. 1985. *Bear Attacks—Their Causes and Avoidance*. New York: Nick Lyons Books.

Houck, O. A. 1993. The Endangered Species Act and its implementation by the U.S. Departments of Interior and Commerce. *University of Colorado Law Review* 64:277–370.

Ingram, H. M. 1973. Information channels and environmental decision making. *Natural Resources Journal* 13:150–169.

Keiter, R. B. 1991. Observations on the future debate over "delisting" the grizzly bear in the greater Yellowstone ecosystem. *Environmental Professionalism* 13:248–253.

Kuehl, B. L. 1993. Conservation obligations under the Endangered Species Act: A case study of the Yellowstone grizzly bear. *University of Colorado Law Review* 64:607–643.

Leopold, A. S., S. A. Cain, C. M. Cottam, I. N. Gabrielson, and T. L. Kimball. 1963. Wildlife management in the national parks. *American Forests* 69:32–35, 61–63.

Lichtman, P., and T. W. Clark. In press. Improving strategic coordination for management of the Greater Yellowstone Ecosystem: Learning from the "vision" exercise. Society and Natural Resources.

Lindblom, C. E. 1980. *The Policy-Making Process*. Englewood Cliffs, N.J.: Prentice-Hall.

Mattson, D. J. 1990. Human impacts on bear habitat use. *International Conference on Bear Research and Management* 8:35–56.

Mattson, D. J., and R. R. Knight. 1989. Evaluation of grizzly bear habitat using habitat and cover type classifications. In *Proceedings—Land Classifications Based on Vegetation: Applications for Resource Management*, D. E. Ferguson, P. Morgan and F. D. Johnson (eds.). U.S. Forest Service General Technical Report INT-257. Washington, DC: Government Printing Office.

———. 1991. *Application of Cumulative Effects Analysis to the Yellowstone Grizzly Bear Population*. Interagency Grizzly Bear Study Team Report 1991C. Washington, DC: U.S. Department of Interior, National Park Service.

Mattson, D. J., and M. M. Reid. 1991. Conservation of the Yellowstone grizzly bear. *Conservation Biology* 5:364–372.

Mattson, D. J., and D. P. Reinhart. 1993. Bear use of whitebark pine seeds in North America. In *Proceedings—International Workshop on Subalpine Stone Pines and Their Environments: The Status of Our Knowledge,* W. C. Schmidt and F.-K. Holtmeier (eds.). U.S. Forest Service General Technical Report INT. Ogden, UT: Intermountain Research Station.

Mattson, D. J., B. M. Blanchard, and R. R. Knight. 1991. Food habits of Yellowstone grizzly bears, 1977–1987. *Canadian Journal of Zoology* 69:1619–1629.

———. 1992. Yellowstone grizzly bear mortality, human habituation, and whitebark pine seed crops. *Journal of Wildlife Management* 56:432–442.

Mazmanian, D. A., and P. A. Sabatier. 1983. *Implementation and Public Policy.* Glenview, Ill.: Scott, Foresman.

Mealey, S. P. (ed.). 1986. *Interagency Grizzly Bear Guidelines.* Denver, Colo.: Interagency Grizzly Bear Committee.

Morgan, G. 1986. *Images of Organization.* Beverly Hills: Sage.

Murie, A. 1940. *Ecology of the Coyote in the Yellowstone.* Fauna of the National Parks of the United States, Bulletin 4. Washington, DC: U.S. Department of the Interior, National Park Service.

O'Reilly III, C. A. 1980. The intentional distortion of information in organizational communication: A laboratory and field investigation. In *The Study of Organizations,* D. Katz, R. L. Kahn, and J. S. Adams (eds.). San Francisco: Jossey-Bass.

Panem, S. (ed). 1983. *Public Policy, Science, and Environmental Risk: Brookings Dialogues in Public Policy.* Washington, DC: Brookings Institution.

Pfeffer, J., G. R. Salancik, and H. Leblebici. 1984. Uncertainty and social influence in organizational decision making. In *Environments and Organizations,* M. W. Meyer, et al. (eds.). San Francisco: Jossey-Bass.

Pimm, S. L., H. L. Jones, and J. Diamond. 1988. On the risk of extinction. *American Naturalist* 132:757–785.

Primm, S. A. 1993. Grizzly conservation in Greater Yellowstone. MA thesis, University of Colorado.

Rees, W. E., and M. Wackernagel. 1992. Appropriated carrying capacity: Measuring the natural capital requirements of the human economy. Paper presented at the Second Meeting of the International Society of Ecology and Economics, Vancouver, British Columbia.

Sabatier, P. 1978. The acquisition and utilization of technical information by administrative agencies. *Administrative Science Quarterly* 23:396–417.

Servheen, C. 1989. The status and conservation of the bears of the world. *International Conference on Bear Research and Management Monograph Series* 2:1–32.

Shaffer, M. L. 1983. Determining minimum viable population sizes for the grizzly bear. *International Conference on Bear Research and Management* 5:133–139.

————. 1992. *Keeping the Grizzly Bear in The American West: A Strategy for Real Recovery*. Washington, DC: The Wilderness Society.

Storer, T. I., and L. P. Tevis, Jr. 1955. *California Grizzly*. Berkeley: University of California Press.

Thomas, C. D. 1990. What do real population dynamics tell us about minimum viable population sizes? *Conservation Biology* 4:324–327.

Twight, B. W. and F. J. Lyden. 1988. Multiple use vs. organizational commitment. *Forest Science* 34:474–486.

————. 1989. Measuring Forest Service bias. *Journal of Forestry* 87:35–41.

U.S. Congress. 1977. Endangered species act oversight: hearings before the Subcommittee on Resource Protection of the Committee on Environment and Public Works, U.S. Senate, 95th Congress, First Session. Washington, DC: Government Printing Office.

U.S. Fish and Wildlife Service. 1993. *Grizzly Bear Recovery Plan*. Missoula, MT: U.S. Fish and Wildlife Service.

Wackernagel, M., and W. Rees. 1992. Perceptual and structural barriers to investing in natural capital. Paper presented at the Second Meeting of the International Society of Ecology and Economics, Vancouver, British Columbia.

Walters, C. 1986. *Adaptive Management of Renewable Resources*. New York: Macmillan.

Westrum, R. 1986. Management strategies and information failure. Paper presented at the NATO Advanced Research Workshop on Failure Analysis of Information Systems, Bad Winsheim, Germany, August 1986.

————. 1988. Organizational and inter-organizational thought. Paper presented at the World Bank Conference on Safety Control and Risk Management, Washington, DC: October 1988.

Yaffee, S. L. 1982. *Prohibitive Policy: Implementing the Federal Endangered Species Act*. Cambridge: MIT Press.

6

The Florida Manatee Recovery Program
Organizational Learning and a Model for Improving Recovery Programs

Richard L. Wallace

The endangered West Indian manatee, *Trichechus manatus,* order Sirenia, family Trichechidae, is a herbivorous marine mammal found in coastal areas of Florida and the southeastern United States (Figure 6-1). Although the species is patchily distributed in the Caribbean, Gulf of Mexico, and northeastern South America, the U.S. population is a distinct subspecies, *T. m. latirostris* (Lefebvre et al. 1989). The Florida manatee is one of the most endangered marine mammals found in U.S. waters and is federally protected under the Marine Mammal Protection Act of 1972 and the Endangered Species Act of 1973 (MMC 1993a).

In response to increasing pressures on the manatee population due to human activities, a federal recovery program for the Florida manatee was initiated by the U.S. Fish and Wildlife Service (FWS) in 1976. The Florida Manatee Recovery Plan, originally adopted in 1980 and revised in 1989, states that its goal is "to restore the Florida manatee as a viable self-sustaining element of its ecosystem" (USFWS 1989:i). This chapter reviews organizational aspects of the program throughout its history. In examining specific periods in the program's development, the chapter analyzes whether organizational learning has taken place to improve the manatee program's implementation and describes a model that may have led to improvements.

The Problem

Since 1976 the manatee recovery program has received several million dollars in personnel and funding support from federal and state agencies, nongovernmental organizations, and private industry. (The

FIGURE 6-1. The Florida manatee. (Photo copyright 1993 Sea World of Florida. All rights reserved. Reproduced by permission.)

combined fiscal year 1993 manatee program appropriations for FWS
and the Florida Department of Natural Resources (FDNR) totaled approximately $4 million according to MMC 1993a.) The program cur-
rently encompasses cooperative efforts by more than seventy federal,
state, and local government agencies and nongovernmental, academic, and private organizations. Yet through 1989 the observed annual manatee mortality in the southeastern United States displayed
an almost constant upward trend. With the exception of an abnor-
mally high number of deaths in 1990 caused by an extreme cold spell,
annual manatee mortality has remained constant since 1989.

Participants have expressed mixed views of the program. Speaking
at a 1992 meeting of the Marine Mammal Commission (MMC) in
Florida devoted largely to a review of the manatee program, the
FWS's assistant director for fisheries stated that the Florida Manatee
Recovery Plan "is probably [FWS's] best recovery plan" and that it
"has served as a model for other plans" being developed or revised
by FWS (MMC 1992a:42). At the same meeting, the director of
the Division of Marine Resources at FDNR, the lead state agency in the
manatee program, addressed the interorganizational cooperation inherent in the recovery plan and manifested in program implementation by stating that "this is a classic example of how to do environ-

mental and ecological management anywhere. It could be picked up and put somewhere else" (MMC 1992a:30). Yet the MMC chairman at the time, a Florida manatee biologist with more than fifteen years' involvement in the recovery program, warned that "despite all the good, there is really no indication that either manatee mortality or the effects of a burgeoning human population in Florida are adequately under control" (MMC 1992a:7). He noted, however, that "the lack of resolution of the problem reflects the size of the problem, not the inadequacy of efforts to resolve it" (MMC 1992a:26).

Throughout its seventeen-year history, the Florida manatee recovery program has encountered numerous organizational difficulties in carrying out its stated goal (MMC 1978a, 1978b, 1978c). Since record keeping began in 1974, observed manatee mortality in Florida has grown from a low of 7 that year to a high of 206 in 1990 (see Table 6-1). Most aspects of the current manatee program are relatively well funded, well staffed, and actively implemented (in terms of the number of recovery tasks being undertaken). As the following discussion illustrates, the program's organization has improved substantially since its troubled start in the mid-1970s. The program now operates by setting reachable implementation goals—such as specific research programs, the development and distribution of educational materials, and the creation and enforcement of waterway speed zones and protected areas—and attempting to meet them with adequate funding and personnel. Many of the specific research and management tasks outlined in the recovery plan appear to be achievable with current levels of funding and personnel.

The problem or, more accurately, the question is this: Is the organization of the manatee program sufficient to reduce manatee mortality, protect necessary habitat, and, ultimately, restore the population to an ecologically viable level within its ecosystem? For most of its seventeen years, it obviously has not been sufficient. Although there are indications (such as no increase in observed annual manatee mortality since 1990) that the program's implementation may be having the desired effects, it is still too early to tell whether the program's organizational changes will ensure effective implementation.

The Florida Manatee

The following paragraphs review manatee ecology, life history, and human-related mortality. These factors constitute the background for the manatee's gravely endangered status.

TABLE 6.-1.

Manatee Mortality in Florida by Cause of Death 1974–1993

Cause of Death	1974	1975	1976	1977	1978	1979	1980	1981	1982	1983	1984	1985	1986	1987	1988	1989	1990	1991	1992	1993	Total
Watercraft collision	3	6	10	13	21	24	16	24	20	15	34	33	33	39	43	50	47	53	38	35	522
Floodgate/canal lock		1	4	6	9	8	8	2	3	7	3	3	3	5	7	3	3	9	5	5	89
Other human-related	2	1		5	1	9	2	4	1	5	1	3	1	2	4	5	4	6	6	6	62
Perinatal		7	14	9	10	9	13	13	14	18	25	23	27	30	30	38	44	53	49	39	426
Other natural		1	2	1	3	4	5	9	41	6	24	19	13	16	24	32	67	14	19	24	300
Undetermined	2	13	32	80	40	23	19	64	35	30	41	38	45	22	25	40	41	39	46	36	675
Total	7	29	62	114	84	77	63	116	114	81	128	119	122	114	133	168	206	174	163	145	2074

Source: FDNR (1994).

Geography and Habitat

The Florida manatee is found year-round throughout the coastal and inland waterways of peninsular Florida and the southeastern coast of Georgia. During the summer months, manatees may range as far north as the Chesapeake Bay in Virginia, east into Louisiana, and, rarely, to the Bahamas (Reynolds and Odell 1991).

Manatees, which move freely between fresh and salt water, usually occupy estuarine and riverine habitat (Hartman 1979). They seek warm water, normally above 15 to 20°C, and are susceptible to mortality due to unseasonably cold temperatures below 10°C (Husar 1977; Hartman 1979; Powell and Rathbun 1984). Therefore, they can be found throughout most of their known range during the warmer months (March through October), but in the winter they congregate in the southern regions of their habitat, where they seek refuge in warm water, including natural warm water springs and the artificial discharges created by coastal power plants (Hartman 1979; Lefebvre et al. 1989; Reynolds and Odell 1991).

Life History

Manatees are largely solitary animals, excepting sexual behavior and mother/calf interactions (Hartman 1979; Caldwell and Caldwell 1985). They become sexually mature between three and ten years of age; although the exact age at which males mature is not known, females are thought to mature between three and four years of age (Brownell et al. 1978; Hartman 1979; Caldwell and Caldwell 1985; Marmontel 1992; Rathbun et al. 1992). Manatees are sexually active year-round (although possibly less so in winter), and females have been known to bear young in all seasons (Hartman 1979; Caldwell and Caldwell 1985).

Gestation periods and calving intervals are not conclusively known (Brownell et al. 1978; Caldwell and Caldwell 1985). Gestation may last approximately thirteen months and females may mate as often as every two and a half years (Hartman 1979; Odell et al. 1992; Rathbun et al. 1992), although one recent study suggests a calving interval of 5.2 years (Marmontel 1992). Calves, which are raised by the female, are weaned by the end of their second year (Hartman 1979; Caldwell and Caldwell 1985; Reynolds and Odell 1991). It is not known how long manatees live in the wild, although individuals may live longer than fifty years (Marmontel 1992, 1993). The manatee's low reproductive rate and relatively long period preceding sexual maturity increase the population's susceptibility to outside pressures that may lead to mortality.

Mortality in Florida

There are no historical estimates of the size of the manatee population in Florida. It has been suggested that several thousand West Indian manatees existed at the time that European settlers first arrived in the United States (Caldwell and Caldwell 1985). Statewide synoptic surveys conducted in 1992 indicate that at least 1856 manatees exist (Ackerman 1992). Manatee mortality has increased dramatically in recent years (see Table 6-1). Since 1976, manatee mortality in Florida has averaged 121 animals per year, reaching a high of 206 in 1990 (FDNR 1994).

The most serious known cause of human-related manatee mortality has been collisions with boats. Perinatal mortality—death related to the loss of dependent calves—is a serious and growing concern as well, and may be related to human impacts. Since 1976, an average of thirty manatees per year have died of boat-related injuries, and in six of the past ten years (through 1993) manatee mortality due to collisions with boats has reached record levels (FDNR 1994). In 1992, boat-related manatee deaths declined by 24 percent from 1991 levels, although overall mortality declined by just less than 5 percent over the same period and boat-related deaths in 1992 still comprised about 23 percent of all observed manatee mortality (FDNR 1992b). Causes of perinatal mortality are largely unknown, but deaths may be due to cold temperatures, premature births, and other natural factors or human-related causes such as pollution and death of nursing mothers due to collisions with boats. Perinatal mortality is the second leading known category of manatee mortality in Florida, surpassed only by boat-related deaths. By far the largest number of observed manatee deaths are due to unknown causes (see Table 6-1).

Habitat degradation, although not a direct cause of manatee mortality, contributes to pressures on Florida's manatee population. The growth of Florida's human population (estimated at 13.2 million people for 1992, based on projections from the 1990 census, and increasing at a rate of 2.7 percent a year; Ackerman et al. 1992), combined with the loss of viable or undisturbed habitat, is a critical threat to the manatee. In this regard, further research is needed to determine manatee habitat requirements and their role in Florida's coastal ecosystems (Packard and Wetterqvist 1986; Reynolds and Gluckman 1988).

Manatee mortality declined in 1991 and 1992, but the 163 observed deaths in 1992 still equaled about 9 percent of the estimated minimum manatee population—a mortality rate that the population may not be able to sustain. Further, the mortality rate was almost certainly higher than what was observed, as an unknown number of manatee deaths go unobserved. Continued annual surveys of manatees in Florida and monitoring of manatee mortality are essential, then, to determine trends.

Viability

Marmontel (1993) conducted a population viability analysis (PVA) on manatees in Florida. Using the Vortex 5.1 software package, a thousand-year simulation was undertaken to determine the probability of persistence and the mean time to extinction (see Lacy 1993). Marmontel acknowledged certain shortcomings of the PVA. Extensive data on longevity were not available, sampling variation and true variation in basic parameters could not be separated due to possible errors in the data set, and genetics and density dependence were excluded from the model (Marmontel 1993). The initial population size used for the simulation was 2000 based on the count of 1856 resulting from the 1992 surveys (Ackerman 1992). Marmontel noted that this relatively low population size lessens the probability of persistence and increases the chances that the population will be adversely affected by environmental variation.

The basic scenario used for the simulation was based on current knowledge of manatee life history, an arbitrary (due to the lack of information on historical numbers of manatees) carrying capacity level of 5000, and estimates of the frequency of three types of catastrophic events: extreme cold temperatures were estimated at an annual probability of 1.33 percent, class 5 hurricanes (248+ km/hour winds) at 0.5 percent, and outbreak of disease at 2 percent. For all three events, a 95 percent multiplicative effect on reproduction and a 90 percent multiplicative effect on mortality were modeled. These variables resulted in the population's extinction in about 750 years (Marmontel 1993).

Based on an assumption that the manatee population in Florida is not currently increasing, Marmontel (1993) concluded from the simulation that the population cannot endure any increase in mortality and, moreover, that random and potentially catastrophic environmental or human-caused events will probably contribute to decreasing chances of long-term persistence. Marmontel (1993:263) also concluded that "only under constant conditions (no environmental variation), inexistence of catastrophes, or very large population sizes does the manatee population have high chances of surviving in the long term."

The Recovery Program

To realize the recovery of an endangered species or population, a program must successfully address "technically demanding biological tasks and uncertainties, limited resources, numerous participants, and intense public scrutiny and involvement" (Clark and Harvey 1991:147). Recovery programs must therefore be open systems. That is, they "must interact with [and] continuously change and adapt to the environment" in order to operate effectively (Daft 1992:9; see also Butler 1991).

In the mid-1970s, when FWS initiated the manatee program, the program's organization was far from an open system. Rather, the program was characterized by goal displacement—in this case, focusing on control of the program rather than on tasks to reduce manatee mortality—and a "crisp" organizational structure. Crisp structures are inherently bureaucratic: they are inflexible with regard to adaptability to uncertainty and variability, and they inhibit analysis and other essentials of wise decision making (Butler 1991). Such organizations are conservative—that is, preoccupied with maintaining the status quo—and therefore very resistant to change (Yaffee 1982).

As we will see, FWS's behavior in the late 1970s manifests some of these features in the manatee program's organization. The root of the problem at that time was the service's lack of appropriate delegation of responsibility for program implementation and failure to make manatee recovery a priority, even in the face of the species' increasing mortality. This shortcoming was compounded by ineffective interactions between its research and management arms, its Washington, regional, and area offices, its program staff and state agencies, and, in general, its staff and those of other federal agencies, particularly the MMC.

The Program in the 1970s

Despite a wealth of FWS personnel from which to draw for the manatee program during the 1970s (no fewer than nine separate offices and divisions of FWS's research and management arms were in some way responsible for manatees in the late 1970s; USFWS 1978), there was no coordination and, as a result, no cohesive program. Although service research staff in Washington and Florida recognized many of the problems affecting Florida's manatees and researchers had been collecting data on manatee mortality since 1974, FWS made no attempt to coordinate the research and management programs. In this sense, with regard to the work being done by research staff, the drive to maintain control of management aspects of the manatee program appeared to be particularly acute in FWS's management arm. Research staff had collected data on increasing manatee mortality in Florida, yet management staff did not use the data to dictate essential actions. Taking such actions would have meant increasing funding, personnel, and time allocations, measures the service's management arm was apparently unwilling to pursue at that time.

Furthermore, there was little communication between the service's Washington and regional offices. And the Southeast Regional Office and Florida management staff—the offices directly responsible for implementing management actions—exhibited a lack of commitment to the program that resulted in no serious actions being taken. Specifically, the highest-ranking member of the service's management staff in Florida at that time showed little concern for the service's role in manatee recovery efforts (see MMC 1978d). He considered himself "an administrator" whose responsibilities did not include close involvement with all the programs under his jurisdiction. Although the manatee program—particularly the work of the West Indian manatee recovery team and its recovery plan for the species—was being developed by the late 1970s, at the time the FWS manager stated publicly that he had "no role in the process of developing, reviewing, or approving the recovery plan . . . , no knowledge of the actions of the . . . recovery team . . . , no authority to initiate action without approval from higher authority . . . , [did not] know how much research or administrative money was available for manatee efforts, and had no control over the initiation of research and development of efforts to protect manatees" (MMC 1978d:30–31).

The case of the FWS manager was not an isolated one. Key staff in the FWS regional office and in several divisions of the agency's Washington headquarters in the late 1970s, despite obvious problems, lacked the initiative to make the organizational changes necessary to improve program performance. The single most important improvement needed at the time was the coordination of recovery activities and corresponding improvement in communication among participants in recovery activities at the federal, state, and local level. To FWS at the time, however, this would have meant, in addition to increasing budgets and staff time for manatee activities, allowing the delegation of responsibility away from bureaucratic centers of control. It eventually took meetings between the executive director of the MMC and the director of FWS to accomplish the needed changes. Meetings in the summer and fall of 1978 to discuss the importance of measures to reduce manatee mortality (which had increased by an order of magnitude in four years) led FWS to restructure the manatee program from the top down and to centralize and coordinate efforts.

Among the most significant changes were those in FWS's staff in charge of manatee program management. By the end of 1980, FWS's top manager in Florida had left, the service had hired a manager, highly regarded by people outside FWS, as manatee coordinator in Florida to oversee research and management activities, and communication had improved between the service's Washington office and the state with the appointment of a new FWS associate director for federal assistance. Following a lengthy public and agency review process, the West Indian manatee recovery plan was completed in April 1980 (USFWS 1980). Although coordinated implementation had yet to be achieved, the plan provided a guide for implementing the service's and the state's recovery activities and set the stage for FWS to create a manatee recovery activities coordinator role in March 1980. The creation of this position laid the groundwork for the current recovery program. The person chosen for the position had several years' experience working with the manatee program, was well known in Florida, and, most important, was well respected by staff in FWS's area, regional, and Washington offices, the state agencies and legislature, Florida industry, and the general scientific and conservation communities. (See, for example, FDNR 1982.) Though new to the formal role of "coordinator," this person had the skills and broad knowledge necessary to undertake the role—credibility, impartiality combined with a sense of duty to the program at hand, and a good understanding of the differing interests involved in manatee issues in Florida. (See also Galbraith 1977 and Clark et al. 1989.)

The recovery coordinator was given the responsibility of developing a comprehensive work plan that would provide a detailed outline of the tasks necessary to implement the recovery plan: what actions were needed to implement them, who the lead and cooperating agencies were, the timetables that had to be met, and the necessary funding levels and sources (see Rose et al. 1981). The coordinator would then act as liaison with all participants in the recovery program in order to implement the recovery plan. Initially the position was funded contractually for one year by MMC with the intent that the coordinator would eventually become a full-time FWS employee. Although the service did not create a permanent staff position for the coordinator at the time due to funding considerations, it supported the position for two years before it was incorporated into FDNR's Division of Marine Resources. The service subsequently created a position of manatee coordinator to oversee FWS management activities and monitor the program in cooperation with the Florida manatee recovery team.

The Program in the 1980s

The creation of the manatee coordinator role and adoption of the comprehensive work plan by FWS in 1982 had a major effect: it elucidated all the research and management tasks expected of the service and related them to all the other actions deemed necessary to the recovery program. Thus program participants both inside and outside the FWS now had an overall context for the work that was being done. Perhaps most important, the many organizations that had in some fashion been involved in manatee research and conservation but were outside the federal sphere—state agencies, nongovernmental organizations, academic institutions, oceanaria, utility companies—became constructively integrated into the program.

By the middle of 1983, FDNR assumed lead agency status for several key management aspects of the recovery program, while FWS continued to focus its resources on manatee research, habitat protection, and general monitoring. The department's funding had been augmented to address increased program responsibilities and staffing needs, and the Save the Manatee Trust Fund had been set up by the state to support FDNR's manatee management activities. Initially endowed through increased annual state boater registration fees that provided $250,000 in the trust fund's first fiscal year (1986–1987), the trust fund in fiscal year 1991–1992 provided more than $2.3 million to the state's manatee program (FDNR 1992a).

By 1985, some progress had been realized with regard to improved implementation of recovery tasks. There were several significant results: evaluation, acquisition, and protection of important manatee habitat; improved biological data collection and population studies; increased education programs aimed at boaters and the general public; increased enforcement of waterway slow-speed zones and manatee protected areas; and the continuation and expansion of a salvage and necropsy program created in 1974 to study manatee mortality. Even so, it became apparent that further improvements were needed. By early 1986, three years into a full-time position with FDNR, the state's manatee recovery activities coordinator had become both geographically and hierarchically distant from the office of FDNR's executive director, thereby reducing his impact on high-level decision making in the state's program. A new executive director of FDNR had attempted to forge a more productive relationship with recreational boating interests in the hopes that this would increase opportunities for manatee conservation efforts. The new executive director was concerned that representatives of the boating industry would be alienated by the manatee coordinator, who was perceived by the boaters as strongly protectionist. Yet the net result of moving the coordinator out of Tallahassee—where FDNR's headquarters are located and executive decision making takes place—was to slow program implementation overall by reducing the manatee coordinator's influence on state-level implementation. At the same time, FWS research and management programs were receiving diminished funds and less staff attention. The effect was serious: manatee mortality, particularly due to manatee/boat interactions, and habitat degradation were still outpacing recovery actions.

The program's organization had experienced a revitalization in the early 1980s resulting in the coordination of recovery planning, improved staff support at the federal and state levels, and the delineation of recovery tasks in the recovery plan and comprehensive work plan. By 1986, however, organizational complacency appeared to have settled into the program. Perhaps it resulted from a tacit belief that since the right actions (coordination, development of a recovery plan) had been taken to address the problems, vigilant efforts and augmented funding became less of a priority. Yet no one had really addressed the increased mortality and habitat-related pressures on the manatee population. Although manatee mortality had declined slightly from 1981 through 1983, by 1988 it was again on the rise, and in fact was now more than double the 1980 levels. Although manatee conservation efforts had increased significantly since the mid-1970s (particularly with regard to the creation of manatee protection zones and the development of education programs), manatee mortality was still outpacing recovery efforts. Although commitment to the program at both the state and federal level throughout the 1980s was vastly improved over efforts in the 1970s, the program still fell short of accomplishing the actions necessary to keep pace with pressures on the manatee population.

The MMC, therefore, critically evaluated the program's organization and made recommendations to FWS and FDNR for improving the program's implementation. The commission noted that attention should focus especially on protecting habitat and controlling development in coastal areas, improving the effectiveness of law enforcement efforts, identifying and sufficiently funding priority research needs, and augmenting education programs (MMC 1987b and 1987c; see also MMC 1987a and Reynolds and Gluckman 1988). Consultations in 1987 between MMC, FDNR, FWS, and other federal and state program participants resulted in FWS's reconstituting the recovery team, which had not met since it had completed the original recovery plan in 1980. The new recovery team, composed of an interdisciplinary membership representing the various political and organizational interests in the manatee program (federal and state governments, nongovernmental organizations, oceanaria, academic institutions, the boating industry, and utility companies), was charged with revising and updating the recovery plan and accompanying comprehensive work plan. As part of program monitoring, the recovery team also produces, in cooperation with FWS's manatee coordinator's office and management staff, an annual status report of all actions taken under the recovery plan (see USFWS 1993). The new recovery team includes people with explicit knowledge of sociopolitical considerations as well as the biological knowledge necessary to implement recovery actions.

The Program Since 1989

The bureaucratic structure of the manatee program has changed little since its reorganization in the late 1970s and early 1980s. Yet by the late 1980s several critical aspects of the recovery program, most notably basic research by FWS and local and regional state education and boat-speed regulatory programs, were operating efficiently and dependably toward the stated recovery goals. In this regard, the revised Florida Manatee Recovery Plan, completed in 1989, succeeded in focusing efforts on task-related issues. It did so by updating the comprehensive work plan and in the process securing the agreement of the primary participants in the recovery program (all of whom signed the recovery plan as a symbol of their support; see USFWS 1989). Due largely to the willingness of FWS, state, and nongovernmental program participants to work together to address the program's implementation, there has been little organizational conflict in the program. The service's research and management staffs have fostered a constructive relationship, and communication among FWS participants in different divisions and between the service and the state agencies remains productive. By the late 1980s, however, the program had yet to bring manatee mortality under control.

In 1989 and 1990, MMC made a number of recommendations to FWS—again emphasizing the need to improve implementation of recovery tasks in order to address, among other things, habitat protection and boat-related mortality (MMC 1989a, 1989b, 1989c). These recommendations culminated in a detailed evaluation of manatee research and management funding needs for the FWS manatee program in fiscal years 1991 through 1995 (MMC 1990). The recommendations were based on the minimal funding needed to meet task objectives outlined in the revised recovery plan. Beginning in fiscal year 1991, funding at levels roughly equal to, or exceeding, those recommended by MMC were allocated by FWS for its research and management programs (USFWS 1991).

State funding improved as well, as the annual budget of the Save the Manatee Trust Fund continued to grow (FDNR 1992a). In October 1989, Florida's governor and cabinet adopted a rigorous manatee protection framework that required thirteen key Florida counties to cooperate with the Florida Departments of Natural Resources and Community Affairs to prepare and implement countywide manatee protection plans to address habitat protection needs and direct human impacts. This action delegated responsibility to the local level, where support of enforcement measures is critical to implementation of recovery tasks. The thirteen counties have cooperated with the two departments in developing these plans. By the end of 1992, one plan had been completed and the other twelve were in various stages of development (MMC 1993a).

Private participation—from organizations such as the Save the Manatee Club, the Florida Power & Light Company, Sea World of Florida, Eckerd College, and numerous others—has also contributed significantly to the accomplishment of key tasks outlined in the revised recovery plan. Among other things, manatee rescue and rehabilitation work, education programs, and research facilities crucial to specific recovery tasks have been carried out and funded by these program participants. The involvement of the public has also had a positive effect. In the past five years, the manatee has become a popular symbol, both in Florida and internationally, of endangered species problems (MMC 1992a). The manatee was designated the official Florida state mammal, and special automobile license plates now depict a manatee motif. Sales of these license plates in conjunction with automobile registrations contributed more than $1.2 million to the Save the Manatee Trust Fund in 1992 (FDNR 1992a).

The FWS, FDNR, and others charged with implementing the recovery plan have continued their efforts to meet recovery needs as outlined in the revised plan. Funding for FWS programs has been adequate to meet most of its recovery responsibilities (see MMC 1992b, 1992c; USFWS 1992), and the service has continued to work closely with the state on protection of crucial manatee habitat. The Florida Department of Natural Resources continues to focus maximum attention on habitat protection and boat speed regulations, as well as on the development of county manatee protection plans. The commission has remained involved in the manatee program, providing periodic seed money for research projects and advising FWS and FDNR on the implementation of specific recovery tasks, particularly those pertaining to habitat protection and boat speed regulations (MMC 1993a).

In 1992 MMC held another manatee program review in Florida to evaluate implementation needs with program participants and review progress since 1989. The meeting resulted in updated recommendations for funding for federal manatee research and management efforts, which the service again indicated that it would meet, and for updating the recovery plan to address needs beyond fiscal year 1994 (MMC 1992b, 1992c; USFWS 1992). As this book goes to press, an updated recovery plan is being readied for release.

Is the Recovery Program Finally Working?

Since 1991, observed manatee mortality in Florida has decreased by about 17 percent (see Table 6-1). Over the same period boat-related deaths declined by one-third and perinatal mortality declined by about one-quarter (FDNR 1994).

It is too early to tell whether stable mortality rates are due to the effects of the revised recovery plan. In a species that breeds slowly and reaches maturity relatively late in its life history it is difficult to measure success in only a few years. But using mortality levels as a criterion for measuring success in the manatee program, at least interim success can be claimed in that the mortality rate has stopped increasing. (There are insufficient data to indicate whether declining manatee numbers are responsible for this perceived trend.)

Organizationally, the manatee program is in better shape now than at any point in its history. Lines of communication between branches of state and federal agencies, especially FWS and FDNR, have remained open, leading to improvements in cooperation between governmental program participants. While the program has become more complex (in terms of the number and diversity of participants), its effectiveness has not diminished. Further, the active participation of nongovernmental organizations has been embraced by FWS and FDNR, and the involvement of these organizations has led to broader program implementation. Both FWS and FDNR now have dedicated personnel at all levels who are willing to make the necessary effort to achieve the goals of the manatee recovery plan. As well, on an organizational level, both agencies should be credited for heeding outside advice—particularly that of MMC—toward improving the manatee program.

Is There Organizational Learning in the Program?

"Organizational learning occurs when members of the organization act as learning agents for the organization, responding to changes in the internal and external environments," say Argyris and Schön (1978:29). But organizational learning can only take place when information on an organization's performance is noted and applied to the set of responsibilities that the organization is designed to carry out. The FWS and, to a lesser degree, FDNR, at one point or another in the recovery program, were hindered by organizational dilemmas that were not readily identified or addressed by changes from within. "To learn to solve dilemmas, an organization must learn to change itself by learning how to acquire new goals, norms, and policies for itself" (Clark et al. 1989:167). The service did not exhibit this behavior at critical junctures in implementation where program and agency organization needed to be refined in order to further manatee recovery goals. In these cases, outside evaluation was necessary to initiate a process that would result in improvements to the program. In both cases, the evaluation and subsequent recommendations for resolution of the dilemmas were provided largely by MMC.

Can MMC be considered the major learning agent for the manatee program, even though its responsibilities are largely advisory and it is not directly involved in day-to-day implementation? To answer this question we first must define the manatee program as an organizational system, look at its constituent parts, and then determine whether or not learning is occurring. According to Butler (1991:1 after Daft 1989), an organization can be defined as a social entity that has a purpose, has a boundary (some participants are considered inside while others are considered outside), and patterns the activities of participants into a recognizable structure. This definition fits the model of recovery programs if we take a macro view of the program.

Is normative learning actually taking place in the recovery program? Or are changes being implemented solely on the basis of revising implementation strategies in the short term without an inherent appreciation by the agencies of the necessary change to their values and norms? For example, has implementation of the manatee program by FWS advanced only by incremental steps specific to the Florida manatee population (that is, one task or problem at a time—a very small-scale view)? Or has FWS, as an organization, realized certain shortcomings inherent in its approach to the program and attempted to revise its overall strategies for program organization and implementation in order to improve the manatee program? The difference between these two approaches is the difference between "single-loop" learning and "double-loop" learning. Single-loop learning is characterized by changes to an organization's strategies and assumptions—frequently in response to a problem—that do not challenge the underlying values and norms of the organization (Argyris 1992). Double-loop learning demands that the organization's values or norms are themselves modified in the process of taking action (Argyris and Schön 1978).

In the manatee program, MMC has demonstrated the ability to question the underlying values and norms of the program, which, in the late 1970s, paid scant attention to the increasing manatee mortality observed in Florida. The commission encouraged changes in the program's organizational structure, particularly concerning leadership, coordination, and communication, that led to a basic restructuring of the norms and values directing the program's implementation. It did this by influencing FWS to elevate the priority of program implementation, which it did, and by assisting in the creation of a Manatee Technical Advisory Committee (MTAC) to provide advice and information on issues including evaluation and coordination of recovery efforts and habitat protection needs. The underlying values and norms of FWS's manatee program today are virtually opposite of those of fifteen years ago. Whereas in the 1970s FWS appeared to disregard many of its management responsibilities concerning the protection of manatees under the Endangered Species Act, today FWS is an active leader in manatee conservation efforts. Its research program has led to vast improvements in the understanding of manatee biology and life history, and, in stark contrast to the agency's actions in the 1970s, its research and management arms work closely together to put that information to use in developing and implementing recovery activities.

In the 1970s, FWS's underlying values and norms, at least with regard to manatee management in Florida, reflected a neglect of its responsibilities under the Endangered Species Act. Since that time, however, FWS has developed a trend toward active and continuing support of the many aspects of the manatee program for which it maintains primary control. The need for MMC's watchdog role has declined significantly as FWS has developed a self-perpetuating and highly functional manatee program. Although this transition gives the appearance of double-loop learning in the manatee program, it is not. According to Argyris and Schön (1978:29): "In double-loop learning, response to detected error takes the form of joint inquiry into organizational norms themselves, so as to resolve their inconsistency and make the new norms more effectively realizable." This has been MMC's role in the manatee program overall, but the results do not in fact demonstrate that FWS and FDNR have learned as organizations.

If the manatee program as a whole can be considered a supraorganization, it may undergo changes in values and norms due to the actions of its constituent agencies. But if we ask whether FWS would have exhibited double-loop learning had MMC not been a participant in the program, a different conclusion may be reached. Considering the inadequate organization of FWS's manatee program in the 1970s, it is likely that the service would have had difficulty undertaking even single-loop learning in the interests of manatee recovery. The organizational development manifested by FDNR with regard to manatees in the early stages of the recovery program, meanwhile, was so contingent upon actions taken by the federal agencies that it is impossible to speculate on the changes that might have taken place at that time without their involvement.

Both FWS and FDNR now operate according to organizational values and norms that encourage them to undertake reflexive single-loop learning in response to the daily problems facing manatee recovery. Indeed, says Yaffee (1982:106), "organizations develop traditions about how things are done and what is important; these become norms to guide . . . behavior." In the case of both agencies, the implementation of a specific set of organizational responsibilities, as outlined in the recovery plan, has become second nature. Budgets and personnel requirements have been developed and institutionalized on the basis of the recovery plan and its implementation schedule, with the assistance and guidance of the annual status reports developed by FWS and the recovery team and periodic advice from the MMC and MTAC. An understanding of what needs to be accomplished in order to achieve recovery tasks (an understanding that FWS's management arm seemingly lacked in the 1970s) has been achieved. In this regard, implementation of recovery tasks, as well as allocation of appropriate funds and personnel to this end, has become the norm. Single-loop learning has in this way also become institutionalized, as necessary changes in short-term recovery activities are undertaken naturally—that is, without bureaucratic inertia—within the context of the recovery plan and implementation schedule.

The program, as directed by the two agencies, may now be having the desired, if incremental, effect of helping to ameliorate adverse impacts on the Florida manatee population. As it is too early to tell whether the current values and norms are faulty (with regard to accomplishing manatee recovery), there is no basis on which to judge whether double-loop learning can yet occur within these two organizations. Should any crises develop, for instance, within the organization of either FWS's or FDNR's manatee programs that disrupt the implementation of the recovery program, it will be interesting to observe whether remedies are initiated from within the organizations or whether the evaluative role of an outside organization (such as MMC, the recovery team, or MTAC) is necessary to steer the program back on course.

Lessons

In gleaning lessons from the preceding case, we must recognize what makes the manatee program distinctive from other recovery programs. The MMC provides an independent review function for the recovery program that contributes evaluative insight into its operations. The commission is directly involved in the implementation of very few research and management tasks necessary to meet program objectives. Yet MMC is statutorily compelled to become involved in the program by the Marine Mammal Protection Act as an advisor/evaluator (although commissioners and members of MMC's Committee of Scientific Advisors, who serve part-time appointments and hold full-time jobs elsewhere, may be directly involved in the manatee program). As it has no regulatory power, MMC "must rely on its powers of persuasion and the weight of the evidence that it can marshal to sustain its [evaluations of and] recommendations to other agencies" (Tobin 1990:221).

The MMC is an official participant in the manatee recovery program, yet its advisory role allows it to evaluate the program's performance. A crucial part of the process of implementing a program is the ability of the implementing organization to evaluate itself and the effectiveness of the job it is doing. Rarely, however, will an organization evaluate itself harshly when implementation is problematic or unsuccessful. (See Chapters 3–11, 16, 17.) Honest self-evaluation might upset the organization's values and norms and weaken its control over the program. The benefit of outside evaluation lies in the fact that it is not controlled by the implementing agency and thus provides a new perspective and stands free of the self-legitimizing biases inherent in self-evaluation.

As an outside observer/evaluator, the MMC and its Committee of Scientific Advisors has been able to maintain a clear view of the program's overall goal. Thus it has avoided the bureaucratic conservatism inherent in most implementing organizations (see Yaffee 1982). The "inside participant/outside evaluator" role therefore gives the manatee program unbiased evaluation and an opportunity for organizational learning not found in most recovery programs. It is the official, formal status of MMC in the context of the recovery program, combined with its penchant for relying on the strongest scientific evidence to support its arguments (whether they are organizational or biological in nature), that has so benefited the manatee program. Although this evaluative role is ostensibly also the function of endangered species recovery teams, MMC's status as an independent federal agency sets it apart from recovery teams in the structure of recovery programs because recovery teams are convened by and ultimately fall under the jurisdiction of a program's lead federal agency, either FWS or the National Marine Fisheries Service.

It might be useful to incorporate the MMC model in other recovery programs. This could be done by creating an independent federal "Endangered Species Recovery Committee" or individual committees staffed with the expertise to advise on certain groups of species or genera and across a range of organizational circumstances. To differentiate them from recovery teams, these committees would have to be conferred with appropriate autonomy and formal status by the federal government (analogous perhaps to MMC an independent agency of the executive branch) so they would not become captives of implementing agencies. They would require some level of authority—akin perhaps to the Marine Mammal Protection Act's statutory requirement that federal agencies must respond to MMC recommendations within 120 days or explain in detail why the recommendations were not followed or adopted (MMC 1993b). Such authority could be created, perhaps, under the Endangered Species Act (although a committee of this nature would no doubt be answerable to the act's Endangered Species Committee, the "God Squad"). Finally, they would require a level of scientific and policy expertise analogous to the MMC and its Committee of Scientific Advisors. The net effect of such a committee would be to provide independent oversight and advice for a recovery program from an organizational standpoint with the capacity to recommend changes to improve the program's organization and operation.

The key to this model is the creation of a review organization as one facet of the overall recovery effort. Its purpose is to frame "tasks in such a way that a larger social context moves to the foreground and technical problem solving becomes a piece of the larger social puzzle" (Schön 1983:274). Such a committee would encourage the program's participants to adopt an analytic view in which the design, evaluation, and modification of organizational structures for species recovery are of primary concern and would instill in them a propensity toward double-loop learning.

Acknowledgments

I am indebted to David W. Laist and John E. Reynolds III for numerous critical comments on this chapter and to John R. Twiss, Jr., for insight on various stages of the manatee program. I am grateful to Tim W. Clark and Richard P. Reading for their advice and assistance.

References

Ackerman, B. B. 1992. Ongoing manatee aerial survey programs—a progress report. Abstract. In *Interim Report of the Technical Workshop on Manatee Population Biology*, T. J. O'Shea, B. B. Ackerman, and H. F. Percival (eds.). Manatee Population Research Report no. 10. Gainesville: University of Florida, Florida Cooperative Fish and Wildlife Research Unit.

Ackerman, B. B., S. D. Wright, R. K. Bonde, D. K. Odell, and D. J. Banowetz. 1992. Trends and patterns in manatee mortality in Florida, 1974–1991. Abstract. In *Interim Report of the Technical Workshop on Manatee Population Biology*, T. J. O'Shea, B. B. Ackerman, and H. F. Percival (eds.). Manatee Population Research Report no. 10. Gainesville: University of Florida, Florida Cooperative Fish and Wildlife Research Unit.

Aldrich, H. 1979. *Organizations and Environments*. Englewood Cliffs, N.J.: Prentice-Hall.

Argyris, C. 1992. *On Organizational Learning*. Cambridge, Mass.: Blackwell.

Argyris, C., and D. A. Schön. 1978. *Organizational Learning: A Theory of Action Perspective*. Reading, Mass.: Addison-Wesley.

Bedeian, A. G., and R. F. Zammuto. 1991. *Organizations: Theory and Design*. Chicago: Dryden Press.

Brewer, G. D., and P. deLeon. 1983. *The Foundations of Policy Analysis*. Homewood, Ill.: Dorsey Press.

Brownell, R. L., Jr., K. Ralls, and R. R. Reeves (eds.). 1978. *Report of the West Indian Manatee Workshop, Orlando, Florida, 27–29 March 1978*. Cosponsored by Florida Audubon Society, Florida Department of Natural Resources, U.S. Fish and Wildlife Service, and Sea World of Florida. Unpublished report. Atlanta: U.S. Fish and Wildlife Service.

Butler, R. 1991. *Designing Organizations—A Decision-Making Perspective*. London: Routledge.

Caldwell, D. K., and M. C. Caldwell. 1985. Manatees—*Trichechus manatus* Linnaeus, 1758; *Trichechus senegalensis* Link, 1795; and *Trichechus inunguis* (Natterer, 1883). In *Handbook of Marine Mammals*. Vol. 3: *The Sirenians and Baleen Whales*, S. H. Ridgway and R. Harrison (eds.). London: Academic Press.

Clark, T. W., and A. H. Harvey. 1991. Implementing recovery policy: Learning as we go? In *Balancing on the Brink of Extinction: The Endangered Species Act and Lessons for the Future*, K. A. Kohm (ed.). Washington, DC: Island Press.

Clark, T. W., R. Crete, and J. Cada. 1989. Designing and managing successful endangered species recovery programs. *Environmental Management* 13(2):159–170.

Daft, R. L. 1989. *Organization Theory and Design*. 3rd ed. St. Paul: West.

Florida Department of Natural Resources (FDNR). 1982. Letter from Casey J. Gluckman, director, Division of Resource Management, to Ronald E. Lambertson, associate director for federal assistance, U.S. Fish and Wildlife Service, 5 March.

————. 1992a. *Fiscal Year 1991–92 Annual Report: Save the Manatee Trust Fund.* Prepared for the Florida State Senate and the Florida House of Representatives. Tallahassee: Florida Department of Natural Resources.

————. 1992b. *Manatee Salvage Data Base: Summary Report.* St. Petersburg: Florida Marine Research Institute, Florida Department of Natural Resources.

————. 1994. *Manatee Salvage Data Base: Summary Report.* St. Petersburg: Florida Marine Research Institute, Florida Department of Natural Resources.

Galbraith, J. R. 1977. *Organizational Design.* Reading, Mass.: Addison-Wesley.

Hartman, D. S. 1979. *Ecology and Behavior of the Manatee* (Trichechus manatus) *in Florida.* Special Publication no. 5. Pittsburgh: American Society of Mammalogists.

Husar, S. L. 1977. *The West Indian Manatee* (Trichechus manatus). U.S. Fish and Wildlife Service, Wildlife Research Report 7. Washington, DC: Government Printing Office.

Lacy, R.C. 1993. Vortex: A computer simulation model for population viability analysis. *Wildlife Research* 20:45–65.

Lefebvre, L. W., T. J. O'Shea, G. B. Rathbun, and R. C. Best. 1989. Distribution, status, and biogeography of the West Indian manatee. In *Biogeography of the West Indies,* C. A. Woods (ed.). Gainesville, Fla.: Sandhill Crane Press.

Marine Mammal Commisssion (MMC). 1978a. Letter from John R. Twiss, Jr., executive director, to Lynn A. Greenwalt, director, U.S. Fish and Wildlife Service, 8 March.

————. 1978b. Letter from John R. Twiss, Jr., executive director, to Lynn A. Greenwalt, director, U.S. Fish and Wildlife Service, 23 August.

————. 1978c. Letter from John R. Twiss, Jr., executive director, to the Honorable Robert L. Herbst, assistant secretary for fish, wildlife, and parks, U.S. Department of the Interior, 9 November.

————. 1978d. Minutes of the public session of the Seventeenth Meeting of the MMC and Thirteenth Meeting of the Committee of Scientific Advisors on Marine Mammals.

————. 1987a. Letter from John R. Twiss, Jr., executive director, to Tom E. Gardner, executive director, Florida Department of Natural Resources, 9 October.

————. 1987b. Letter from John R. Twiss, Jr., executive director, to the Honorable Frank H. Dunkle, director, U.S. Fish and Wildlife Service, 19 November.

————. 1987c. Letter from John R. Twiss, Jr., executive director, to the Honorable Frank H. Dunkle, director, U.S. Fish and Wildlife Service, 29 December.

————. 1989a. Letter from John R. Twiss, Jr., executive director, to Susan Recce Lamson, acting director, U.S. Fish and Wildlife Service, 26 May.

————. 1989b. Letter from John R. Twiss, Jr., executive director, to Maryanne S. Bach, deputy assistant secretary for fish, wildlife, and parks, U.S. Department of the Interior, 11 July.

————. 1989c. Letter from John R. Twiss, Jr., executive director, to the Honorable John F. Turner, director, U.S. Fish and Wildlife Service, 1 August.

————. 1990. Letter from John R. Twiss, Jr., executive director, to the Honorable John F. Turner, director, U.S. Fish and Wildlife Service, 2 March.

————. 1992a. *Proceedings of the Thirty-Second Meeting of the MMC and the Twenty-Sixth Meeting of the Committee of Scientific Advisors on Marine Mammals, Tallahassee, Florida, April 30–May 2, 1992.* Vol 1. Washington, DC: MMC.

————. 1992b. Letter from John R. Twiss, Jr., executive director, to the Honorable John F. Turner, director, U.S. Fish and Wildlife Service, 17 June.

————. 1992c. Letter from John R. Twiss, Jr., executive director, to the Honorable John F. Turner, director, U.S. Fish and Wildlife Service, 16 October.

————. 1993a. *Annual Report to Congress, Calendar Year 1992.* Washington, DC: MMC.

————. 1993b. *The Marine Mammal Protection Act of 1972 as Amended.* Washington, DC: MMC.

Marmontel, M. 1992. Age and reproductive parameter estimates in female Florida manatees. Abstract. In *Interim Report of the Technical Workshop on Manatee Population Biology*, T. J. O'Shea, B. B. Ackerman, and H. F. Percival (eds.). Manatee Population Research Report no. 10. Gainesville: University of Florida, Florida Cooperative Fish and Wildlife Research Unit.

————. 1993. Age determination and population biology of the Florida manatee, *Trichechus manatus latirostris*. Ph.D. dissertation, University of Florida.

Odell, D. K., G. D. Bossert, M. T. Lowe, and T. D. Hopkins. 1992. Reproduction of the West Indian manatee (*Trichechus manatus*) in captivity. Abstract. In *Interim Report of the Technical Workshop on Manatee Population Biology*, T. J. O'Shea, B. B. Ackerman, and H. F. Percival (eds.). Manatee Population Research Report no. 10. Gainesville: University of Florida, Florida Cooperative Fish and Wildlife Research Unit.

Packard, J. M., and O. F. Wetterqvist. 1986. Evaluation of manatee habitat systems on the northwestern Florida coast. *Coastal Zone Management Journal* 14(4):279–310.

Powell, J. A., and G. B. Rathbun. 1984. Distribution and abundance of manatees along the northern coast of the Gulf of Mexico. *Northeast Gulf Science* 7:1–28.

Rathbun, G. B., J. P. Reid, R. K. Bonde, and J. A. Powell. 1992. Reproduction in free-ranging West Indian manatees. Abstract. In *Interim Report of the Technical Workshop on Manatee Population Biology*, T. J. O'Shea, B. B. Ackerman, and H. F. Percival (eds.). Manatee Population Research Report no. 10. Gainesville: University of Florida, Florida Cooperative Fish and Wildlife Research Unit.

Reynolds, III, J. E., and C. J. Gluckman. 1988. *Protection of West Indian Manatees* (Trichechus manatus) *in Florida*. Final report to the Marine Mammal Commission, Washington, DC: MMC.

Reynolds, III, J. E., and D. K. Odell. 1991. *Manatees and Dugongs*. New York: Facts on File.

Rose, P., J. Baker, and D. Peterson. 1981. *Comprehensive Work Plan for the West Indian Manatee* (Trichechus manatus). Washington, DC: U.S. Fish and Wildlife Service.

Schön, D. A. 1983. *The Reflective Practitioner*. New York: Basic Books.

Tobin, R. J. 1990. *The Expendable Future: U.S. Politics and the Protection of Biological Diversity*. Durham, N.C.: Duke University Press.

U.S. Fish and Wildlife Service (USFWS). 1978. Letter from Lynn A. Greenwalt, director, to John R. Twiss, Jr., executive director, MMC, 1 May.

————. 1980. *West Indian Manatee Recovery Plan*. Prepared by the U.S. Fish and Wildlife Service in cooperation with the West Indian Manatee Recovery Team. Washington, DC: U.S. Fish and Wildlife Service.

————. 1989. *Florida Manatee* (Trichechus manatus latirostris) *Recovery Plan*. Report prepared by the Florida Manatee Recovery Team. Atlanta: U.S. Fish and Wildlife Service.

————. 1991. Letter from John F. Turner, director, to John R. Twiss, Jr., executive director, MMC, 12 March.

————. 1992. Letter from Richard N. Smith, deputy director, to John R. Twiss, Jr., executive director, MMC, 15 July.

————. 1993. *1992 Annual Progress Report for the Florida Manatee Recovery Plan*. Prepared by the Jacksonville Field Office, U.S. Fish and Wildlife Service, with cooperation of the Florida Manatee Recovery Team.

Yaffee, S. L. 1982. *Prohibitive Policy: Implementing the Endangered Species Act*. Cambridge: MIT Press.

7

The Red-Cockaded Woodpecker Recovery Program

Professional Obstacles to Cooperation

Jerome A. Jackson

The endangered red-cockaded woodpecker (*Picoides borealis*) is a non-migratory species that is endemic to mature, open pine forest ecosystems of the southeastern United States (Figure 7-1). It is a high-profile endangered species and the subject of prolonged conservation attention and a number of important legal cases. Although this woodpecker's endangerment is due directly to habitat loss, the fundamental causes are ultimately rooted in the socioeconomic context of the region. The major problem facing recovery is the insistence by government agencies that management must fall within the constraints of "desired" management practices imposed by the forest industry, i.e., short-rotation, even-aged management. There are many well-documented weaknesses in implementing the Endangered Species Act (see, for example, King et al. 1977; Yaffee 1982; Jackson et al. 1983), and some of these shortcomings are evident in efforts to restore this species (see Ray and Guzzo 1993:89). At the center of these efforts are several federal and state agencies, universities, nongovernmental conservation organizations, private enterprises, and individuals. Conservation efforts have been variously "professional" and "cooperative" over the decades.

This chapter focuses on my professional interactions with the agencies and their staffs responsible for conserving the red-cockaded woodpecker and the ecosystems upon which it depends. Although this is a complex subject that can only be examined briefly here, problems are clearly evident in the organizational and professional systems involved in endangered species conservation. This chapter presents my assessments of some of these problems and suggests what can be done to address them positively and constructively.

FIGURE 7-1.   The red-cockaded woodpecker. (Photo by Jerome A. Jackson.)

The view presented here is a personal one based on over twenty-five years of research on the species throughout its range as well as involvement with recovery efforts—as leader of the species' recovery team for eight years, as principal author of the first recovery plan for the species, and as an expert witness in several legal actions. My research has been supported by the National Science Foundation, Fish and Wildlife Service, Department of Energy, U.S. Army, National Park Service, Forest Service, Bureau of Land Management, the State of Mississippi, and Georgia-Pacific Corporation. My employment at a state university has made it possible for me to work somewhat free of the constraints felt by biologists employed within the various bureaucracies. Nevertheless, as we shall see, opposing the bureaucratic line can have its costs even for nongovernmental employees. I must emphasize that this chapter is from my perspective and that the examples have been selected from my personal experience. Although similar experiences have been shared by others, I recognize that my vantage point is unique and that others may have differing perspectives and draw different conclusions. I do not wish to add to the continued conflict, but present and document my perspective as a basis for comparing, discussing, and hopefully improving and developing a consensus on conservation of the species.

Endangerment and Current Status

Endemic to the mature, open pine forests of the southeastern United States, the red-cockaded woodpecker evolved with the southern pines and was once common within its range. For a summary of the natural and political history of the species, see Jackson (1971, 1987, 1994). Its life history is intimately tied to the forests and to the frequent lightning-started fires that characterize the region.

The red-cockaded woodpecker copes with its fire-dominated ecosystems by excavating nest and roost cavities in living pines rather than in more vulnerable dead trees. To make use of these live trees, however, means that excavating a cavity can take months or years rather than the two weeks woodpeckers typically need for dead trees. The problem has been made easier for the species as a result of development of a strong sociality such that groups of birds, called clans, work together to excavate and maintain cavities that are then handed down from generation to generation through male offspring. Cavity excavation is further facilitated by the selection of trees that are infected with the red heart fungus (*Phellinus pini*), which softens the dead heartwood of older pines. The disease enters longleaf pine (*Pinus palustris*) at about the age of eighty years and other pines at about sixty years and may take two decades to spread enough to be of value to the birds. The forest industry is aware of the disease and trees are generally harvested before the fungus sets in. And therein lies a basic conflict between the birds and humans: the birds need the fungus; the logger wants rot-free wood.

Fire suppression and the resulting crowding and replacement of pines with hardwoods, the clearing of southern forests for nonforest uses, the woodpecker's rigid social system, and fragmentation and isolation of populations have further reduced the bird's numbers and chances for survival. The red-cockaded woodpecker has been officially listed as endangered since 1970, and today more than 80 percent of the populations are on federal lands (Jackson 1978; Lennartz et al. 1983). Yet even these have declined precipitously in recent decades (Conner and Rudolph 1989).

Recovery Efforts

Decline of red-cockaded woodpecker populations was recognized long before the species was officially declared endangered (Steirly 1957), and at least token efforts were made by federal agencies to protect cavity trees. The acceptance of clearcutting as a standard harvesting and management approach for southern pines in the 1960s and early 1970s, however, resulted in very rapid decline throughout the species' range. Although recovery efforts recognize the importance of old-growth forest for the birds, they have yet to come to grips with the basic problems of habitat loss and fragmentation due to clearcutting and short-rotation forestry.

The Agencies

The U.S. Fish and Wildlife Service (FWS) has responsibility for enforcing the Endangered Species Act (ESA) and managing relatively small red-cockaded woodpecker populations in several national wildlife refuges (NWRs). Although the species was historically known from three FWS regions (Southeast, Southwest, and Central), its only population in the Central Region (in southern Missouri) was extirpated in the 1940s (Jackson 1971). Most remaining populations are in the Southeast Region, which takes the lead responsibility for the species.

Most of the woodpecker's populations occur on lands managed by the U.S. Forest Service. All populations on Forest Service lands today fall within Region 8, though the service's research biologists are split between two subregions. Although the Forest Service supports extensive management-oriented research on the woodpecker, the split between management and research branches is such that the service's researchers can only make recommendations and managers have the freedom to decide what to include in forest management plans. Perhaps the second largest number of red-cockaded woodpeckers are found on Department of Defense (DOD) lands located throughout the species range. These include several different management jurisdictions—Army, Air Force, Navy, Marine Corps—and further subdivisions within the services. Other federal agencies with responsibility for small numbers of red-cockaded woodpeckers include the National Park Service (NPS), Department of Energy (DOE), and Bureau of Land Management (BLM). Various state lands, including state forests, wildlife management areas, parks, and highway right-of-ways, also support the species. As well, there are many scattered active cavity tree clusters on private and industrial lands.

The range of government responsibilities and the forest habitat needs of the red-cockaded woodpecker have also attracted the attention of numerous nongovernmental conservation and forest industry organizations (NGOs). Legal actions relating to the species and implementation of the ESA have come from both sides of the NGO fence. With such a web of ownership and management authorities, it is no wonder that coordination of management and recovery efforts has been difficult and demonstrably successful management elusive.

Brief History of Management

Early management for the species was initiated by the Forest Service in some areas even before it was listed as an endangered species. This management consisted of leaving cavity trees stand when the remainder of the forest was cut—a practice, however, that did little to stem the species' decline. Management progressed to leaving cavity trees and a 200-foot buffer of trees around them—another measure that proved inadequate. The role of fire in controlling hardwood understory and promoting the open, parklike conditions the birds seem to favor has long been known but generally ignored. The management recommended for the red-cockaded woodpecker in the first recovery plan (Jackson et al. 1979) might have stemmed the precipitous decline of many populations, but the plan was never really implemented (Freeman 1984; Bigony 1991). Following legal actions by conservation groups, recommendations for management provided in the second recovery plan (Lennartz and Henry 1985) began to be widely implemented during the 1980s. Two major breakthroughs—the development of translocation (DeFazio et al. 1987) and artificial cavity excavation techniques (Copeyon 1990)—coupled with intensive, positive forest management since the late 1980s have produced the first evidence of stability or even growth in some populations (Jackson 1994). Clearcutting, however, a practice opposed by conservation groups and many nongovernmental biologists (see Larmer 1989 and McFarlane 1992), remains the centerpiece of management "tools" included in the recovery plan.

Professional and Organizational Obstacles

A number of obstacles to professional and cooperative efforts have impeded conservation of this species. A few have significantly affected the restoration program in complex ways and are examined here: agency conservatism, decentralized administrative authority, the FWS permit process for research, Section 7 implementation, "consensus building" methods of the agencies, commitment to public education, education of professional ESA implementers, and agency commitment and roles. These considerations form the basis of the general lessons I offer at the end of the chapter.

Conservatism in Federal Agencies

The FWS is a conservative agency (see Yaffee 1982). This conservatism manifests itself in several ways: hesitating to take action until a crisis arises, stifling employee initiatives, and supporting business as usual. Much of the agency's behavior appears to be a reaction to outside pressure.

First, crisis seemed to be the best motivator of the FWS (and the Forest Service and DOD as well) to action on behalf of the red-cockaded woodpecker. The trend was clear: several major federal actions on behalf of the species were the result of political/legal crises brought to a head by people outside the federal bureaucracy. Some examples: it took the threat of legal action by the Sierra Club Legal Defense Fund, the Wilderness Society, and the National Audubon Society to get the FWS to respond to mismanagement of the red-cockaded woodpecker at Fort Benning, Georgia (Freeman 1984, Turner 1991); it took legal action by the Sierra Club Legal Defense Fund, the Texas Committee on Natural Resources, and others to get the Forest Service to provide more positive management for the species in national forests (McFarlane 1992); and it took Hurricane Hugo to get the FWS and the Forest Service to implement translocation and artificial cavity construction techniques on a wide scale. Even on FWS lands—national wildlife refuges—appropriate management did not follow the first recovery plan. Prescribed fire was to be used to control hardwood understory at three-to-five-year intervals, yet budget constraints and other priorities often resulted in only one prescribed burn in ten years (Heinrichs and Heinrichs 1984). In Noxubee NWR, the species plummeted from thirty two active colonies to sixteen as a result of the birds abandoning sites overgrown with hardwoods (personal observation).

Second, the bureaucracy has stifled internally initiated action (see Wilson 1980 and Yaffee 1982). On several occasions agency employees blew the whistle on agency activities that were contrary to the species' conservation, presumably because they felt they could not remedy the problems by working internally. An example: three civilian employees of the army at Fort Benning, Georgia, were indicted in 1992 for destroying red-cockaded woodpecker colonies. The situation was brought to light by civilian employees of the army who reported the infractions in a clandestine fashion with late night calls and parking lot meetings. (I was one of the recipients of the calls and a participant in the meetings.) In another example, information that led to successful litigation against the Forest Service came from within the agency. (For details see McFarlane 1992.) Forest Service employees provided a steady flow of data but were unable to bring about in-house management improvements because of the agency's authority relationships and culture (see Kennedy and Thomas 1992). Morrison (1989) provides a similar example (although not related to the red-cockaded woodpecker) in which a Forest Service employee was reportedly given a choice of demotion or transfer for opposing a clearcut operation. I do not know that these examples are common occurrences, but they illustrate that some employees believe they cannot work solely within the federal bureaucracy to achieve formal policy goals.

The third professional obstacle to cooperative efforts is the support of business as usual. One example illustrates this problem well. The case deals with drilling gas wells on D'Arbonne NWR, Louisiana. D'Arbonne NWR was established as mitigation for a Corps of Engineers project, but mineral rights were not purchased with the surface ownership. Although most of the refuge is not suitable habitat for red-cockaded woodpeckers, five clans persisted in a small pine forest. When owners of the mineral rights decided to drill one gas well every 8 acres, the refuge manager objected, citing noncompliance with ESA, and enlisted local conservation-minded citizens and me to help protect the birds. The Sierra Club and Defenders of Wildlife filed suit and the National Audubon Society assisted. At the trial the coordinator for the FWS testified that the loss of five colonies at D'Arbonne would not jeopardize the continued existence of the species. Not surprisingly, the judge decided in favor of drilling (Turner 1987).

These three forms of agency conservatism—hesitating to act without a crisis, stifling of employee initiatives, and supporting business as usual—combine to produce less conservation management for the red-cockaded woodpecker than might otherwise occur. These bureaucratic features are themselves part of the overall species conservation problem. The next obstacle is a problem in authority management.

Decentralized Authority in the FWS

The FWS holds ultimate power to implement the ESA. That power was formerly centralized in the Office of Endangered Species in Washington. In 1981, however, authority for implementing the act (for all species) was transferred to the regional FWS offices (J. Sheppard, Office of Endangered Species, Washington, DC, pers. comm.). This transfer resulted in some dramatic differences in ESA implementation, as is illustrated in the way recovery plans were produced. The Southeast Region, unlike all other regions, abolished endangered species recovery teams established to advise the regional director and prepare recovery plans. The former recovery team was established in 1975 and existed until 1982 before being terminated. At that time FWS contracted with the Forest Service to revise the recovery plan for the species. During this revision, the red-cockaded woodpecker continued to decline because the approved plan was not being implemented (Bigony 1991).

The FWS's coordinator for the species, a former Forest Service employee, was given responsibility for overseeing the revision process and in the end he coauthored the revision (Lennartz and Henry 1985). When the revision was sent out for external review, however, extensive comments called for major revisions, yet only minimal changes were made. The "revised plan" is in fact a completely new plan bearing no resemblance to the plan put together by the recovery team. Many people in both the academic and conservation communities, including a committee of scientists appointed by the U.S. Section of the International Council for Bird Preservation and the American Ornithologists Union, objected to the biased and inadequate nature of the second recovery plan. For example, Ligon et al. (1986) charged that home range requirements were based on breeding season requirements of a single South Carolina population, rather than yearly needs of the species in various parts of its range. Essentially the plan presents a "cookbook" approach—one management strategy—even though the species is widespread and locally adapted to forest habitats of several pine species in a wide range of climatic regimes. Management needs in the flatwoods of Florida are quite different from those in the mountains of Kentucky. Nevertheless, the plan remains unchanged from the form adopted in 1985 by the FWS.

In the case of the red-cockaded woodpecker, decentralization of endangered species responsibility led to abandonment of the recovery team concept in favor of recovery plan preparation by a single person—in this case, an employee of an agency that had been found in violation of Section 7 of the ESA for its management of the species. This was likened in conservation circles to the fox being placed in charge of the chicken house. As we shall see, authority management problems constrain research as well as management.

The FWS Permit Process

A major professional obstacle to nonagency professionals is red tape—a multiplicity of permits and annual reports required for scientists wishing to work with an endangered species. Without these permits, work cannot be done. To band red-cockaded woodpeckers, for example, a scientist must first have a federal bird-banding permit. Next, appropriate state permits are also required. Moreover, the FWS requires an additional, separate endangered species permit. If the scientist wishes to work in a national wildlife refuge, a national forest, or other federal lands, further permits are required. Obtaining the multiplicity of permits can be a task that varies from routine issuance to near impossibility.

In many instances there seems to be no established process for appeal if applications are denied. A case in point involves my 1990 requests for permits to capture, band, and color-band red-cockaded woodpeckers for population ecology studies in D'Arbonne (Louisiana) and Noxubee (Mississippi) NWRs. A permit to work in D'Arbonne was essentially issued by return mail, good for a period of three years, with minor restrictions. An eight-page request for a permit to continue nearly twenty years of work in Noxubee NWR, however, was turned down ostensibly because I was not following the scientific method, my literature survey was inadequate and cited too many of my own papers, there were questions about the statistics I proposed using, and past annual reports included information that was "incomplete and questionable," although no annual report was ever previously questioned. When I appealed the rejection to the regional director, I was informed that the refuge manager had final say. Ultimately the Sierra Club came to my assistance and, combined with a letter of support from the attorney general of Mississippi, I was issued a permit. That permit, however, had many restrictions that essentially precluded the conservation work required—for example, it required a 24-hour advance notice to work in colonies near refuge headquarters for which I had the greatest history, no banding of nestlings, and no work after dark (which is necessary to capture birds on the roost). I learned subsequently that the assistant refuge manager later applied for and received a permit to do woodpecker banding in the refuge following my methods (Jackson 1982).

The process for obtaining permits under the current FWS permit system seems open to many influences. Moreover, it appears that permits requested for FWS personnel may be issued in preference to (and under much more lenient guidelines than) those requested by nongovernmental researchers and with little regard for experience with the species. The current system, then, seems biased against the nongovernmental researcher. With the red-cockaded woodpecker becoming increasingly restricted to federal lands and research on the species becoming increasingly dominated by government biologists, information about its status is becoming increasingly controlled by the agencies. As we shall see, actions between agencies with regard to the species often leave much to be desired.

Section 7: Woodpeckers, the Military, and the Forest Service

Section 7 of the ESA requires federal agencies to ensure that their activities do not jeopardize the existence of threatened and endangered species. The FWS oversees government activities through an informal and formal consultation process authorized by Section 7. Through this process the service issues biological opinions about proposed federal projects and their effects on listed species.

The U.S. Army has been through several difficult Section 7 consultations for red-cockaded woodpeckers on southern military installations (see Jackson and Parris 1994; Nickens 1993). The support the army gives this woodpecker and other endangered species varies and is sometimes questionable. For example, the recently published "Commander's Guide to Environmental Management" (Pringle 1991) only briefly mentions endangered species and the ESA. While commanders in the field may be poorly informed, DOD biologists have generally received good training (DOD 1991).

In cooperation with the FWS and the army, I have repeatedly been involved in red-cockaded woodpecker management on military bases. Some management resulted in Section 7 consultation. In one instance it resulted in grand jury indictments of civilian employees of the army for ESA violations (Dailey 1992). Although support within the military has improved—from open hostility to genuine interest and concern—FWS support of endangered species conservation under ESA has lagged behind essential management. The military has never used national defense to exempt one of its projects from the ESA, yet the FWS has almost always yielded to military requests—almost automatically. This near "acquiescence" has resulted in the destruction of dozens of red-cockaded woodpecker colonies and thousands of acres of foraging habitat (Jackson and Parris 1994; Nickens 1993).

Several examples illustrate the weaknesses in Section 7 application. At Fort Polk, Louisiana, the attitude of army biologists and commanders has been to do what it takes to protect the woodpecker. Interest has been high and sincere. When the army wanted to construct a multipurpose range at Fort Polk, it presented the FWS with three alternative sites, all with active cavity tree cluster of the species on them (Jackson and Parris 1994). The army's preferred site harbored the largest number of active cavity tree clusters. Although the FWS recommended selection of a different site, it approved the army's choice. Their biological opinion stated that:

> The Army's selected alternative (2A) represents the most severe impact to the Fort Polk red-cockaded population and over the long term will likely result in loss of most if not all clans occupying the intensive use area. . . .
>
> Therefore, it is the Biological Opinion of the U.S. Fish and Wildlife Service that the selected alternative (2A) for siting of the MPRC and its cumulative effects is not likely to jeopardize the continued existence of the red-cockaded woodpecker. . . .
>
> [The] amount of incidental take that is possible and would not be a violation of the "taking" prohibitions of Section 4(d) and 9 of the Act is estimated at up to 52 birds . . . a loss of 20 clans. . . . [Jordan 1984]

In the end, the army selected a different alternative for reasons other than the red-cockaded woodpecker. Monitoring has thus far revealed no birds killed and only one active cavity tree cluster within the range abandoned as a result of the project (Jackson and Parris 1994).

Fort Benning, Georgia, presents a substantial contrast in attitudes of civilian and military personnel and examples of inaction and inadequate Section 7 implementation. In 1976, I consulted on a possible conflict involving an active cavity tree cluster of red-cockaded woodpeckers at the center of a site where a new barracks complex was being built (Freeman 1984). Upon arrival, I was ushered into a colonel's office and his first words to me will never be forgotten: "Dr. Jackson, I ain't never seen a red-cockadoodled woodpecker and I never want to see one. You do what it takes to move them." The army had already destroyed all trees at the colony site except for four cavity trees. The FWS gave permission for the army to continue its activities, but the army had to fund an effort to move the birds and provide positive management for a nearby colony. In 1980, I returned to see how the mitigation colony was faring and was shocked to find an obstacle course in the middle of the active cavity tree cluster (Freeman 1984). The FWS had required mitigation, but there had been no follow-up to assure that it was done.

Section 7 consultations with the Forest Service demonstrated similar patterns. For example, a biological opinion regarding management of red-cockaded woodpecker habitat in Texas national forests cited several major shortcomings (Nelson 1977). Once issued, however, both the FWS and the Forest Service essentially ignored the opinion. In 1988, after a decline of more than 70 percent in some populations (Conner and Rudolph 1989) and following suits filed by the Sierra Club Legal Defense Fund, the Texas Committee on Natural Resources, and the Wilderness Society, Judge Robert Parker mandated management that the FWS should have required and enforced more than a decade earlier (McFarlane 1992; Nickens 1993). The ruling has changed the species' management in national forests across the Southeast (Nickens 1993).

Section 7 consultation is intended as a mechanism to assure that the actions of federal agencies do not further endanger endangered species. Although the concept is a good one, its implementation has at times seemed more a routine seal of approval for agency actions than a real effort to protect endangered species. When biological opinions have concluded that actions would jeopardize the woodpecker, the jeopardy opinion has not always been followed by appropriate action. Often the goal seems to be the defusing of difficult problems within the bureaucracy rather than species recovery. Consensus-building efforts within the bureaucracies appear to be another manifestation of the same problem.

Consensus Building: Agency Models

As restoring endangered species is a prolonged, multiparty activity, having consensus on goals, timetables, and tasks is helpful. Consensus is best achieved through forums and participation processes that blend the views of interested parties. But these are not always the methods used to find consensus. Two examples of "consensus" building are examined here: national forest management planning and a special summit meeting on the red-cockaded woodpecker.

National Forest Management Plans. Management plans for southern national forests go through a long development and approval process including public hearings and preparation of environmental impact statements. The process is tedious and involves the public much less than is represented. The 1976 National Forest Management Act and the 1969 National Environmental Policy Act require the Forest Service to plan forest management and compare alternative management programs. The public is then asked to comment on alternatives framed by the service—but often there is no good environmental choice. Until the Texas court decision on national forests, for example, the only forest management approach for managing red-cockaded woodpecker habitat was clearcutting. But, in fact, there are many problems with the approach (Jackson 1986; McFarlane 1992).

The way in which national forest planning fails this species' conservation is illustrated in the Final Environmental Impact Statement for the Kisatchie National Forest Land and Resource Management Plan, Louisiana (USFS 1986). Of eight alternatives, the Forest Service packaged Alternative D as the "red-cockaded woodpecker" alternative. Although it was management's preferred alternative, it was not the best alternative for the woodpecker. Alternative D called for eighty-year rotations for longleaf pine and sixty-year rotations for other pine species—far shorter rotations than the recovery plan's recommendations of a hundred years and eighty years, respectively. Alternative E (initially labeled the "amenity" alternative) provides the recommended longer rotations, but that fact is not mentioned in the narrative, only in a table (USFS 1986:table 11-2, p. II-21). All the alternatives called for clearcuts of up to 80 acres, a narrow range of possible options. In one sense, the public had virtually no real say in the overall management of their national forest. To the degree that this is true, the plan fails the American public and it fails the red-cockaded woodpecker.

The Red-Cockaded Woodpecker Summit. In March 1990, I received an invitation (Rosen 1990a) from the National Wildlife Federation to attend "a summit meeting of experts on the biology and management of the red-cockaded woodpecker." The invitation stated:

> The summit is intended to provide a neutral forum for open discussion among experts on key points regarding the biology and management of the species. . . . We hope the result of the summit will be a consensus among participants, based on scientific evidence and expert opinion, on as many key points as possible about the biology of the species as well as specific forest management practices that will lead to protection and recovery of the bird. . . . No minutes of the meeting will be kept and no "outsiders," including staff of the National Wildlife Federation will participate in the actual discussion sessions. . . . All participants are expected to participate as independent professionals. Agency and organization

policy regarding the species should not be allowed to bias discussion. . . .

A second letter a week later (Rosen 1990b) said: "We were able to choose only 24 participants . . . from over 60 candidate scientists. . . . The process we used to select participants was intended solely to identify the top 24 researchers and managers." When I arrived at the summit, I learned that the "top 24" included key administrators within the Forest Service and FWS, as well as people with whom I was unfamiliar. Nine of the "top 24" were from the Forest Service, and at least three additional service employees ("outsiders") sat in on sessions and conferred with service participants. Missing were several researchers with distinguished publication records. Thus the composition of the forum brought the purpose and prospects for the whole meeting into question.

It appeared, from the very beginning, that the summit had purposes other than what had been represented. The Forest Service largely controlled the agenda and discussions and seemed to want a "scientific" consensus for their red-cockaded woodpecker environmental impact statement and perhaps an endorsement for their role in ongoing legal actions against them. During sessions we were divided into three groups, and each was instructed to reach a consensus on key management issues for the species (no minority opinion allowed). Those of us nonagency experts who shared similar ideas had been split up and our voices submerged. Those with decades of experience with the species were given equal voice with those having very limited experience. The meeting appeared set up to co-opt many of us and enhance the Forest Service's legitimacy.

Despite the early assurance that minutes would not be taken, a 36-page summary report was published (Southeast Negotiation Network 1990). In it were identified "numerous areas of consensus" that were reached by the twenty-four "widely respected experts on the RCW" (p. 1–2). There were some truly worthwhile points of consensus–such as the agreement that summer prescribed burns were needed. But there was also endorsement of issues—such as even-aged management—that some of us strongly oppose. Gene Wood, a participant in the summit and a professor at Clemson University, submitted a detailed "disclaimer" that was circulated with the summary (Wood 1990:5). He noted, among other things, that: "While it was frequent that general agreement was reached on many general statements, substantial disagreement existed on many specific statements. Representing all of this as consensus is misleading as to the complexion of discussion and array of points of view."

Neither national forest planning mechanisms nor the Forest Service/National Wildlife Federation summit did little to build consensus. To the contrary, they polarized the community of interests around the red-cockaded woodpecker and harmed the prospects for its conservation. Publication of the summary report on the summit was an ill-conceived attempt to "educate"—an endeavor at which federal agencies have failed.

Public Education: Whose Responsibility?

In the first recovery plan, educating the public about the species was considered an integral part of the recovery effort in recognition that the ultimate causes of the species' decline are human attitudes and behavior (Jackson et al. 1979). In fact, it was considered so important that it was part of the first item of the step-down plan. If the species is to be saved, these attitudes and behavior must be changed through an active, well-led educational campaign implemented by the FWS and other responsible agencies.

The second recovery plan, however, deletes the education portion (Lennartz and Henry 1985). A strong case can be made that this omission makes the work of the FWS even harder. Education about the woodpecker's habitat requirements, reasons behind its endangered status, links with other components of southeastern pine forest ecosystems, and how woodpecker management benefits other components of the southern pine forests should be an essential part of the recovery effort. Emphasis should be placed on the species' positive role in forest ecosystems (control of insect pests, provision of nest sites for other species) and the positive effects of the species' management on game species such as northern bobwhite (*Colinus virginianus*; Brennan and Fuller 1993) and fox squirrels (*Sciurus niger*).

Lack of public knowledge creates misconceptions and negative attitudes about recovery. Education should respond quickly to erroneous or misleading statements advanced in lobbying against conservation. For example, some forest industries claim that woodpecker management will reduce funds for schools and roads, that wood and paper products could escalate in price, and that "the woodpecker" could have "a serious impact on hunting, not only on national forest lands but on all forest lands, both public and private" (Anon. 1989:1). The putative economic costs of endangered species management are the typical focus of news stories (as in Rivas 1989). Some truly are outrageous. The Texas Farm Bureau, for example, published an article titled "Red-Cockaded Woodpecker—Endangered Species Issues Threaten All of Agriculture" (Barnett 1988). Even employees of federal agencies have used their positions to lobby the public against red-cockaded woodpeckers (see Savelle 1993) without response from the FWS.

There is a great need, therefore, for strong public education on the recovery effort, the species, and its many beneficial aspects in regional ecosystems. Such a program is lacking today. And education programs should not end with the public but should extend to those charged with implementing ESA policy.

Educating the Professionals

The range of knowledge and skills needed for effective, efficient, and equitable ESA implementation is substantial. Most of those entrusted with its implementation are well motivated and hardworking. But it takes more than well-meaning, hardworking professionals to bring about a program's success. Many other variables are essential.

In practice, the many areas of knowledge needed for successful programs are often absent or underused. Traditionally, the wildlife profession takes a narrow focus on problems and emphasizes management of game species. This tendency reflects the history of the discipline and professional training. For decades, students in university wildlife departments, especially in the South, wrote theses and dissertations on game species, furbearers, and sometimes a pest species. The FWS, Forest Service, and other federal agencies turned to these departments for their employees. It is therefore not surprising that, following passage of the ESA, the middle to top levels of federal agencies were staffed by people with game management backgrounds. For these agencies, a quick "retooling" and "retraining" was needed to administer the ESA. This process began with personnel at the bottom of the hierarchy. Despite the need and current efforts, the overall management system remains largely dominated by traditional thinking.

Finding new employees with the right knowledge and skills requires new university training. But the ESA caught university wildlife departments by surprise. Introducing endangered species management into wildlife departments is still a slow process. The study of endangered species was more often done in zoology or biological sciences departments. Graduates of these departments were not always given equal footing in competition for wildlife jobs with the agencies. But since the 1970s, a number of red-cockaded woodpecker biologists with zoology/biology backgrounds have been hired by the federal agencies.

Getting knowledgeable endangered species biologists into agency systems has been hindered by a traditional university education, especially in the southeastern United States. Most universities have a school of forest resources within which wildlife management is a subprogram. At Mississippi State University, for example, the wildlife student must major in forestry with a "wildlife option." This arrangement promotes a view that foresters are central to management and wildlife biologists are a subset of management. Professionals are taught that forests should generate profits, even if other forest values are harmed. This is even true in national wildlife refuges, where even-aged management (a euphemism for clearcutting) has prevailed in southern refuges (Braun et al. 1978). Although clearcutting remains a dominant management practice in southern refuges, the situation is slowly changing. For example, vigorous management of the species was instituted in Noxubee NWR in the late 1980s following a 50 percent decline in active colonies in about fifteen years.

Just as the wildlife profession focused on game management, the forestry profession continues to focus on maximum-economic-return forestry. Dan Lay, a pioneer of research on the red-cockaded woodpecker, said it succinctly: "Modern silviculture is based on getting your money as early and as fast as you can before something happens. The longer you leave a tree out there, the more risk you have for insect damage or lightning damage" (quoted in Bigony 1991:14). True multiple use of forest lands, including special uses such as endangered species management, run against the established modus operandi and violate basic principles that have been taught for decades.

Antagonistic sentiments prevail in some forestry classrooms even today. A former forestry professor, whose course was required for wildlife students, would take students to local lumber mills and other forest industries and, before the proprietor could begin, the professor would ask him: "What problems are the damned environmentalists causing you?" We do not know how widespread this practice is, but clearly a profession that has been educated to respect endangered species and understand their role in ecosystems would do much to promote implementation of the ESA and recovery of species. An agency's commitment will only be as strong as the commitment of the people within it.

Agency Commitment and Roles

Wildlands—and as a result endangered species—are increasingly limited to those lands under federal ownership, particularly in the eastern United States. It is likely that more than 80 percent of red-cockaded woodpecker colonies are on federal lands (Jackson 1978; Lennartz et al. 1983), and this percentage increases each year. Federal land managers and agency administrators increasingly seem to want "their own people" to do all endangered species work on "their" lands. With nonagency biologists out of the way, the presentation and interpretation of data are theirs. As a result, there is either no accountability or accountability comes so late that harmful impacts affecting the species are beyond correction.

Science requires data verification. In some cases, this does not come easy on federal lands. A case in point is the status of red-cockaded woodpeckers at the Savannah River Plant (SRP), South Carolina, a 300-square-mile nuclear facility operated by DOE whose forests are managed by the Forest Service. From the mid-1970s through the early 1980s, under DOE contracts, I studied the behavioral ecology of the species at SRP and monitored its decline from eighteen active colonies to as few as five birds including only one breeding pair (McFarlane 1992). The cause for decline was clear: older trees were being removed through clearcutting to establish even-aged stands grown on short rotations. In addition, the woodpecker's remaining habitat was not being properly managed. Long absence of fire resulted in dense understory growth, a condition conducive to colony abandonment.

Inquiries and comments on the Forest Service's inadequate management fell on deaf ears or, worse, antagonistic ears. I had long argued against the sixty-year rotations on which longleaf pine was being managed, and finally the forest supervisor informed me that "just for your birds" the longleaf pine rotation would be increased to eighty years. I thanked him for the information but suggested that even eighty years was inadequate since the average cavity tree age in longleaf pines is about a hundred years. About a month later, I found most of the remaining foraging range of one clan clearcut. When I confronted the forest supervisor and reminded him of his eighty-year rotation, he calmly said "Yes, we have an eighty-year rotation there, but we have to start at zero." Lennartz and Stangel (1989) also describe the plight of the woodpecker at SRP, noting that the Forest Service was "fully committed to restoring the red-cockaded population." They neglect to state that the commitment finally came in 1985, after the species had declined from eighteen colonies to five birds.

When my contract came up for renewal in 1984, the Forest Service successfully argued that since they were responsible for managing the forest and habitat, they should also be responsible for research on the red-cockaded woodpecker. The renewal proposal I submitted to DOE was turned down and a Forest Service proposal to do the same work was funded at several times the cost of my proposed budget. Ultimately a subcontract from the Forest Service, for more money than I had originally requested, was given to me and my colleagues to do part of the work. Following the rationale I proposed, young females of the species were introduced to males-only colonies. The procedure worked (DeFazio et al. 1987). The SRP population had a new lease on life, if only it could hang on until the habitat improved.

My involvement at SRP ended in 1986. At the third red-cockaded woodpecker symposium, a Forest Service employee told of the wonderful success story at SRP and how the woodpecker's population had swelled to nearly forty birds (Gaines et al. 1994). The tale, however, is far from over. The habitat available at SRP is still marginal, and Lennartz and Stangel (1989:15) acknowledge that "decades will have to pass before the trees mature into 85–95-year-old nesting habitat." The Forest Service has now introduced nearly forty birds to the population; the population, in other words, has grown by about as many birds as have been introduced. As Griffith et al. (1989:479) note: "Without high habitat quality, translocations have low chances of success regardless of how many organisms are released or how well they are prepared for the release."

These examples raise serious questions about agency commitment and the roles they assume in endangered species conservation. Independent monitoring and evaluation of agency programs for endangered species are sorely needed.

Lessons

The experiences cited throughout the chapter form the basis for a number of professional lessons useful to both agency and nonagency people involved in endangered species restoration. Overall, the lessons focus on the individual—the practicing conservation biologist. Essentially there are six lessons:

- *Know thyself and know when to act.* To be most effective, professionals must be committed and maintain high professional and ethical standards. They must also appreciate their own limitations. Saying "I don't know" should be part of professional practice. Sacrifice is essential at times. Learning must be a central part of the work, and a good understanding of people is vital. Know when and how to speak out effectively and when not to try. Professionals should be careful of what they put into print. If it turns out to be incorrect years later, it can come back to haunt future efforts.

- *Know thy species.* Knowing the species' biology is absolutely essential. Learn the literature first. Learn all aspects of the species' life history and community interactions and interdependencies. This knowledge gives a biologist a firm foundation for action and credibility. Management recommendations must be based on this knowledge, as lack of reliable knowledge hinders conservation.

- *Know thy colleagues.* To be effective, professionals should appreciate subtle differences in people's personality, knowledge, cognitive ability, and social skills. This knowledge is absolutely essential to enable them to interact effectively with others. Lines of communication should be developed, fostered, and maintained. Be open to advice and criticism from colleagues. Bridging communication gaps and promoting open dialogues are ongoing challenges.

- *Know thy organization.* In-depth knowledge of organizational systems is also absolutely essential for effective practice. Organizational pathologies are common; knowing how to recognize and solve them is crucial. Just as endangered species need management, there are many structural and cultural features of organizations that need explicit management attention. Mounting evidence suggests that a large bureaucratic system is not the best organization to be in charge of species restoration.

- *Know thy opposition.* Not all parties in a complex program will necessarily agree, nor is it necessary that they do. Nevertheless large, prolonged programs like the red-cockaded woodpecker recovery effort, which involve hundreds of people and scores of organized interests, are prone to divisive interpersonal and interorganizational problems. The scale of these problems makes them difficult to ameliorate. In such circumstances it is vital to understand the perspectives of the opposition. This knowledge can be essential in finding opportunities to increase communication and cooperation.

- *Regroup and evaluate.* In working in complex endangered species projects over many years it is necessary to maintain a distant perspective. This may even require withdrawing for periods from frontline action, but never fully giving up participation. Cooperation is essential, but it is sometimes painful. Maintain your perspective, energy levels, and commitment. Never lose sight of the overarching goal: Save The Species.

When the Endangered Species Act became law, the FWS was woefully unprepared for the task of implementing it. Over the last twenty years the service has shown itself to be very conservative indeed in the act's implementation. Moreover, FWS biologists were not trained for endangered species conservation, having been educated in university programs that focused on game species management. Other federal agencies, such as the Forest Service and DOD, revealed similar professional shortcomings.

The federal bureaucracy is learning about the needs of endangered species management, albeit very slowly. The twenty years since passage of the ESA have seen the retirement of many old-guard wildlife careerists. Academia seems to be changing more slowly to meet the training needs of frontline conservation biologists involved in endangered species management—probably because of the combined effects of declining enrollments, shrinking budgets, and long careers of tenured faculty. Nonetheless, some highly motivated, well-trained biologists with endangered species backgrounds are finding places within federal agencies. The pace of species' decline, however, may make this a race that some species will lose.

A number of basic changes are needed in the endangered species bureaucracy (Chapter 16). The FWS should lead the way as a staunch advocate for endangered species conservation, a strong and exemplary protector of species, and an uncompromising enforcer of the ESA. The service needs to educate the public about the plight of endangered species and be responsive to other organizations and individuals committed to restoration. Only time will tell whether the FWS, along with other federal and state agencies, can rise to the challenge.

Acknowledgments

I especially thank Tim Clark, who organized this collective effort as a course taught at the University of Michigan in the fall of 1992 and who brought us together again in the winter of 1993 to discuss the problems of implementing the ESA. My wife, Bette, who shared much of my fieldwork and many of the triumphs and frustrations of working with the red-cockaded woodpecker, provided helpful comments. Richard Conner of the U.S. Forest Service and Robert Hole at Mississippi State University read earlier drafts of the chapter and offered useful comments.

References

Anonymous. 1989. Woodpecker could impact hunting. *Willamette Wildlife* 3(3):1, 3.

Barnett, M. 1988. Red-cockade woodpecker—Endangered species issues threaten all of agriculture. *Texas Agriculture* (Texas Farm Bureau), 21 October, p. 6.

Bigony, M. L. 1991. Controversy in the pines. *Texas Parks & Wildlife* 49(5):12–17.

Braun, C. E., K. W. Harmon, J. A. Jackson, and C. D. Littlefield. 1978. Management of national wildlife refuges in the United States: Its impacts on birds. *Wilson Bulletin* 90:309–321.

Brennan, L. A., and R. S. Fuller. 1993. Bobwhite quail and red-cockaded woodpeckers: What's the connection? *Mississippi Wildlife* 5(1):7–8.

Conner, R. N., and D. C. Rudolph. 1989. Red-cockaded woodpecker colony status and trends on the Angelina, Davy Crockett, and Sabine National Forests. USDA Forest Service Research Paper SO-250.

Copeyon, C. K. 1990. A technique for constructing cavities for the red-cockaded woodpecker. *Wildlife Society Bulletin* 18:303–311.

Dailey, L. B. 1992. Benning workers charged in red-cockaded woodpecker case. *Columbus Ledger-Enquirer* (Georgia), 30 January, p. 1.

DeFazio, J. T., Jr., M. A. Hunnicutt, M. R. Lennartz, G. L. Chapman, and J. A. Jackson. 1987. Red-cockaded woodpecker translocation experiments in South Carolina. *Proceedings of the Annual Conference of the Southeastern Association of Fish and Wildlife Agencies* 41:311–317.

Department of Defense (DOD). 1991. *Proceedings of the Red-Cockaded Woodpecker Management Workshop*. Marine Corps Base Camp LeJeune, North Carolina.

Freeman, J. T. 1984. Woodsman, spare that woodpecker. *Defenders* 59:4–13.

Gaines, G. D., W. L. Jarvis, K. Laves, et al. 1994. Red-cockaded woodpecker management on the Savannah River site: A management/research success story. In *The Red-Cockaded Woodpecker: Species Recovery, Ecology and Management*, D. L. Kulhavy, R. Costa, and R. G. Hooper (eds.). Nacogdoches, Texas: Center for Applied Studies, College of Forestry, Stephen F. Austin State University.

Griffith, B., J. M. Scott, J. W. Carpenter, and C. Reed. 1989. Translocation as a species conservation tool: Status and strategy. *Science* 245:477–480.

Heinrichs, J., and D. B. Heinrichs. 1984. Rare birds and big trees: East—the woodpecker and the pines. *American Forests* 90(3):24–49.

Jackson, J. A. 1971. The evolution, taxonomy, distribution, past populations and current status of the red-cockaded woodpecker. In *The Ecology and Management of the Red-Cockaded Woodpecker*, R. L. Thompson (ed.). Tallahassee: Bureau of Sport Fisheries and Wildlife, U.S. Department of Interior, and Tall Timbers Research Station.

———. 1978. Analysis of the distribution and population status of the red-cockaded woodpecker. In *Proceedings of the Rare and Endangered Wildlife Symposium*, R. R. Odom and L. Landers (eds.). Technical Bulletin WL4. Atlanta: Georgia Department of Natural Resources, Game and Fish Division.

———. 1982. Capturing woodpecker nestlings with a noose—A technique and its limitations. *North American Bird Bander* 7:90–92.

———. 1986. Biopolitics, management of federal lands, and the conservation of the red-cockaded woodpecker. *American Birds* 40:1162–1168.

———. 1987. The red-cockaded woodpecker. In *Audubon Wildlife Report 1987*, R. L. DiSilvestro (ed.). New York: Academic Press.

———. 1994. Red-cockaded woodpecker (*Picoides borealis*). In *The Birds of North America*, No. 85. A. Poole, and F. Gill (eds.). Philadelphia: The Academy of Natural Sciences, and Washington, DC: American Ornithologists' Union.

Jackson, J. A., and S. D. Parris. 1994. The ecology of red-cockaded woodpeckers associated with construction and use of a multi-purpose range complex at Fort Polk, Louisiana. In *The Red-Cockaded Woodpecker: Species Recovery, Ecology and Management*, D. L. Kulhavy, R. Costa, and R. G. Hooper (eds.). Nacogdoches, Texas: Center for Applied Studies, College of Forestry, Stephen F. Austin State University.

Jackson, J. A., W. W. Baker, V. Carter, T. Cherry, and M. L. Hopkins. 1979. *Recovery Plan for the Red-Cockaded Woodpecker*. Atlanta: U.S. Fish and Wildlife Service.

Jackson, J. A., B. J. Schardien, and P. R. Miller. 1983. Moving red-cockaded woodpecker colonies: Relocation or phased destruction? *Wildlife Society Bulletin* 11:59–62.

Jordan, D. B. 1984. Letter of 20 November 1984 to Col. A. D. Smith, Ft. Polk, La.

Kennedy, J. J., and J. W. Thomas. 1992. Exit, voice, and loyalty of wildlife biologists in public natural resource/environmental agencies. In *American Fish and Wildlife Policy: The Human Dimensions*, W. R. Mangun (ed.). Carbondale: Southern Illinois University Press.

King, W. B., J. A. Jackson, H. W. Kale II, H. F. Mayfield, R. L. Plunkett, Jr., J. M. Scott, P. F. Springer, S. A. Temple, and S. R. Wilbur. 1977. The recovery team–recovery plan approach to conservation of endangered species: A status summary and appraisal. Report of the Committee on Conservation 1976–1977. *Auk* 94 (supplement):1DD–19DD.

Larmer, P. 1989. A clearcutting ban for the birds. *Sierra* 74(2):28–30.

Lennartz, M. R., and V. G. Henry. 1985. *Red-Cockaded Woodpecker Recovery Plan*. Region 4. Atlanta: U.S. Fish and Wildlife Service.

Lennartz, M. R., and P. W. Stangel. 1989. Few and far between. *Living Bird Quarterly* 8(4):15–20.

Lennartz, M. R., P. H. Geissler, R. F. Harlow, R. C. Long, K. M. Chitwood, and J. A. Jackson. 1983. Status of red-cockaded woodpecker populations on federal lands in the South. In *Red-Cockaded Woodpecker Symposium II Proceedings*, D. A. Wood (ed.). Tallahassee: Florida Game and Fresh Water Fish Commission.

Ligon, J. D., P. B. Stacey, R. N. Conner, C. E. Bock, and C. S. Adkisson. 1986. Report of the American Ornithologists Union Committee for the Conservation of the Red-Cockaded Woodpecker. *Auk* 103:848–855.

McFarlane, R. W. 1992. *A Stillness in the Pines*. New York: Norton.

Morrison, C. 1989. Engineer: Forest Service punished him for opposition to clear-cut. *Atlanta Constitution*, 4 April, pp. 1A, 8A.

Nelson, W. O., Jr. 1977. Final biological opinion: Red-cockaded woodpecker Section 7 consultation on Texas National Forests. Letter to James E. Webb, regional forester, U.S. Forest Service, Atlanta.

Nickens, E. 1993. Woodpecker wars. *American Forests* (January/February): 28–55.

Pringle, W.J.B. 1991. *Commander's Guide to Environmental Management*. Aberdeen Proving Grounds, Md.: U.S. Army Corps of Engineers.

Ray, D. L., and L. Guzzo. 1993. *Environmental Overkill*. Washington, DC: Regnery Gateway.

Rivas, M. 1989. Fight over woodpecker habitat continues. *Dallas Morning News*, 1 January, pp. 41A–42A.

Rosen, R. A. 1990a. Letter of 2 March 1990 to Jerry Jackson, Mississippi State.

———. 1990b. Letter of 9 March 1990 to Jerry Jackson, Mississippi State.

Savelle, W. 1993. Have a voice in forestry issues. *Starkville Daily News*, 7 January, p. 10.

Southeast Negotiation Network. 1990. *Summary Report: Scientific Summit on the Red-Cockaded Woodpecker*. Atlanta: Georgia Institute of Technology.

Steirly, C. C. 1957. Nesting ecology of the red-cockaded woodpecker in Virginia. *Atlantic Naturalist* 12:280–292.

Turner, T. 1987. Trouble on the split estate. *Defenders* 62(1):28–36.

———. 1991. Ground zero: The American military vs. the American land. *Wilderness* 55(194):10–36.

USDA Forest Service (USFS). 1986. *Final Environmental Impact Statement: Kisatchie National Forest Land and Resource Management Plan.* Atlanta: Forest Service, Southern Region.

Walters, M. J. 1992. *A Shadow and a Song.* Post Mills, Vt.: Chelsea Green.

Wilson, J. Q. 1980. *Bureaucracy: What Government Agencies Do and Why They Do It.* New York: Basic Books.

Wood, G. 1990. Memorandum to the file, 26 June 1990, circulated to Red-Cockaded Woodpecker Summit participants with the summary report: *Scientific Summit on the Red-Cockaded Woodpecker.* Atlanta: Georgia Institute of Technology.

Yaffee, S. L. 1982. *Prohibitive Policy: Implementing the Endangered Species Act.* Cambridge: MIT Press.

8

The California Condor Recovery Program

Problems in Organization and Execution

Noel F. R. Snyder

The California condor (*Gymnogyps californianus*, Figure 8-1) is one of the most critically endangered species in the United States and has been the subject of public concern for over a century. An intensive program to study and conserve the condor began in the late 1930s under the leadership of Carl Koford (1953). Koford was followed by a succession of other workers. Of these, Miller et al. (1965), Wilbur (1978), and Snyder and Snyder (1989) have provided detailed summaries of conservation endeavors. Many of the issues in condor conservation have also been discussed, with varying degrees of accuracy, in a steady stream of magazine articles, books, and book chapters by authors not directly involved in the conservation program. (See, for example, Phillips and Nash 1981, Darlington 1987, DiSilvestro 1989, Tober 1989.)

The principal agencies involved in condor conservation have been the U.S. Fish and Wildlife Service (FWS), the U.S. Forest Service, the California Department of Fish and Game (CDFG), the California Fish and Game Commission (CFGC), the Bureau of Land Management (BLM), the National Audubon Society (NAS), the Los Angeles Zoo (LAZ), and the Zoological Society of San Diego (ZSSD). Field research and conservation activities have been conducted primarily under the leadership of the FWS and NAS; the Forest Service and BLM have been the principal habitat managers for the species; and the LAZ and ZSSD have had primary responsibilities for the captive breeding program.

My direct involvement in the condor program extended from early 1980 to mid-1986, primarily as the FWS's biologist in charge of field research and conservation activities. In this capacity, I had an opportunity to observe closely many of the strengths and weaknesses in

FIGURE 8-1.   The California condor. (Photo by N.F.R. Snyder.)

how efforts were organized and implemented for this species. Here I summarize what appear to be some of the most important generalizations that can be drawn from that experience. As background, I begin with a brief look at the recent organization of condor recovery efforts. This section is followed by a review of a period of intense controversy—the crisis of 1985—which illustrates many of the problems that have affected not only the condor program but other endangered species programs as well. After offering some potential solutions to these problems, I turn to basic recommendations on the role of recovery plans in recovery efforts and on the makeup and functioning of recovery teams.

Organization of Recovery Efforts

Like many endangered species efforts, the California condor program has been a multiagency effort coordinated by means of a recovery team. A condor recovery team has existed since the early 1970s, and prior to that time there was a condor advisory committee that served, in effect, as a recovery team. The recovery team has had several incarnations, and each has included a broad representation of involved agencies and biologists. Like other recovery teams, the condor teams have been established by appropriate regional directors of the FWS and have served in an advisory capacity to these regional directors. The only exception to this pattern was the condor recovery team of 1980 to 1985, which was established by FWS officials in Washington, D.C., and made recommendations directly to these officials.

My overall view of the functioning of the various condor recovery teams through 1986 is strongly positive. Members of the teams were dedicated and competent and quite consistently strove to design and recommend optimal conservation actions for the species. Given the information that was available to them, these teams did an excellent job in formulating strategies. The recovery plan written by the first team for the condor in 1974 was the first recovery plan written for any endangered species, although it has since been rewritten several times and remains in a state of flux. Most of the problems experienced during the years I was in the program stemmed not from any mal-functioning of the recovery team but from more fundamental flaws in the way efforts were organized and funded.

During the years of my participation, principal funding for the pro-gram came from the FWS, but much of this money was actually lob-bied annually from Congress by the NAS as "add-on" appropriations and was in part transferred to the NAS for its role in the program. This influence over the budget gave NAS considerably more clout with the FWS than was true of other cooperators in the program, and was one of the major considerations in the disputes that developed. The overall budget for the program by the mid-1980s was over $1 million annually, and the very size of this budget was a major factor influencing decision making. Perhaps not surprisingly, NAS was more successful than FWS biologists and the recovery team in getting its strategies adopted by the FWS in a number of instances.

The principal disputes during the early 1980s, however, arose from a lack of consensus between the FWS and the CFGC regarding the proper role of intensive approaches, such as radio telemetry and cap-tive breeding, in conservation efforts. Since both these agencies had to grant permits for such activities, their continued disagreement effec-tively delayed intensive research and conservation efforts for several years. At that time, those of us participating in the field program felt uniformly frustrated by the level of caution exhibited by the CFGC. By 1982, NAS was threatening to leave the program if the CFGC did not begin to allow recommended actions. By 1983, however, the CFGC had in many respects become the most progressive agency in the pro-gram and my perspective on the state/federal conflict had changed considerably. It had now become apparent that the balance of au-thority represented by the dual state and federal permitting system was capable of preventing some highly detrimental actions proposed by one agency or the other. This was sometimes of crucial benefit for the program.

While the complexities in the development of consensus among cooperators often slowed things down considerably, this at least generally ensured a more careful examination of strategies than would have otherwise taken place. The sorts of mistakes that can occur in endangered species programs include moving too fast on issues as well as moving too slowly—depending on circumstances, a relatively cumbersome process of consensus development can sometimes be highly beneficial. Overall, a fairly strong argument can be made that crises are often less crucial than they seem at the time. Often there really is time for a process of consensus development that involves all parties having a full opportunity to be heard. Nevertheless, it is well to recognize that when the last few black-footed ferrets (*Mustela nigripes*) are about to die of disease, one needs quick action, not endless debate. Similar situations demanding quick action occurred in the condor program as well.

In general, the various condor recovery teams have functioned well in offering a forum for debate by diverse interests, and I believe this has been highly beneficial. Conservation strategies are often uncertain in their risks and values, and it is entirely appropriate that they be examined closely before implementation. If consensus can be achieved by a diverse recovery team, this has tremendous value in enabling government agencies to implement policies that are both biologically and politically sound. Government agencies often act with such extreme caution on controversial issues that they are unwilling to allow anything to happen that could possibly backfire. Recovery team consensus can be the mechanism to break an impasse of inaction. This was demonstrated repeatedly with the condor recovery team.

Nevertheless, the advantages in relying on recommendations of the condor recovery team have not always been recognized by all participants in the program. This was particularly true in 1985, when team recommendations ran counter to the goals of certain powerful program participants. During this period the team did not command enough political power to retain a primary role in charting conservation strategies for the species. And even though the strategies of the team did eventually prevail, the team itself wound up as a casualty of the controversies.

The Condor Crisis of 1985

By the end of 1984, knowledge about the factors limiting the remnant wild population was increasing steadily, and the condor program was proceeding rapidly in establishing an "insurance" captive population. The wild population included only fifteen individuals at that point and was still declining, although the conservation program had achieved a recent increase in total numbers of condors through multiple-clutching wild pairs (of which there were then five in existence). Fourteen young had been taken captive from 1982 to 1984, mostly as eggs, and good genetic representation had been gained from several pairs.

As a consequence of this success, the recovery team developed a plan whereby multiple-clutching would continue, but when five progeny were gained in captivity from any pair, additional progeny from the pair would then be returned to the wild in a release program designed to sustain and bolster the wild population. This strategy was accepted by all cooperators in the program, including the FWS, the CFGC, and the NAS, and was official policy by early 1985. Since two pairs were already represented by five young apiece in the captive flock, it appeared that if these pairs produced additional young in 1985, some of their progeny could be returned to the wild. Preliminary plans for such releases were under way.

Unfortunately, a basic assumption of these plans was that survival of wild pairs would continue to be relatively good, as in the years just preceding. The winter of 1984–1985 revealed this assumption to be grossly unrealistic. Of the five pairs active in 1984, only one survived to breed in 1985, and this pair was not one of the two pairs already represented by five young apiece in captivity. Clearly the accepted criteria by which birds could be released to the wild would not allow any releases of birds in 1985.

In fact, with the loss of six of the fifteen birds left in the wild (40 percent of the population) during the winter of 1984–1985, the wisdom of the basic strategy of doing releases became highly questionable. It was becoming obvious that overall mortality rates were much too high to allow any realistic hope of sustaining the wild population, even by expending the "capital" that had been achieved by then in captivity. The known mortality factors—especially lead poisoning (which was first identified as a major threat in 1984 and 1985)—were not ones that could be effectively countered in the near term. The only defensible near-term strategy left was to defer releases to the wild and concentrate on attempting to achieve a viable captive population by bringing in the last wild birds. Although this goal was eventually achieved, the process by which it was accomplished was tortuous (see Snyder and Snyder 1989).

Responses of the Recovery Team and Agencies Involved

The recovery team met in April 1985 to assess the implications of the heavy mortality of the previous winter and to develop recommendations for the future. At that point it was clear that the two pairs on which releases would depend no longer existed and that some other birds had also disappeared, although there were still some doubters as to whether six birds had truly been lost over the winter. In any event, the team voted unanimously that there should be no near-term releases of captives to the wild, consistent with policies adopted the previous year. On the question of whether all remaining wild birds should be brought into captivity there was no consensus as yet. About half the team (including myself) urged this course of action; the other half (including the NAS representative) was in opposition.

The team did develop a consensus, however, to solicit recommendations from the most prominent population geneticists in the country as to whether the existing captive flock could be considered adequate. The team also agreed to reconvene once the recommendations were obtained in order to determine whether consensus on what to do with the remaining wild birds would then be possible.

The recommendations of population geneticists were promptly solicited, and responses were received within a few weeks. Their responses were entirely uniform: the existing captive population was much too deficient in genetic breadth to allow confidence that it could serve as a viable base for preserving the species. The only way to achieve such a base would be to bring the remaining wild birds into captivity.

However, although these responses were received and circulated among team members, no meeting of the recovery team followed. The FWS would not allow such a meeting. Instead it came out with its own recommendations in May 1985 (at the urging of NAS and without any consultation with the recovery team or its own field biologists in California). These recommendations were that three birds in the wild flock should be taken captive and replaced with three birds released from captivity to the wild. The logic of these recommendations was neither clearly presented nor obvious. There was no recognition of the evident inviability of the wild population and the futility of throwing away any captives in trying to prolong its existence.

Shortly thereafter the CFGC came out with its own recommendations. Aware of the previous deliberations of the recovery team and the recommendations of the population geneticists, the CFGC authorized the trapping of all wild birds and concurred with the recovery team that no birds should be released to the wild in the near term. The only point in common between the FWS position and the CFGC position was that three wild birds could be trapped into captivity, and this was accomplished during the summer of 1985.

Debates over whether the rest of the birds should be taken captive, and whether there should be releases, continued through the rest of the year. The recovery team was allowed only one more meeting—a session closely supervised by FWS officials in August 1985. The team at this point reaffirmed its opposition to near-term releases of captives and was able to achieve a consensus that at least three of the six remaining wild birds should be taken captive. By the fall of the year, there was a consensus on the team that all remaining wild birds should be taken captive, although no formal meeting of the team to express this consensus was ever permitted. This same consensus was nevertheless stated as a resolution of participants in the Third International Vulture Symposium held in Sacramento in November 1985.

Meanwhile, the FWS continued to seek a compromise position acceptable to both the NAS and the CFGC that would still allow the possibility of some birds remaining in the wild and releases to the wild in the near future. In October, the CFGC finally agreed to a FWS position of three more birds coming into captivity and a possibility of releases in 1986 if agreement could be reached on a site for releases and if there were three releasable birds available. While the CFGC still favored all wild birds coming into captivity and no near-term releases of captives to the wild, it took the position that it was better to get at least three more of the wild birds into captivity in the short term than to risk getting none of them into captivity in a continued stalemate. The CFGC was still left in a position where it could veto releases because of the conditions attached.

In late November, the female of the last remaining wild breeding pair, a bird not yet authorized for captivity, was captured to replace her defective radio transmitters. A blood sample taken at this time, but not analyzed until after the bird was released, proved to be heavily contaminated with lead. This finding led the veterinarians and toxicologists advising the program to recommend that the bird be immediately recaptured for treatment. The detection of high lead levels in this condor (she had had low lead levels when captured earlier in the fall) proved to be a crucial development in convincing the FWS that the remaining wild condors were still at high risk—especially since this particular bird was one that the NAS and some personnel in the

FWS had claimed was at very low risk from such poisoning because of her supposedly "safe" foraging habits and range.

In mid-December 1985, Richard Smith, an associate director of the FWS, sought and obtained a reversal of FWS policy. The FWS announced that it now supported bringing all remaining wild condors into captivity and refraining from near-term releases of captives to the wild—the long-standing position of the CFGC and the consensus position of the moribund recovery team. The reversal was attributed primarily to the high lead levels in the November-trapped condor, but also to an absence of any captive birds behaviorally suitable for release in 1986 as well as the fact that a new pair was forming in the wild between one of the birds slated for capture and one of the birds to be left in the wild. At the same time as it announced its intention to take the last wild condors into captivity, the FWS also declared its intention to follow through with the establishment of Hudson Ranch as a condor refuge, an acquisition strongly advocated by the NAS.

Before trapping could begin, however, the NAS filed suit to obtain an injunction against capture of the last wild birds, alleging that this decision was arbitrary and capricious and that taking the last birds captive would doom habitat preservation efforts. The NAS won the first round in the courts but was ultimately defeated on appeal. The last surviving wild condors were finally all trapped into captivity by Easter Sunday 1987.

Failures in Information Transfer and Basic Causes of Controversy

During the NAS lawsuit, documents came to light revealing that information transfer was being significantly distorted between levels in the FWS. In fact, the recovery team's position rejecting near-term releases of condors in 1985 was turned around 180 degrees by middle-level administrators of FWS and presented as an endorsement of the FWS position recommending releases. The recovery team's position was also misrepresented to the NAS board by NAS administrators anxious to gain board support for their position. Even my own position opposing near-term releases and favoring trapping all birds into captivity was at times misstated at higher levels in the FWS.

One of the most important forces driving these and other events during the 1985 crisis was a fear among certain middle-level administrators, both in FWS and NAS, that complete loss of the wild condor population would threaten the continuation of the condor field program. Loss of the program was especially worrisome because of all the spin-off benefits ("overhead monies") that would be affected. (Of the funds being generated by the program only a minority were actually getting to the field level to support research and conservation for the condor.) Other program participants were more concerned about other potential adverse consequences of removal of all birds from the wild. In their view:

- It would be impossible to save condor habitat if all birds were brought into captivity. The NAS was particularly concerned that its campaign to persuade the FWS to purchase Hudson Ranch as a condor refuge would ultimately fail with no birds left in the wild.

- Valuable wild traditions of the condor population would be forever lost if all birds were taken captive.

- Invaluable trained personnel would be lost from the field program to other jobs if the program was even temporarily discontinued.

- It would be nearly impossible to reestablish a field program in the future if all birds were taken captive.

Nevertheless, in the view of many of us in California, none of these concerns was nearly as compelling as the need for salvaging a biological entity—the California condor—under desperate conditions where every bird added to the captive flock might be the difference between success and failure. By 1985 the species had still not been bred in captivity, and it was speculative whether most individuals would reproduce under captive conditions and whether enough genetic diversity remained among the last few individuals to allow recovery. The wild population was clearly not salvageable. To worry about saving habitat, condor traditions, and condor programs and personnel under this circumstance was to miss the forest for the trees.

Furthermore, when the foregoing concerns were examined closely they all had fundamental weaknesses. We had no evidence that the condors were in any way limited by habitat availability. And in any event, the Forest Service had guaranteed protection of all condor use areas under its control into the indefinite future even if all condors were taken captive. The importance of Hudson Ranch to future recovery of the species was far from clear, even though the ranch had been used regularly by condors for foraging. While it was legitimate to

argue that certain condor traditions might be important for future survival of the species in the wild, the assumption that all condor traditions were worth saving was highly questionable. Evidently some of
these traditions were leading the birds into high vulnerability to un-
controllable mortality factors. Leaving birds in the wild would preserve "lethal" as well as "valuable" traditions. Further, if one wanted
to save condor traditions in general, they could probably best be saved
by taking birds into captivity with potential release back into the wild
at some future date, rather than by watching the last birds perish in
the wild. That termination of the field program would lead to skilled
personnel taking other jobs was true enough. But the argument that
a field program could not be revived with skilled personnel at some
future date was not persuasive.

To suggest, in effect, that individuals of critically endangered species
must be sacrificed to save programs, personnel, or habitat is one of the
most dubious positions that can be advanced in endangered species
conservation and should always be viewed with great suspicion when
it surfaces. In any event, during the crisis a number of us proposed a
viable way to save the program that would not entail sacrificing any
California condors. This was a proposal to initiate surrogate release
studies with Andean condors (*Vultur gryphus*) in potential future release areas for California condors during the period when California
condors would all be in captivity. While the surrogate release proposal
was made primarily for biological reasons—especially to develop locally appropriate release techniques—we were all fully aware that it
would solve programmatic problems for the FWS as well. Nevertheless, the proposal was initially rejected by the FWS and some other
agencies in part because they felt it would be politically inviable. The
proposal proved to be both politically and biologically sound, was ultimately implemented, and did indeed save the field program.

Some Problems and Solutions
While the motives of various individuals and organizations during the
crisis of 1985 were undoubtedly complex and diverse, and while
events of that period often seemed more confusing than enlightening
at the time, certain generalities emerged from the crisis and are worth
emphasizing:

- Accurate and insightful field data on the factors causing species
  endangerment are absolutely essential for development of successful conservation strategies. Attempts to base strategies on self-

serving rationalizations and philosophical preferences have a negligible probability of success.

• Tight control over the flow of information, coupled with centralized decision making, often leads to faulty decisions.

• Parties with vested interests in particular recovery decisions should be free to voice their opinions, but should be excluded from a primary role in making those decisions.

• The FWS's financial dependence on nongovernmental organizations (NGOs) can lead to strong influences of the NGOs on public policy.

• The FWS's interference in the continuous functioning of the condor recovery team and disregard for its recommendations greatly impeded a proper resolution to controversies. The condor team needed much more independence from the FWS.

• The restrictions placed on public disclosure of the recovery team's recommendations allowed misrepresentations of these recommendations.

Many of these generalities are not unique to the condor program but apply as well to conservation efforts with many other species.

How can these problems be combated? I can suggest a number of means by which such difficulties might be reduced or averted, although some of the solutions would demand very difficult to achieve restructuring of organizations:

1. Top administrators should make efforts to receive information from a wide variety of sources in addition to immediate subordinates and should be mandated to discuss key issues directly and confidentially with field personnel and recovery teams.

2. Chains of command should be kept as short and simple as possible. Programs overburdened with administrators are especially likely to use endangered species to further their own preservation as a higher priority than the goal of aiding the species.

3. Government agencies must fully recognize the logic in creating independent recovery teams, of high quality, in allowing open communications by the teams with the public, and in following team recommendations. Recovery teams of poor quality will rarely be able to guide species recovery efforts successfully.

4. Government agencies must strive to avoid financial dependence on NGOs, as tempting as this option may be in the short term.

5. Primary authority for field decisions should be vested in field biologists—within the framework of general policy recommendations proposed by recovery teams and accepted by responsible agencies. Centralization of control is demoralizing to fieldworkers, especially when centralized decisions are unsound and undermine the abilities of fieldworkers to succeed in their primary responsibilities. Centralized control leads ultimately to a loss of high-quality fieldworkers in addition to jeopardizing the success of programs.

6. Emphasis needs to be placed on the hiring of top-quality personnel at the field level. Attempting to run an endangered species program without well-motivated and highly skilled personnel is a betrayal of the public trust. In contrast, programs with highly qualified personnel need a minimum of supervision and provide the best results for minimum cost. As obvious as it may seem, this point is sometimes forgotten by administrators eager to hire personnel mainly on the basis of their presumed or demonstrated willingness to support agency policies, right or wrong.

7. The system of rewards and benefits in government organizations and many NGOs needs restructuring to favor efficiency and decentralization of authority and to punish proliferation of bureaucracy. Above all, actual performance in achieving species recovery, not loyalty to the organization, must become the primary basis for rewarding employees.

8. To ensure maximum objectivity and efficiency in recovery actions, research on behalf of endangered species should be performed as much as possible by nonagency personnel. A tremendous resource for high-quality research exists, for instance, in university professors and in graduate student degree programs at universities. This resource should be utilized to a much greater extent than it has been. Maximum use of this resource would go a long way toward solving many of the control and expense problems that have developed in endangered species conservation.

Admittedly, some of these recommendations are easy to state but not so easy to implement. Once bureaucracies are in place, they are almost impossible to dislodge because of their instincts for survival and the rigidity inherent in government employment practices. Few administrators think of themselves as bureaucrats or recognize the harm caused by administrative actions that lead to steadily declining authority and efficiency at the field level. In fact, many of the most burdensome actions are undertaken in the belief that they will ensure "better science," enhance "accountability," and foster apparently desirable goals such as "long-term planning." Even when it becomes common knowledge that certain programs have become impossibly bureaucratic and ineffective, with researchers hobbled by paperwork and restrictions, the solutions proposed usually tend only to be cosmetic reorganizations of bureaucracy that offer no fundamental improvement.

Those of us who participated in the FWS Endangered Wildlife Research Program when it was first established can remember how efficiently things worked and how much less expensive it was for the taxpayer. In those rosy years, when the program was small and had virtually no middle-level administrators, authority and control of field operations were largely delegated to field personnel, morale in the field could not have been higher, and progress in many species efforts was outstanding.

Unfortunately since those early years, the bureaucracy administering the program and the overall budget for the program have expanded greatly while the number of field stations has actually shrunk from nine to just two (and part of another). At the same time, authority for field actions has progressively shifted from the field stations to higher levels of the FWS. Morale and effectiveness of field biologists have declined greatly, and most have left the program voluntarily.

These regrettable trends are not confined to the FWS. Bureaucracies seem to evolve inexorably in all organizations, whether governmental or private, and the process is incremental: each step leads to the next in a gradual, almost unnoticeable, fashion. Unfortunately, the stable asymptotic state is one of near paralysis at the field level, with enormous budgets are dedicated almost entirely to middle levels of administration and spent almost completely on nearly useless committee meetings, training sessions, pointless paperwork, and salaries, travel, and benefits for administrators.

A basic problem seems to be this: most organizations tend to become too large for top administrators to know directly what is happening at the field level and to be able to make intelligent and timely decisions on the basis of such knowledge. When top administrators are not in direct contact with field personnel, they become vulnerable to manipulation by middle-level administrators. Further, they become reluctant to delegate authority to the field level and find it difficult to evaluate the performance of field personnel accurately. Once organizations grow to a point where middle-level administrators are perceived as essential, a dangerous chain of consequences is set into motion. Ultimately the goal shifts from maximizing field accomplishments to maximizing the power and benefits accruing to middle-level administrators. Perhaps the only way to avoid this shift is for conservationists to recognize that small is not only beautiful, it is absolutely necessary if there is to be long-term effectiveness in organizations.

If organizations cannot be kept small, perhaps the most that can be hoped for is the continuous inception of new programs to replace those that have gone through the normal bureaucratic evolution and become impotent. For a number of years, these new programs may be quite highly motivated and effective—until they too build up administrative structure in the middle and fade into senescence.

Recovery Teams vs. Recovery Plans

While properly constituted recovery teams have proved to be a valuable component of recovery efforts, official recovery plans have generally been a disappointment. The fundamental assumption of traditional recovery plans is that detailed long-range planning is both feasible and worthwhile. With most endangered species work, however, there is insufficient knowledge to plan intelligently very far into the future, and recovery plans have had a discouraging tendency to become obsolete within months or even weeks of the time they have been finalized. Meanwhile they have demanded a tremendous amount of time, energy, and resources to develop and review. I have been personally involved in the writing of recovery plans for three endangered species—the Puerto Rican parrot (*Amazona vittata*), the Puerto Rican plain pigeon (*Columba inornata wetmorei*), and the California condor—and I do not believe that these plans have had anywhere near the value that was once envisioned for them.

Research on endangered species has repeatedly demonstrated a high degree of unpredictability in what is learned as programs progress. (See Snyder et al. 1987 and Snyder and Snyder 1989.) Yet the advances in understanding achieved by research frequently overturn previously held assumptions on how best to conserve species.

With only small populations available for study, efforts to identify the principal limiting factors are plagued by large temporal fluctuations in the importance of these factors. Conservation strategies have to change rapidly with the intrinsic instability of the information base being developed for the species. The process is basically inductive— rather than deductive ("generative" rather than "calculative" in the terminology of Westrum 1986)—and cannot be successfully encapsulated in a traditional recovery plan format because commonly many of the elements important for recovery are often unknown at the start. The recovery plan process is not presently designed to accommodate rapid changes in understanding, for it is a laborious and time-consuming task to continually redraft plans to keep them consistent with current information and then gain approval for these changes. Although recovery plans are supposed to be reviewed annually and continuously updated, this almost never takes place (see GAO 1988), because the task is so overwhelming. The energy that would be entailed is clearly much better spent performing recovery actions under a much less ambitious planning arrangement.

In general, recovery plans are much too detailed and are burdened with many recommendations that are based on too little information. Commonly a substantial fraction of actions listed in plans are never funded or implemented or even taken very seriously (see GAO 1988). Often many actions are of questionable utility from the start and are only included for the sake of "completeness." There is no real value in tabulating these low-priority actions and little value in general in planning ahead in a detailed manner on more than an annual basis. Things usually change too rapidly.

I would much prefer to see a different mode of recovery team action adopted and implemented—one that would emphasize interim and ad hoc documents addressing specific parts of the recovery process and specific time periods such as the breeding season of a given year. In effect, this is the way the California condor recovery program operated during much of the 1980s. The recovery team was involved not nearly so much in trying to keep the official recovery plan updated as in preparing position documents that were smaller in scale and much more timely and focused in impact. These documents, like the recovery plan, were directed primarily toward the federal and state agencies in charge of administering the program, but they had much more influence than did the recovery plan itself. There was neither the time nor the staff to be constantly revising the full plan in an attempt to keep it current. The recovery plan was generally viewed as an "after the fact" document that would have to be fixed up some day rather than as a "predictive" document.

Because recovery plans become outdated rapidly and include many recommendations of questionable value, they should not generally be trusted for guidance in any long-term sense. Trust should instead be placed in continuously functioning quality recovery teams and their most current ad hoc recommendations. Unfortunately, the finalization of a recovery plan for a species has often been the pretext for dissolving the recovery team (e.g., the Puerto Rican parrot recovery team)—as if once a plan is generated, there is no further need for expert input to the FWS. Nothing could be more biologically short-sighted and politically counterproductive.

Recovery plans have had the very objectionable tendency to become surrogate endpoints of recovery efforts—perhaps because they are tangible results that agencies can point to as accomplishments whereas actual progress toward recovery goals is often much harder to define and evaluate. Even the Government Accounting Office review of recovery efforts (GAO 1988) used the number of recovery plans produced as a measure of progress. Yet the existence of a recovery plan tells us virtually nothing about whether progress toward recovery is taking place. In reality, substantial progress often occurs in the absence of an approved plan and, conversely, can fail to occur even when an approved plan exists. Recovery plans are at best a minor component of the functioning of an effective recovery effort. While they are not completely without value, their merits often do not justify their costs. Unfortunately, they have gained a status far beyond their true worth.

Despite their many weaknesses, I would not recommend that recovery plans be abolished completely. Instead, I believe they should be scaled back to brief and general narrative accounts of the state of knowledge about the factors limiting species populations, broad research needs, and potential conservation strategies. More focused interim and ad hoc plans, including detailed budgets and assignments of responsibilities, should form annual appendixes to the general plan. The highly complex step-down outlines and multiyear budgets that form the bulk of present-day recovery plans should be abandoned for lack of utility.

Makeup of Recovery Teams

Despite my enthusiasm for recovery teams, as opposed to traditional recovery plans, it is important to recognize that recovery teams can only function well if they are properly constituted and empowered.

This has not always been the case. Clark and Westrum (1989) give a valuable general discussion of the characteristics of high-performance teams in wildlife conservation. To expand on their observations, I want to emphasize four characteristics that appear to be especially important in the constitution and functioning of high-quality recovery teams.

First: Teams must include the principal biologists who have worked with and are working with the species in question—representing the full range of biological perspectives relating to the species.

Second: Teams should also receive input (but not necessarily voting representatives) from all major agencies involved in the conservation of the species, whether federal, state, or private. When involved agencies have biologists or other relevant technical specialists working with the species, these personnel should be candidates for full membership on the team. When involved agencies lack such personnel, they should provide only nonvoting consultants to the team. Above all, the voting membership of teams should not include personnel whose only function is to represent agencies—biological and pertinent conservation perspectives must remain the paramount focus in recovery team recommendations. Further, agencies and organizations with serious conflicts of interest over conservation of the species should not be represented with voting membership on recovery teams, although teams should certainly seek comments from such agencies and organizations.

Third: Teams should also include members or consultants with general expertise in ecology and small population biology, as well as individuals with real expertise in public education and outreach. A large fraction of endangered species problems involve conflicts without economic or political interests that cannot be solved with such expertise. Moreover, depending on particular circumstances, recovery teams should be flexible enough to add personnel with other special talents as needs arise.

Fourth: Teams must be given a measure of independence from government agencies so that they can give the best possible advice to the agencies on a continuing basis. There are several components of this independence:

- Deciding whether or not a recovery team should be formed or terminated for a particular endangered species, and who should be on this team, should not simply be decisions of the regional FWS director. This concentration of power in the hands of regional directors has led to many abuses (see Miller et al. 1994). A much better arrangement would be for such decisions to be made by the FWS in consensus consultation with appropriate professional organizations, such as the National Academy of Sciences, the Society for Conservation Biology, and the American Ornithologists Union (in the case of endangered birds). For example, determining whether a recovery team is needed for a particular bird species and who should be on the team could be decided by a panel composed of single representatives of the FWS, the American Ornithologists Union, and the Society for Conservation Biology.

- Recovery teams should be encouraged to communicate with the public—not forbidden to do so, as is currently the practice of the FWS. This communication, of course, must not apply to specific information that could result in jeopardizing species (divulging locations of breeding pairs, for example), but it should emphatically apply to general policy recommendations. The public has a right to know what the most knowledgeable experts are recommending for a species—and a right to view that information uncensored by agency biases. This idea does not in any way threaten the final authority of the FWS and state agencies to make decisions regarding endangered species, but it will signal when they are implementing actions that are not recommended by recovery teams.

- It is especially important to recognize that recovery team chairpersons should be unaffiliated with the FWS or state resource agencies. Government employees face fundamental limits in how independently they can act. For example, during much of the condor crisis of 1985 there was no way that the condor recovery team could meet because the chairperson was an employee of the FWS who was instructed by the service not to allow the team to meet (obviously because the FWS intended to pursue actions it knew the recovery team would not endorse). Several months later, this incarnation of the condor recovery team was dissolved. These were fundamental mistakes on the part of the FWS that blinded the agency to the very expertise it so desperately needed in the crisis. Once the FWS's intended course of action proved both politically and biologically inviable, a new recovery team was formed and things returned to a more reasonable operational basis.

Under current practices, recovery teams for the FWS are made up largely of government personnel (approximately 77 percent of team members in a recent survey of recovery teams by Miller et al. 1994). Further, approximately 89 percent of recovery team chairpersons in the same survey were either FWS or state government personnel. Many of these personnel are vulnerable to reprisals if they deviate significantly from the agency's goals and objectives in their recommendations.

In general, recovery teams should have much greater proportional representation of nongovernment personnel if the agencies are to get the most objective and valuable advice. If implemented properly, the involvement of professional organizations in selecting the recovery team personnel should ensure a good balance of nonagency and agency personnel and should also correct the regrettably frequent exclusion from teams of some of the best experts on particular endangered species.

A properly constituted recovery team can usually offer federal and state agencies a course of action that represents a consensus of involved agencies and is both biologically and politically viable. In my experience, the political sophistication of well-formed recovery teams has often far surpassed that of federal and state agencies. This point has frequently been misunderstood. Federal and state agencies have sometimes found themselves at odds with the recommendations of recovery teams for very parochial reasons—such as exaggerated fear of special interests—and have often failed to appreciate the biological and political wisdom underpinning the recommendations.

Unfortunately, a number of recovery teams have not been well constituted—teams formed almost solely of administrators with no direct knowledge of the species involved, for example, or teams formed with members whose agencies have serious conflicts of interest with conservation of the species. Such recovery teams cannot be expected to offer anything very perceptive regarding conservation efforts. They seem to be formed primarily when the FWS officials involved are not truly interested in expert advice and seek only to retain maximum control over programs or to mollify special interests while giving the public an impression that their policies represent a consensus of expert opinion.

Further recovery teams do not currently even exist for a number of crucially endangered species, such as the Puerto Rican parrot and the red-cockaded woodpecker (*Picoides borealis*), because of arbitrary decisions of regional directors. These situations cry out for remedy.

Lessons

The California condor program has experienced many of the problems in organization that have afflicted other endangered species programs. In many respects the condor program has functioned better than many other programs, though at tremendous financial and personal costs. Maintaining an open, task-oriented program has been a challenge in the face of agency pressures for control, self-perpetuation, and self-aggrandizement. The major challenges for the future include maintaining high-quality personnel at the field level, giving them enough authority to function effectively, and maintaining an open forum for debate of strategies with the widest possible input from informed experts. As these goals are not automatic outcomes of the way endangered species efforts are currently organized, I have tried to suggest certain ways in which they can be more surely reached in the future.

I find the recovery team concept to be an excellent idea, but it has been badly abused in many cases. Recovery teams need to be much more consistently composed of high-quality personnel, to be much more independent of agencies, and to be empowered in various ways that do not usurp final authority from responsible agencies but do encourage agencies to act in the most responsible manner. A recovery team's recommendations should be open to the public and divorced from the tyranny of recovery plans (which must be fundamentally revised in concept). Further, recovery team actions should be expanded in such realms as recommending specific personnel and agencies to perform essential recovery actions. Administrators need to be educated in the overwhelming advantages of trusting high-quality recovery teams and employing high-quality, as opposed to merely obedient, field researchers. While no recovery program, however well designed and staffed, will be free from mistakes, top-down efforts characterized by low-quality recovery teams, pedestrian fieldworkers, and tightly controlled flow of information are almost guaranteed to fail.

The basic problems in administering endangered species programs are no different from those found in other programs (Chapters 14 and 16). The same mistakes in organization and implementation seem to recur endlessly, despite considerable discussion of these mistakes. (See Yaffee 1982; Clark et al. 1989; Tobin 1990; Kohm 1991.) The amount of taxpayer money that is wasted in accomplishing very little is often phenomenal, and there often seems to be no sure method to achieve true efficiency and productivity. While the most promising way to make progress in this regard is no doubt through wide discussion of the problems, so long as proposed solutions do not come to grips with basic weaknesses in human nature, progress may continue to be limited.

Recent trends are not encouraging. Instead of a deemphasis of the importance of traditional recovery plans, we are starting to see state resource agencies imitating the federal government by expending resources and staff in creating "state recovery plans" for the very same species for which federal plans already exist. Instead of adequate support for the field studies necessary to establish what factors are limiting wild populations of many endangered species, we often see enormous sums spent on "population viability analysis workshops" even though enough demographic data to make such workshops fruitful are lacking for almost all species. We see enormously expensive captive breeding programs being proposed for species even though studies to establish the need for captive breeding have not been made. And, sadly, we see government agencies moving progressively toward monopoly control of information by their control over who is allowed to do research on endangered species and by their increasingly channeling research activities to in-house researchers (see Miller et al. 1994). The sum total of such trends is steadily declining effectiveness in recovery efforts in the face of steadily mounting financial costs. The situation can be changed—but probably not without a major effort to recognize the fundamental problems causing current difficulties and not without a major reorientation of organizations involved in endangered species conservation.

Acknowledgments

Early drafts of this chapter were reviewed by Scott Derrickson of the National Zoological Park, Jim Wiley and Herb Raffaele of the FWS, and Vicky Meretsky of the University of Arizona. I greatly appreciate the many useful improvements they suggested.

References

Clark, T. W., and R. Westrum. 1989. High-performance teams in wildlife conservation: A species reintroduction and recovery example. *Environmental Management* 13:663–670.

Clark, T. W., R. Crete, and J. Cada. 1989. Designing and managing successful endangered species recovery programs. *Environmental Management* 13:159–170.

Darlington, D. 1987. *In Condor Country*. Boston: Houghton Mifflin.

DiSilvestro, R. L. 1989. *The Endangered Kingdom*. New York: Wiley.

General Accounting Office (GAO). 1988. *Endangered Species: Management Improvements Could Enhance Recovery Programs*. GAO/RCED-89-5. Washington, DC: Resources, Community, and Economic Development Division.

Koford, C. B. 1953. *The California Condor*. National Audubon Society Research Report no. 4. New York: National Audubon Society.

Kohm, K. A. (ed.). 1991. *Balancing on the Brink of Extinction: The Endangered Species Act and Lessons for the Future*. Washington, DC: Island Press.

Miller, A. H., I. McMillan, and E. McMillan. 1965. *The Current Status and Welfare of the California Condor*. National Audubon Society Research Report no. 6. New York: National Audubon Society.

Miller, B., R. Reading, C. Conway, J. A. Jackson, M. Hutchins, N. Snyder, S. Forrest, J. Frazier, and S. Derrickson. 1994. Improving endangered species recovery programs: Avoiding organizational pitfalls, tapping the resources, and adding accountability. *Environmental Management* 18(3).

Phillips, D., and H. Nash (eds.). 1981. *The Condor Question: Captive or Forever Free?* San Francisco: Friends of the Earth.

Snyder, N.F.R., and H. A. Snyder. 1989. Biology and conservation of the California condor. *Current Ornithology* 6:175–267.

Snyder, N.F.R., J. W. Wiley, and C. B. Kepler. 1987. *The Parrots of Luquillo: Natural History and Conservation of the Puerto Rican Parrot*. Los Angeles: Western Foundation of Vertebrate Zoology.

Tober, J. A. 1989. *Wildlife and the Public Interest*. New York: Praeger.

Tobin, R. J. 1990. *The Expendable Future: U.S. Politics and the Protection of Biological Diversity*. Durham, N.C.: Duke University Press.

Westrum, R. 1986. Management strategies and information failures. Paper presented at the NATO Advanced Research Workshop on Failure Analysis of Information Systems, Bad Winsheim, Germany, August 1986.

Wilbur, S. R. 1978. The California condor, 1966–76: A look at its past and future. *U.S. Fish and Wildlife Service North American Fauna* 72:1–136.

Yaffee, S. L. 1982. *Prohibitive Policy: Implementing the Federal Endangered Species Act*. Cambridge: MIT Press.

9

The Florida Panther Recovery Program
An Organizational Failure of the Endangered Species Act

Ken Alvarez

The Endangered Species Act of 1973 (ESA) has become a centerpiece of American conservation and a prominent conduit for its energies. It is an ambitious federal law designed to draw the separate and diverse public agencies of land stewardship into accord on measures needed to halt the decline of imperiled organisms under their jurisdictions and to restore diminished populations to a safe number within a protected setting. These organizations, the federal and state land management agencies, have in common the management of land—whether it be for the harvest of wildlife, timber, or other resources or to provide for cattle grazing or outdoor recreation. But the two decades since passage of the act have been marked by glaring examples of dysfunction in the branches of government charged with its implementation (Chase 1986; McFarlane 1992; Alvarez 1993). Stolid resistance from the land management bureaucracy to the new thinking and novel operations that must be employed in species recovery has weakened the intent of the act. This malaise is well illustrated in the Florida panther recovery program.

This chapter describes the performance of four government agencies—the U.S. Fish and Wildlife Service (FWS), the National Park Service (NPS), the Florida Game and Fresh Water Fish Commission (Florida GFC), and the Florida Department of Natural Resources (Florida DNR). They were all experienced in wildlife matters, each in its own way, and represented the best instruments the government could put forward. The panther challenge, however, was a new type that was not amenable to resolution by the practices upon which they had come to rely. Where urgency was required, they delayed (in

FIGURE 9-1.   The Florida panther. (Photo courtesy of Florida Game and Fresh Water Fish Commission.)

captive breeding, in examining the outbreeding question, in planning a refuge). Where unified management of public land was important, each went its own way. Where educating the public was essential, they sequestered information. Through their collective behavior, they have illuminated a central failing in the popular enterprise to save endangered species toward which the nation has laid a statutory foundation and committed considerable resources. Now let us review the record.

Background and Organization

The Florida panther (*Felis concolor coryi*, Figure 9-1) is one of twenty-seven subspecies of a widespread cat of the Western Hemisphere. In Florida it is called a panther. Elsewhere in North America it is commonly called puma, cougar, or mountain lion. Prey of large size, usually deer, are required to satisfy its caloric requirements, especially in the case of pregnant or lactating females. Consequently, individual cats must have large home ranges to assure ample opportunities for securing food. The big cat was largely extirpated from eastern North America by 1900. It survives today in the interior of southern Florida where poor soils and a seasonally flooded landscape discouraged human settlement well into the twentieth century.

In 1976 the FWS, prompted by the Florida Audubon Society, formed a Florida panther recovery team and charged it with drafting a recovery plan. The team was made up of four government biologists and two nongovernment appointees, one of which was an experienced breeder of large felines. The Florida GFC was designated the lead agency (Pritchard 1976). So negligible were data on the Florida panther that in the beginning no consensus could be formed by the team on whether a population of panthers remained in the wild. Some thought that the few reliable records represented captives that had escaped or been released. Extensive field searches by observers trained to detect signs of the cat, however, soon produced plaster casts of tracks from widely dispersed sites south of Lake Okeechobee. Evidence showed that the Florida panther, however precariously, still roamed the swampy terrain of southern Florida. In 1981 two panthers were captured and affixed with radio collars.

Upon FWS approval of the recovery plan in 1982, the recovery team was disbanded. Radio telemetry research then made its way forward to probe the size and distribution of the population. In 1983 a panther died during capture, raising a public outcry against the use of radio transmitters. The controversy had two important outcomes: veterinarians were sent to the field to safeguard the animals during captures, thereby adding a wealth of physiological data to the expanding base of information, and a legislative act mandated a Florida Panther Technical Advisory Council to advise Florida GFC on its administration of the recovery program. Members were appointed by the governor. This council of five included one official each from a federal and state agency charged with endangered species management: Oron P. (Sonny) Bass, from NPS, and myself, a biologist with the Florida DNR, were named. Two positions were for persons from universities, or private research institutions, who were experienced in big cat research: these were filled by John Eisenberg and Melvin Sunquist from the Florida Museum of Natural History. The remaining slot was for an appointee from the public. It was occupied by Robert Baudy, a breeder of rare felines and other animals.

The council promptly began making recommendations that were not always received with enthusiasm by the state game agency, which was extremely sensitive to criticism of recreational hunting. But the council could not avoid questioning the impact of this activity in the Big Cypress National Preserve. There, in 1983, hunting was a loosely regulated affair lasting 222 days of the year. (The general gun season encompassed sixty days.) Dogs were used to pursue white-tailed deer

(*Odocoileus virginianus*) and feral hogs (*Sus scrofa*), and six thousand off-road vehicles were registered for use (council minutes, 3 Nov. 1983). Still, few deer were being harvested.

Some recommendations made by the council applied to other agencies and were not viewed congenially by them. In one instance in 1986, the council proposed that all contiguous public lands, without regard for the administrative boundaries dividing them, be managed in concert to support Florida panthers. Given the large home ranges of individual cats, relatively few could inhabit the land controlled by any single agency, and the different management doctrines that prevailed might not provide the special measures, where needed, to sustain a viable population. It would be important for these separate jurisdictions to work together, unified in purpose by a comprehensive plan (council recommendations, 19 Nov. 1986). A collaborative design of this kind was never adopted, however.

To take another example, in 1985 the council made a case for NPS to modify its doctrine opposing special management of a single species. The council's recommendation was to increase white-tailed deer above the natural carrying capacity of NPS lands that did not produce many deer (council recommendations, 4 Oct. 1985). The agency responded by founding a new entity to oversee recovery (letter, Moorehead to Brantly, 2 Aug. 1985). Called the Florida Panther Interagency Committee (FPIC), it was modeled on the Interagency Grizzly Bear Committee (IGBC), a mechanism previously established by the NPS when controversy enveloped its grizzly bear management policy in Yellowstone National Park (Chapter 5). The FPIC included the FWS, the Florida GFC, and the Florida DNR. This bureaucratic combination has been in charge of the Florida panther recovery program ever since.

Delegates from these four agencies gather periodically in an assembly called the Technical Subcommittee. Ostensibly this unit works independently, framing positions on technical issues (in the manner of the Florida Panther Technical Advisory Council) and then submitting them to the judgment of the FPIC. However, the representatives comprising the Technical Subcommittee have not distanced themselves from partisan interests. Consequently, it has usually functioned as a secretariat brokering the different, and often self-interested, viewpoints of the parent agencies into a mutually acceptable position—as can be seen by reviewing the Technical Subcommittee's review of the Florida panther recovery plan in 1987. The plan was drafted during a series of meetings that lasted a year. The FWS urged the inclusion of several important objectives: naming a target number for a minimum viable population, setting up a peer review process using scientists independent of the bureaucracy, and expanding the panther's prey base (white-tailed deer) to the desired plane on lands administered by NPS (Panther INFO 1986:4). None of these objectives were written into the recovery plan.

The Mechanism of Implementation

Implementation of the ESA falls to a land management bureaucracy of many parts. Several agencies, with varying ideas on land and wildlife stewardship, and with habits of thinking and procedure embedded in their respective traditions, administer public land for different purposes. Their sometimes incompatible prescriptions for land and wildlife management do not always foster harmony if they are required to manage a resource together. A good place to observe this disharmony at work is at the southern tip of the Florida peninsula where six contiguous administrative units are under the jurisdiction of five different government agencies. Four of them—FWS, NPS, Florida GFC, and Florida DNR—are state and federal instruments having a clear responsibility for the survival of the Florida panther. A fifth government entity—a regional state agency, the South Florida Water Management District—controls hundreds of thousands of acres that might be used in an ancillary way to help sustain the animal.

This vast instrument of land custody, with its variant aims, arose out of historical events that are uniquely American. A retrospective glance at its origins will assist in understanding the present difficulties with endangered species management. A strong conservation sentiment matured in the United States while much natural land was still unclaimed and could be put aside for general use. The government soon set up agencies to administer these continental reserves: the U.S. Forest Service was authorized near the turn of the century to supply timber on a sustained-yield basis; in 1916 the National Park Service was founded to protect and preserve the nation's spectacular landscapes. To the original lands the federal government withdrew from the public domain others have since been added by purchase. States too have acquired land, and they too set up commissions to protect wildlife resources and regulate their harvest. In time, strong constituencies grew to back all these causes and defend the agencies that embodied them against threats from noncompatible interests: timber companies and loggers to ensure the harvest of trees as a principal use of national forests, a National Parks and Conservation Association to parry threats to the parks and to lobby for appropriations, and organized sportsmen to give political support to the game and fish commissions. These organized supporters exert a strong influence on prevailing policies and agency behavior and constitute a barrier to change.

The FWS, which is ostensibly in charge of recovery programs, lacks a similar base of political power. Its functions are diffuse. The FWS did not, like the other entities, enter the world on a specific date to follow a defined purpose. It is descended from several nineteenth-century government offices set up to aid commercial fish resources, in one case, and agricultural production in another. Since then other assignments have been added (and sometimes removed), and the FWS has come to be lodged in the Department of Interior. Today, the national wildlife refuges form one of the most visible of the agency's several arms.

Oversight of the ESA has been assigned to the FWS. This endeavor has suffered in recent years under the dominion of unsympathetic political appointees at the Department of Interior. James Watt and Manuel Lujan, for example, were known for their lack of sympathy for endangered species. By tradition, the secretary of interior is chosen from the West, a region rife with feelings against federal "meddling" in local affairs. During the 1970s the appointed heads of the Fish and Wildlife Service were men formerly associated with the game departments of western states. It will be apparent that such arrangements prevented FWS from influencing recalcitrant state wildlife agencies that were participating in endangered species programs. A study of operations pertaining to the Florida panther, the black-footed ferret (*Mustela nigripes*; Chapter 4), and the California condor (*Gymnogyps californianus*; Chapter 8) will bear this out.

Did the ESA Implementation Work?

Have the policies of the FPIC met the demands of panther recovery? An opinion can be formed by asking what the responsible parties might reasonably have done and then reviewing their performance. By 1985 research had revealed a picture sufficient to identify the main features of a recovery strategy. Panther numbers were estimated at thirty to fifty. The age structure appeared to be skewed toward older animals, suggesting low recruitment into the population (Robertson et al. 1985). Inbreeding was thought likely, due to decades of isolation at a low number (letter, Alvarez to Brantly, 4 Oct. 1985). Only about a dozen panthers used the public lands that served as a refuge. The other cats were to the north on private lands—mainly cattle ranches (McBride 1985).

These findings indicated that a strategy to recover the Florida panther must develop along certain lines. First, the number of breeding adults needed for security against the combined impact of inbreeding and natural disasters would have to be fixed as an objective. Second, a refuge of a size and character calculated to support the target number would have to be planned. Given the large region needed by a wide-ranging predator of this kind, even the sizable combination of protected public lands dominating the southern extremity of Florida, as they are presently managed, would not be adequate. Large landholdings to the north of these sprawling parks would have to be assessed for inclusion. Moreover, the linking of all potential refuge lands in the south to suitable lands farther north would have to be pursued as a policy (letter, Alvarez to Brantly, 1 Feb. 1987). These two objectives—finding a target number for a panther population and finding a refuge for them to live in—would comprise the irreducible minimum in planning necessary to sustain the cats.

Third and fourth would be captive breeding and a consideration of potential genetic problems. Bringing some individuals into captivity would provide security against a calamity overtaking the wild population of Florida panthers. And since it had apparently been isolated for an unknown period, questions about genetic fitness pressed for attention. Florida's dusky seaside sparrow (*Ammodramus maritimus nigrescens*) offered a lesson in the neglect of a captive breeding project— as well as the discord that could be generated by advocating outbreeding. The last three surviving male sparrows had been captured in 1979. But it was too late. There were no females. A debate ensued, pitting the FWS against conservationists and nongovernment biologists, over the issue of outbreeding with another subspecies. FWS was averse to the proposal.

Fifth and last, the public would have to understand and support the physical scale and administrative complexity of this program. The biological issues, some of them novel, must be understood as well. An educational project of considerable scope and prolonged duration would be necessary (council recommendations, 14 Nov. 1983). These were topics that the Florida Panther Technical Advisory Council, beginning in 1983 and continuing through 1986, highlighted for the Florida GFC and, by extension, the other three agencies in the FPIC. Since space limitations preclude detailed treatment of all these issues here, three have been selected to represent the difficulties of implementing recovery measures: planning a refuge, implementing captive breeding, and educating the public. These and other issues have been treated elsewhere in detail (Alvarez 1993).

Planning a Refuge

What was the FPIC's response to the need for planning a refuge? In 1985 the Florida Panther Technical Advisory Council urged the Florida GFC to begin preliminary mapping of private lands north of the Big Cypress National Preserve to assess their suitability as habitat (council recommendations, 28 May 1985). Nothing came of this. A year later, Dr. Larry Harris, a professor at the University of Florida, appeared at a Technical Subcommittee meeting with a scheme to do the mapping. His offer was well received by those present, including delegates from Florida GFC (TS minutes, 13–14 Jan. 1987). Thus encouraged, he subsequently submitted proposals to the agency. He was finally told, late in 1987, that there was no money for the work (Larry Harris, pers. comm.).

Meanwhile, on another front, in 1985 a private conservation organization took the initiative for making the habitat linkage idea a reality. William Partington, of the Florida Conservation Foundation, sponsored a workshop to which elected officials, the news media, agency representatives (including the Florida GFC), and various private interests came and contributed. More than two hundred people attended, and the event was favorably reported in the press. Based on information gathered at this meeting, Partington drafted a bill to be introduced at the next assembly of the state legislature—where it was killed early in the session by landowning interests (David Gluckman, pers. comm.). Partington held several more workshops and tried to enlist support for the project from the Florida GFC, but none was forthcoming.

Without a motivating force in the FPIC to draft a blueprint for the refuge, the idea languished without form until 1990, when a Gainesville citizen, Holly Jensen, sued the FWS. She had been aroused when the *Federal Register* reported that captive breeding, after years of debate and delay, was at last scheduled to begin (*Federal Register* 1990:32). Panthers were to be withdrawn from the wild to captivity that year. The plan was to take three adults and two kittens the first year, to be followed by two adults and six kittens per year through 1992. Jensen was disturbed at the absence of official thought on when and where captives were to be reintroduced and concerned that the outbreeding option had not been discussed. Her lawsuit produced action. In 1992 the FWS began circulating drafts of a habitat plan for panthers in southern Florida. So it can be seen that expert advice from different quarters produced no substantial results. In the end it took litigation to finally produce a draft plan. The six- to seven-year span between the recommendations and action has been characteristic of the bureaucratic delays in the Florida panther recovery program. Other urgently needed measures have been similarly delayed.

When the comment period for the final draft of the habit protection plan closed on 30 April 1993, the director of the Florida GFC wrote the FWS complaining that the document had been released prematurely, without proper review by the FPIC, and could not be endorsed by his agency unless "major" changes were made (letter, Brantly to Pulliam). Never mind that the Florida Panther Technical Advisory Council had urged the state game agency to start work on a habitat plan in 1985 and 1986, or that the FPIC had been inert for years, or that it was litigation which had compelled the FWS to produce a plan, or that the Florida GFC had been provided a copy for review on as timely a basis as everyone else.

The FPIC has never embraced the proposal for a unified plan of management on public lands aimed at providing more deer to panthers as recommended by the Florida Panther Technical Advisory Council in 1985 (council recommendations, 28 May 1985). The Florida Park Service (a division of Florida DNR) long resisted measures to artificially improve deer habitat. Like its federal counterpart, the Florida Park Service adheres to a philosophy of "natural systems management" that has created a strong bias against "unnatural" measures being employed on behalf of panthers. The NPS, in the same way, will not tolerate what it considers a larger-than-natural number of deer, claiming that it is prohibited by the Organic Act of 1916 (letter, Baker to Branan, 29 May 1985). An environmental organization, Florida Defenders of the Environment, challenged this obstruction by producing a 33-page legal report which found that the organic statutes of the National Park Service did not in fact prohibit taking special steps for more deer as claimed—and, furthermore, that the ESA mandated the agency to act affirmatively to conserve the endangered panther using any and all methods (Hamann and Tucker 1987).

The revised recovery plan of 1987 calls for "separate agencies to evaluate habitat protection and management actions on their respective lands and initiate actions to enhance panther conditions as appropriate" (FWS 1987). In 1987 environmentalist Jerry Gerde wrote to Florida's senators complaining about the absence of a design to unite the federal and state properties at the southern tip of Florida through a single plan for managing panthers. One senator asked the agencies why this was so. Each of the four agencies each gave a different justification. The Florida DNR opposed the initiative by speculating that "some of the agencies might have policies which prevent them from doing it" (letter, Gardner to Parker, 1 Dec. 1987). The Florida GFC answered by quoting the recovery plan (letter, Logan to Parker, 20 Nov. 1987). The NPS declared that the law prohibited it from managing for a single species—adding for effect that since the other three agencies of the FPIC had signed the recovery plan, they obviously agreed with the Park Service's position (letter, Baker to Graham, 14 Sept. 1987). The FWS, in reply to the inquiry, conceded the importance of habitat management by the National Park Service: "We are fully aware that Everglades NP along with the other areas in SW Florida under public control hold the key to the future for the Florida panther." The official said the specifics, however, would be taken care of in the future—after further evaluation of the data (letter, Pulliam to Gerde, 2 July 1987).

There has been no coordinated thrust to provide a secure and nourishing habitat for the Florida panther. Indeed, the work has been frustrated by outright organizational aversion. Another task, that of acquiring a captive population as insurance against disaster striking the wild one, has been characterized by endless delays and irresolution as well.

Captive Breeding

When the recovery team was appointed in 1976, Robert Baudy, the breeder of rare felines, was named to the team. Baudy, without waiting for official sanction, started a captive breeding project at his own expense in 1978. He learned of Florida panthers held in a tourist attraction, Everglades Wonder Gardens, and acquired several for breeding. These cats were referred to as the "Piper stock"—after the owners of the zoo. But their pedigree became suspect when tales circulated that a South American puma had been part of the zoo inventory years before. The rumor was later supported by electrophoretic analysis (Roelke 1986–1987; O'Brien et al. 1990). Thus the Piper stock was not considered to represent "true" Florida panthers, and Baudy was so informed (letter, Belden to Baudy 1980). But in disqualifying these cats for the captive breeding program, a complication should be noted. Years before, captives from this same stock had been released by the National Park Service, two in Everglades National Park in 1957 (Biennial Animal Census Report for Everglades National Park, 20 Jan. 1958), with additional releases of five more cats in 1965 (Superintendent's Monthly Narrative Report for Everglades National Park, Sept. 1965) and 1968 (letter, Allin to Pipers, 25 Mar. 1968).

In 1983 Robert Baudy was appointed to the Florida Panther Technical Advisory Council. In 1984 the Florida GFC commissioned a series of expert reports on the methods and feasibility of breeding the Florida panther in captivity. The fourth and last of these reports stated that inbreeding might prove to be an obstacle to panther recovery; if so, breeding stock from outside the wild Florida population would be needed. The most readily available source for this genetic introgression would be from the Piper stock (Eisenberg 1985).

In 1985 the Florida GFC moved to implement a captive breeding project (but not in cooperation with Baudy) after a young male panther (called "Big Guy") was injured by an automobile and made a captive. The state agency applied to the FWS for a permit to remove two young adult cats from the Big Cypress National Preserve (permit application, 15 April 1986). This initiative was supported by the Florida Panther Technical Advisory Council. Nevertheless, the decade of the 1980s ended with no serious effort having been made to assemble a breeding stock, even though a benefactor, the Gilman Paper Company, came forward to supply the capital, the facilities, and the expertise. This company operates a large complex for endangered wildlife in northern Florida near the Osceola National Forest, a promising reintroduction site for the Florida panther. "Big Guy" was placed in the captive facility, and in 1987 an aged female was captured for the project. She was apparently too old to breed, however, and died in 1988.

In 1989 the FWS breathed life into the project by contracting with the Captive Breeding Specialist Group (CBSG), headed by Dr. Ulysses Seal, of the International Union for Conservation of Nature and

Natural Resources, now known as the World Conservation Union. His team was to apply state-of-the-art expertise to assess the inbreeding threat and recommend appropriate procedures. The CBSG held a population viability workshop in which all data on the genetics and demography of Florida panthers were entered into population models and trends were estimated for the future. The CBSG concluded that the Florida panther was "a declining population with an 85% probability of extinction in 25 years" (Ballou et al. 1989:2). The report warned that survival in the wild would require "a captive and multiple wild populations (a minimum of three) of at least 30–50 adults each. The establishment of a captive population is the only management intervention that can assure survival of the Florida panther for 100 years" (Ballou et al. 1989:3). The CBSG emphasized the importance of losing no more time in beginning the captive arrangements. It urged taking at least six panthers less than eighteen months old to form a captive population before the end of 1989.

Six months passed before the FPIC even convened its Technical Subcommittee to review the recommendations. And then, the Florida GFC declined to act, claiming that Seal had not provided enough information—even though the key participants at the Technical Subcommittee meeting had also been present at the CBSG workshop where everyone had been urged to express opinions and raise concerns so they could be addressed. The CBSG was obliged to repeat the workshop late in 1989.

At last, in 1991, captive breeding was officially set in motion—after a delay occasioned by public resistance and a lawsuit, which, as previously noted, had the positive result of mandating a debate on outbreeding and on the preparation of a habitat protection plan. Two panthers were then being held. To these were now added newborn kittens and juveniles between three and six months of age. (Two of the new arrivals were males, three were females.) Of the three females, one died of a heart defect and another from unknown causes. A third contracted pneumonia but recovered after a lung was removed. During 1992 two males and two females were added to the captive stock. No attempt was made to capture additional panthers during 1993.

It is interesting to compare this prolonged delay in starting a captive breeding project to the parallel progress of Robert Baudy. When told that the cats he had acquired to initiate the captive breeding project were genetically compromised, Baudy began breeding them anyway to sell the offspring. By 1992 he had raised and sold over 100 animals. These animals, from the Piper stock, may harbor genetic material no longer found in the wild Florida panther population (Seal 1991). Despite substantial—and increasing—signs that adverse effects of inbreeding are accumulating rapidly in the wild population (Seal 1992) the FPIC has for years ignored the potential for the Piper stock to be used for genetic introgression.

Educating the Public

Given the complexity of the panther program, with its vast geographic requirements and potential for friction with established interests, gaining public support for the many difficult measures would be a vital element of the recovery strategy. Enormous obstacles face an enterprise on the scale of saving Florida panthers (even aside from the ponderous incorrigibility of multiple bureaucracies). Public opinion is not always receptive to the varied and sometimes novel tasks that must be carried out to save endangered species. Other recovery programs have been burdened by citizen resistance (Ogden 1983). Three times the Florida panther enterprise has been threatened by controversies: once over radio telemetry, later by resistance to removing panthers from the wild for captive breeding, and most recently by opposition from agricultural and other propertied interests in southern Florida to a proposed plan for habitat preservation.

The many opportunities for provoking hostile opinion can only be overcome by concerted labors to build a supportive base for the recovery effort. It is important to specify the large spaces that will be required and to identify land uses compatible with panther survival. As well, the intricacies of small-population biology must be explained to laypeople. The need for research, captive breeding and outbreeding, and all the other potentially fractious realities must be presented.

But here again we encounter the dominant obstacle. It is primarily the agencies that possess the resources and data to educate. In Florida they have not used their assets to advantage. Instead, information has been applied to protect departmental prerogatives. In 1986 the Florida GFC took the lead in publicizing the work of the FPIC. The medium was a newsletter entitled *Coryi*. Each agency contributed articles describing its projects and sharing its ideas with readers. But never did *Coryi* try to impart a whole picture of the big problems or suggest what must be done to solve them. An effective population size, securing adequate private lands for a refuge, the value of managing public lands for more deer—these vital topics to name a few were ignored. Precious years have passed with no broad understanding of the complex requirements of saving the Florida panther ever having been instilled in the public mind.

Yet opportunity has been there, waiting for exploitation. In Florida concern over the loss of wildlife and its habitat constitutes a majority opinion. Polls have found that over 60 percent of Floridians believe that spending for the environment should increase (Duda 1987). In 1985 the Florida GFC commissioned a poll in which opinion was canvassed in northern Florida, where attitudes are more akin to those of the rural South than in the more cosmopolitan lower peninsula. Respondents were questioned about the desirability of reintroducing panthers to the region—the result was a landslide in favor of reintroduction (Cristoffer and Eisenberg 1985). A later poll showed 57 percent of citizens willing to spend more money to save the Florida panther (*Palm Beach Post*, 14 Sept. 1986). Many citizens in the state pay extra to purchase a Florida panther license plate each year—in effect, a voluntary tax. Ironically, the license revenue is supposed to be used to educate citizens about the need to preserve panther habitat. But the FPIC has not tapped the fund or provided guidance on using it for educational purposes.

There is much evidence pointing to a pervasive public sentiment for saving the state animal, sentiment that could be turned to advantage by the FPIC. It is instructive to glance at another similar enterprise in Florida in which education paid off. The West Indian manatee (*Trichechus manatus*) recovery program (Chapter 6) has done a superb job over many years of building support for manatee protection. In 1992 boating speed limits were proposed in coastal areas throughout Florida to protect manatees. Although there was fierce resistance from the boating industry, the social and political forces favoring protection had grown very strong and overcame the opposition. The Save the Manatee Club can be joined by subscription. Members receive a newsletter keeping them informed of issues as they evolve—including advice on when to write letters to elected officials about important issues. The club is an integrated part of the recovery mechanism.

A complex recovery program is unlikely to progress very far unless helped along by an organizational arm directed at public education. That is one organizational lesson to be drawn from the panther affair in Florida. Such lessons are secondary, however, to the need for basic reforms in the government agencies that oversee the whole.

Organizational Lessons

The principal lesson from the Florida panther recovery program is that land-management agencies will not readily depart from established practice when presented with a novel challenge. And if several similarly affected agencies are required to combine their efforts, coherent policies are virtually impossible. The NPS has refused all suggestions that it increase deer numbers. The Florida DNR has resisted also. A few deer feeders and mineral blocks were set out by the latter agency, and forage was planted on a modest scale, but a decided lack of official enthusiasm soon terminated this work. The accomplishments were primarily cosmetic. A serious obstacle to vigorous actions by the Florida GFC is its fear that recreational hunting will be curtailed. FWS, ostensibly in charge of the recovery enterprise, is incapable of bringing the other participants into line. The endangered Florida panther has not, as it should have, stimulated a disinterested official analysis of its plight. Nor has it been a catalyst for meaningful, unified action.

Can flexibility be infused in this unresponsive bureaucracy by structural adjustment? History offers a model in the quasi-independent commissions set up by the states to conserve wildlife resources. In the late nineteenth century, sportsmen's organizations in state after state induced their legislatures to create departments of fish and game, but while the lobbying was effective the resulting protection of wildlife was not. The sportsmen learned that a statute to establish a force of wardens was insufficient. An effective arm of implementation did not automatically accompany a good law because, in this case, positions were filled by political appointment. The sportsmen renewed their campaign, this time working to reform the departments, guided by the idea of placing policy and hiring in the hands of a qualified commission. Ultimately, wildlife policy in all the states came to be vested in commissions, some very able, which were usually made up of people accomplished in the business profession. It was they who forged the effective organizations that restored depleted stocks of game and fish (Reiger 1975; Trefethen 1975).

While experiences of the business world were adequate as a training ground for citizen commissioners to build a law enforcement structure for game protection, the technical knowledge required today for endangered species management is highly specialized and difficult to acquire outside an academic setting or without training in a related discipline. The commissioners of today are not up to the task. A review of the conduct of game and fish departments in the widely separated cases of the Florida panther, black-footed ferret (Chapter 4), and California condor (Chapter 8) shows a similar pattern of dysfunction. Today, experts in the field of conservation biology are present in every state's universities and research institutions. This reservoir of expertise should be tapped for service on game and fish commissions.

The professional commission could also be employed at the national level where the dearth of spirit and clout in the FWS is a conspicuous weakness in the arrangements for recovering endangered species. A national endangered species commission, made up of professionals of experience and repute in the field of conservation biology, implanted in the FWS, could strengthen the agency. This proposal has a useful model to recommend it: the Florida Marine Fisheries Commission made up of appointed experts who recommend policy to the governor and cabinet. This group, for administrative purposes, is attached to the Florida DNR, but its opinions are not subject to department review. They go directly to the state's executive body for a ruling. In the same way, a National Endangered Species Commission in the FWS (see Chapters 6 and 16), being appointed, would not have to conform its thinking to the imperatives of career security. It would, however, have secretarial help, per diem expenses, and access to all the agency resources of information to carry out its charge. Its recommendations to the agency head, unconstrained by departmental frictions and constraints, would be open to review by all.

The Florida Marine Fisheries Commission, while free of agency bias, is not invulnerable to special interests. By law its membership must represent users of marine resources as well as scientists, and policy decisions can be appealed at the executive level of state government. Nonetheless, its recommendations rest on a rigorous and well-argued review of data, and controversial policy decisions are subject to relentless public scrutiny.

Quasi-independent commissions could be the stick to animate a resistant bureaucracy. It should be noted, however, that even a vigorous new leader, or a new system of leadership, can be frustrated by resistance to new directions from within a government bureaucracy. Tenured careerists who believe a proposed change of policy will violate cherished practices can be endlessly inventive in devising ways to foil change. Therefore it would be desirable, in the course of redesigning an organization, to try and alter ingrained attitudes as well. And for that, a new ideal must be promoted.

Organizational behavior often has an idealistic element. The ideals of land stewardship are a potent force in government agencies that manage land and wildlife. Given this psychological component it will be important to reshape not only institutional structures but motivations as well. Organizations are not simply mechanisms. They are human aggregates, and humans need to be part of something larger than themselves. A land management agency can fill this need. Each agency was born of a popular cause that was transmuted through the legislative process into a departmental form and charged with addressing a great problem and hope of the time: the National Park Service to preserve awe-inspiring landscapes; the U.S. Forest Service to use the earth for the good of humans; state game and fish departments to conserve wildlife for public use and enjoyment. Such ideals buttress an agency's sense of mission, usually embodied in a doctrine or a philosophy.

This is the intangible but fervid core of resistance to change in the land management bureaucracy. The idea of mission drives the psychodynamics of agency behavior. It defines ethical conduct. Whatever aids the mission is good; whatever threatens it must be deflected. What might appear unethical to an outsider can often seem perfectly justified in the partisan defense of a government agency. The sense of mission can overpower or undermine any ad hoc collaborative structure. Evidence for this assertion can be seen in the FPIC, where recommendations to benefit the endangered Florida panther—increasing deer on NPS lands, a unified management plan for contiguous public preserves, restrictions on recreational hunting, educating citizens— have been resisted, turned aside, or vitiated by participating agencies who sensed a threat to their respective missions. This is why an alternative ideal can be useful in shifting a paradigm. It is the "carrot" of change. And a compelling new ideal stands waiting to serve: preservation of the earth's biological diversity.

Professional Lessons

Professionals in an organization will rarely venture beyond the strictures of the organizational culture. Organizational imperatives dominate their behavior. This being so, dysfunction in endangered species programs would respond to correction more quickly by reforming the agencies than by training a new breed of professionals. Fresh ideas entering an organization at the bottom can take years to make a strong imprint upon policy, and the rescue of imperiled wildlife populations is intolerant of delay.

This is not to say that a broader training for professionals would be futile. Understanding the entanglements of organizational intransigence that wait to frustrate the rational prosecution of wildlife science can, at the least, alert the professional to the strengths and failings of the organizational environment and the frustrations that may be encountered there. And, on a progressive note, multiplying the liberally trained professionals in an organization supplies it with an organic momentum for a paradigm shift. Not all frustrated professionals are passive. The more energetic work relentlessly for change—contacting forces outside the agency, working with professional societies, writing articles, even books.

Shortcomings observed among professionals in the Florida panther recovery program can be grouped in three categories. First are those who have not kept abreast of developments in the field of conservation biology and are not conversant with its principles. Second are those who do know but nevertheless adjust to agency imperatives, sometimes resignedly, sometimes willingly, not trying to redirect them. In the third category are the professionals who have acquired a vested interest in some part of the program and strive to protect it. An example would be the field researchers who resisted captive breeding and disparaged the growing evidence of inbreeding depression, seeing proposals to capture even a limited number of cats for security as a threat to their projects.

In summary, the nation has not made the proper arrangements to carry out the mandate of the ESA. The government agencies in charge were not formed for that specific purpose, and their operations and thinking have become rigidly fixed by long habit and constituency pressures. As they are presently structured it is unlikely that they will ever adopt the flexibility needed to bring a difficult endangered species task to a successful conclusion.

Seeking the Best Solution

How can these organizational incapacities—the unwillingness to face unpleasant facts, the paralyzing fear of adverse elements of opinion, the aversion to the new discipline of conservation biology—be remedied? These marks of official weakness are recurring in one endangered species program after another (Clark 1984; McFarlane 1992; Alvarez 1993). The best solution would be to reshape the policymaking apparatus so that it incorporates the necessary technical competence and promotes a heathy climate for professional operations.

Short of that, there are makeshift strategies. One approach is to get all pertinent information into the open where it can be aired and probed. A wide scientific debate can at least expose unrealistic policy options (Rosenbaum 1985). Thus we should seek out the means and mechanisms for drawing the organizations and interested factions into the forum where differing opinions can be displayed, challenged, and put on record. Resistance can be anticipated from the bureaucracy. Officials there prefer to operate solely within the established organs of interagency collaboration, like the FPIC in Florida and the IGBC in the Yellowstone region (Chapter 5). But such combinations are unlikely to treat contentious issues seriously or to give unsettling ideas a hearing, much less a thorough examination. In Florida the forum was put to good use. A prominent environmental organization—Florida Defenders of the Environment—sponsored a conference on the Florida panther in 1986. Spokesmen for the four agencies participated, and the key issues were put on display (Branan 1986).

Another method was also used in Florida to push independent ideas and opinions into public view. In 1984 an environmental organization, the Florida Conservation Foundation, published an investigative article on the Florida Game and Fresh Water Fish Commission's resistance to imposing hunting restrictions on lands inhabited by panthers (Grow 1984). The article came to the attention of the governor, Bob Graham, who intervened to pressure the commissioners (for the governor appoints commissioners). They, in response, modified regulations to ease hunting pressure on the deer herd.

Parallel organizations may also be productive. Certainly they have been used to improve the performance of business and industry. The device has the potential to invigorate faltering endangered species programs as well (Clark et al. 1989). The Florida panther recovery program provides an example. In 1989 the FWS invited the CBSG to the recovery program. This body of professionals made a rigorous analysis of panther genetics and published their findings, which recommended a captive breeding program (Ballou et al. 1989). The CBSG later outlined strategies for outbreeding (CBSG 1991). Shadow organizations of this kind can be employed to some advantage in an adventitious role, but bureaucratic resistance to them can be formidable and unyielding at best (Johnston 1993). The CBSG review of genetics in 1989 revealed alarming evidence of inbreeding. Later, in May

1991, the CBSG recommended policies on outbreeding. At a workshop in October, 1992, the CBSG again presented evidence of a worsening genetic situation (Seal 1992). Finally, in June, 1993, the FPIC approved outbreeding, but then another obstacle appeared. The proposal has been unable to win approval from the Washington office of FWS, although authorities there have made no official response to the scientists they employed to recommend policies on outbreeding.

When all else fails to animate a torpid bureaucracy, the law can be brought to bear. In 1991, as related earlier, a Gainesville resident, Holly Jensen, acting in concert with the Fund for Animals, took the FWS to court over its apparent indifference to preparing a panther habitat protection plan and for its apathy in investigating the question of outbreeding. Legal action had forced some progress on both these matters by the summer of 1993.

While the forum, the *ENFO* newsletter, the parallel organizations, and the force of law performed well and each moved the program forward by increments, they have functioned as levers, not an engine. Though useful for moving dead weight, they are incapable of infusing a dynamism into the bureaucracy that would enable it to advance under its own power. Excellence in responding to the challenge of saving imperiled species—or something approaching it—can be attained only by fundamental reforms in America's land management agencies.

Acknowledgments

I would like to thank John Eisenberg, Katherine Ordway Professor of Ecosystem Conservation, and James N. Layne, director of research for Archbold Biological Station, for reviewing this chapter and offering many helpful suggestions and sometimes challenging opinions.

References

Alvarez, K. 1993. *Twilight of the Panther: Biology, Bureaucracy and Failure in an Endangered Species Program.* Sarasota: Myakka River Publishing.

Ballou, J. D., T. J. Foose, R. C. Lacy, and U. S. Seal. 1989. *Florida Panther,* Felis concolor coryi, *Population Viability Analysis and Recommendations.* Report prepared by the Captive Breeding Specialist Group, Species Survival Commission, IUCN. Washington, DC: U.S. Fish and Wildlife Service.

Branan, W. V. 1986. *Survival of the Florida Panther: A Discussion of Issues and Accomplishments.* Conference proceedings. Tallahassee: Florida Defenders of the Environment.

Captive Breeding Specialist Group (CBSG). 1991. Subspecies, populations, hybridization, and conservation of threatened species. CBSG Working Group draft policy. U.S. Fish and Wildlife Service.

Chase, A. 1986. *Playing God in Yellowstone: The Destruction of America's First National Park.* New York: Atlantic Monthly Press.

Clark, T. W. 1984. Biological, sociological, and organizational challenges to endangered species conservation: The black-footed ferret case. *Human Dimensions in Wildlife Newsletter* 3:1–15.

Clark, T. W., R. Crete, and J. Cada. 1989. Designing and managing successful endangered species programs. *Environmental Management* 13:159–170.

Cristoffer, C., and J. F. Eisenberg. 1985. *Report No. 3 on the Captive Breeding and Reintroduction of the Florida Panther in Suitable Habitats.* Tallahassee: GFC.

Duda, M. D. 1987. *Floridians and Wildlife: Sociological Implications for Wildlife Conservation in Florida.* Technical Report no. 2. Tallahassee: GFC.

Eisenberg, J. F. 1985. *Final Report on the Captive Breeding and Reintroduction of the Florida Panther into Suitable Habitats.* Report no. 4. Tallahassee: GFC.

Federal Register. 1990. 55:32 (15 February 1990).

Grow, G. 1984. New threats to the Florida panther. *ENFO* (Florida Conservation Foundation)

Hamann, R., and J. Tucker. 1987. *Legal Responsibilities of the National Park Service in Preserving the Florida Panther.* Gainesville: Center for Governmental Responsibility, University of Florida College of Law.

Johnston, K. 1993. *Busting Bureaucracy: How to Conquer Your Organization's Worst Enemy.* Homewood, Ill.: Kaset International.

McBride, R. 1985. *Population Status of the Florida Panther in Everglades National Park and Big Cypress National Preserve.* Homestead, Fla.: Everglades National Park.

McFarlane, R. W. 1992. *A Stillness in the Pines: The Ecology of the Red-cockaded Woodpecker.* New York: Norton.

O'Brien, S. J., M. E. Roelke, N. Yuhki, K. W. Richards, W. E. Johnson, W. L. Franklin, A. E. Anderson, O. L. Bass, Jr., R. C. Belden, and J. S. Martenson. 1990. Genetic introgression with the Florida panther, *Felis concolor coryi. National Geographic Research* 6(4):485–494.

Ogden, J. 1983. The California condor recovery program. In *Bird Conservation,* S. A. Temple (ed.). Madison: University of Wisconsin Press.

Panther INFO. 1986. Technical Subcommittee minutes, 18 June 1986. U.S. Fish and Wildlife Service.

Pritchard, P.C.H. (ed.). 1976. *Proceedings of the Florida Panther Conference, March 17–18, 1976.* Tallahassee: Florida Audubon Society and Florida Game and Fresh Water Fish Commission.

Reiger, J. F. 1975. *American Sportsmen and the Origins of Conservation*. New York: Winchester Press.

Robertson, W. B., Jr., O. L. Bass, Jr., and R. T. McBride. 1985. Review of existing information on the Florida panther in EVER, BICY and environs with suggestions for needed research. Unpublished technical report. Homestead, Fla.: Everglades National Park.

Roelke, M. E. 1986–1987. *Florida Panther Biomedical Investigation*. Tallahassee: GFC.

Rosenbaum, W. A. 1985. *Environmental Politics and Policy*. Washington, DC: Congressional Quarterly.

Seal, U. S. (ed.). 1991. *Genetic Management Considerations for Threatened Species with a Detailed Analysis of the Florida Panther* (Felis concolor coryi). Workshop report. Washington, DC: U.S. Fish and Wildlife Service.

———. 1992. *Genetic Management Strategies and Population Viability of the Florida Panther* (Felis concolor coryi). Workshop report. Washington, DC: U.S. Fish and Wildlife Service.

Trefethen, J. 1975. *An American Crusade for Wildlife*. New York: Winchester Press.

U.S. Fish and Wildlife Service (USFWS). 1987. *Florida Panther Recovery Plan*. Washington, DC: U.S. Fish and Wildlife Service.

I0

Candidate and
Sensitive Species Programs
Lessons for Cost-Effective Conservation

Craig R. Groves

Much attention has been focused on the ever-lengthening list of threatened and endangered species and problems associated with their recovery. But "candidate" species, an official group of species whose status may warrant listing as threatened or endangered under the Endangered Species Act (ESA), have received much less notice. Yet about 3700 species are officially considered candidates and two or three hundred of these, many of them plant species, may already be extinct (Meese 1989). About 330 of these candidate species are classified as Category 1, meaning that substantial information exists to support their listing as threatened or endangered (*Federal Register* 1991, 1993). The rest of the candidates are designated as Category 2, indicating that conclusive data on biological vulnerability and threat are not available but listing as threatened or endangered may be appropriate in the future.

This chapter presents examples of ongoing conservation programs for two current and one former Category 2 animal species in Idaho: the Coeur d'Alene salamander (*Plethodon idahoensis*, Figure 10-1), harlequin duck (*Histrionicus histrionicus*, Figure 10-2), and wolverine (*Gulo gulo*, Figure 10-3). These conservation programs are cooperative efforts between the U.S. Department of Agriculture's Forest Service and the Conservation Data Center of the Idaho Department of Fish and Game (IDFG). After examining the professional and organizational weaknesses encountered in conducting these programs, the chapter concludes with steps for designing and conducting a model conservation program for candidate species based on the lessons learned from these experiences.

FIGURE 10-1.   The Coeur d'Alene salamander. (Photo by C. R. Groves.)

Candidate and Sensitive Species

Several biologists have pointed to the long list of candidate species as evidence that the ESA is not working well and argue that we need solutions for *preventing* species from becoming endangered (Scott et al. 1991). A recent analysis of the ESA by Wilcove et al. (1993) revealed that most federally listed species, subspecies, and populations did not receive ESA protection until their total population size or total number of populations was very low and the chances for successful recovery were diminished. One alternative strategy for preventing species from becoming endangered appears to be underutilized: designing conservation programs for candidate species, particularly those in Category 2, whose populations have not yet reached a critically low level. These conservation programs can be carried out in an ecological and political environment that has not reached the "emergency room conditions" faced by many endangered species, and at substantially less cost.

Many Category 2 species are thought to be rare, suspected to be declining, and are poorly understood biologically. Thus the overall dilemma facing biologists is how to design a conservation program for species that are often difficult to find, not widely known or appreciated by the general public, and for which little biological information is available. To secure funding to initiate a conservation program for a rare or declining species, it is usually necessary that the species be classified by a federal or state agency in some special category such as a

FIGURE 10-2.   The harlequin duck. (Photo by C. R. Groves.)

"candidate species" under the ESA within the Interior Department's
Fish and Wildlife Service (FWS) or a similar designation in the Forest
Service and Interior Department's Bureau of Land Management
(BLM) called "sensitive species." In fact, there is considerable overlap
between these two designations—many candidate species are also
sensitive species and vice versa.

The FWS publishes separate lists of candidate plant and animal
species, respectively, for review every two years or so (such as the
Federal Register 1991). Both the public and pertinent state and federal
agencies have the opportunity to submit comments to the FWS for
adding a candidate species to the list, removing one, changing the
status of a species from one category to another, documenting threats
to a taxon, or recommending critical habitat designations. The state
FWS office often consults with species experts and state agencies that
have responsibility for rare plants or animals under Section 6 of the
ESA. They seek information on population trends of candidates (improving, declining, stable, unknown) as well as relevant additions,
deletions, or changes to category status. Then state-level recommendations are forwarded to the FWS regional office, which consolidates
comments on candidate species or potential candidates whose distribution occurs over more than one state. Based on its review, the regional office submits recommendations to the national FWS office.

In the three examples discussed here, each species either is or was

FIGURE 10-3.   The wolverine. (Photo by remote camera, Jeff Copeland, Idaho Department of Fish and Game.)

designated a candidate and is currently classified by the Forest Service as a sensitive species in Idaho and Montana (Moseley and Groves 1992). Sensitive species are taxa identified by the regional forester "for which population viability is a concern as evidenced by significant current or predicted downward trends in population number or density or significant current or predicted downward trends in habitat capability that would reduce a species' existing distribution" (Forest Service Manual Title 2670.5 #19). Although the BLM has a similar designation, its program for sensitive species lacks the legal standing that exists under the 1976 National Forest Management Act (NFMA) requiring national forests to maintain viable populations of all native vertebrate species within the planning area (Wilcove 1993). The examples discussed in this chapter focus only on Forest Service, not BLM, sensitive species. It is Forest Service policy to develop and implement management practices to ensure that sensitive species do not become federally listed as threatened and endangered. Because of this policy and the viability requirement of NFMA, the Forest Service's sensitive species program affords an excellent opportunity to conduct conservation programs for candidate and sensitive species, particularly those not on the brink of extinction.

The candidate or sensitive species examples discussed in this chapter involved two Forest Service regional offices (Northern and

Intermountain). Both offices went through similar designation processes to compile an official list of sensitive species. First, inter-agency committees were established by the Forest Service—usually consisting of service staff, state game and fish biologists, Natural Heritage program personnel, and local academic experts. Second, a list of nominated species was screened and scored according to such criteria as distribution, range, reproductive strategy, population trend, habitat trend, and habitat requirements. Third, with this information the committee made its recommendations to the regional forester for official designation. Once the species was so designated, inventory, monitoring, and research projects would be initiated. Both Forest Service regions also published action plans (USFS Intermountain Region 1989; USFS Northern Region 1991) describing the needs, actions, responsibilities, and timetables necessary for implementing threatened, endangered, and sensitive species programs.

When sufficient biological information has been obtained, Forest Service policy calls for a conservation assessment to be written by a person with expertise on a particular sensitive species. In the Northern Region, development of conservation assessments and strategies is being directed by regional wildlife staff. Recently, the Forest Service has instituted interagency and interregional working groups (such as the Western States Carnivore Group) composed of both researchers and managers to develop habitat conservation assessments for taxonomically similar groups of sensitive species such as furbearers.

Sensitive Species and Conservation Efforts

The three species considered here (Coeur d'Alene salamander, harlequin duck, and wolverine) run the gamut from poorly known and underappreciated (salamander) to charismatic megafauna (wolverine). All three occur primarily on public lands, and none are on the edge of extinction. We turn now to the ecology and status of each species, as well as a brief summary of the conservation efforts undertaken to date.

Coeur d'Alene Salamander

The Coeur d'Alene salamander is the only lungless salamander (*Plethodontidae*) known from the northern Rocky Mountains and one of only four salamander species known to occur in the region (Cassirer, Groves, and Genter 1993). Its distribution includes northern

Idaho, western Montana, and southern British Columbia. Its taxonomic status remains uncertain: some biologists consider it a subspecies of the more widely distributed Van Dyke's salamander (*P. vandykei*); others cite biochemical evidence indicating that it is a distinct species.

Only two major biological studies of the Coeur d'Alene salamander have been conducted, making it perhaps the least studied *Plethodon* species. These salamanders are usually only active aboveground during moist weather in spring and fall. They may spend up to seven months underground in moist spaces among fractured rock where they can avoid desiccation in summer and freezing in winter. When aboveground, they can be found in three types of habitats all closely tied to water: springs or seeps, waterfall spray zones, and edges of streams.

The Coeur d'Alene salamander was listed as a Category 2 species in 1983. In 1984, the FWS received a petition to list the species as endangered due to its low population numbers and habitat destruction. The FWS decided in 1986 not to list the species due to information presented by expert herpetologists that the species was more abundant than indicated by distributional records and persistent questions concerning its taxonomic status (*Federal Register* 1986). In 1987, the Forest Service listed the Coeur d'Alene salamander as a sensitive species. That same year, surveys were initiated in Idaho that continued through 1992. (For a summary see Cassirer, Groves, and Genter 1993.) These surveys were designed to locate new populations, determine the status of previously known populations, and collect macrohabitat information at each site where animals were observed. As a result of these surveys, the number of sites where the salamander was known to occur increased from just a few in each state to nearly two hundred. Based on the number of new occurrences resulting from these surveys, the FWS downlisted the species from Category 2 to 3c (more widespread and abundant than previously thought). Subsequently, a conservation assessment was written for the species (Cassirer, Groves, and Genter 1993). This assessment contains information on biological background, management threats and guidelines, monitoring needs, research needs, and an implementation schedule. Its implementation is pending Forest Service approval.

Harlequin Duck

Harlequin ducks are small sea ducks that winter in coastal areas and breed inland on mountain streams. As such, they are the only waterfowl species in North America to nest exclusively on swift-flowing mountain streams. The species occurs in two distinct populations, one in the North Atlantic and one in the North Pacific (Cassirer et al. 1991). The Atlantic population is thought to be declining. The Pacific population is much larger, but there are no reliable estimates of its size.

Compared to most waterfowl species, harlequin ducks have received little scientific attention. Prior to the Idaho investigation, studies on the breeding grounds had been conducted in only four places—Montana, Alaska, Wyoming, and Iceland. (For a summary see Cassirer and Groves 1991.) Due to the species' small population, restricted distribution, and a lack of information, it was designated a sensitive species in the Forest Service's Northern Region in 1987 and in the Intermountain Region in 1990. As a result of work described later and similar efforts in Montana, the FWS listed the harlequin duck as a Category 2 species in 1991 (*Federal Register* 1991).

In 1987, the first field surveys for harlequins in Idaho were initiated. Prior to this time, little was known about the historical distribution or status of the species. From 1987 to 1990, seventy-five streams were surveyed in Idaho for harlequins and information on distribution, population status, breeding biology, and habitat use was gathered. (For a summary see Cassirer and Groves 1991.) Based on these surveys, the breeding population was estimated to be less than one hundred individuals, distributed primarily on twenty-eight streams over about 38,000 km² in northern and north central Idaho (Cassirer et al. 1991). From 1991 to 1993, a more intensive study was conducted on selected streams in northern Idaho (Cassirer and Groves 1992). The objectives of this study were to determine population density, productivity, and habitat use, to investigate factors affecting density and productivity, and to determine appropriate population monitoring techniques. The field aspects of this study have been completed; data analysis and reports will be finalized in 1994.

While conducting these surveys and studies, efforts were made to inform the public and other interested groups. Posters seeking information on sightings were widely distributed, prominent articles were published in major newspapers, an article was published in the state fish and game magazine (Groves et al. 1990), and a report on the study appeared on the IDFG's monthly television show. In addition, a symposium on the species was held in 1992 (IDFG 1992), an international working group was formed, and a presentation on the conservation status of the species was made to the Pacific Flyway Council in the fall of 1992.

Wolverine

Little biological information is available on the wolverine in North America. Only four major ecological investigations have been conducted: one in western Montana, two in Alaska, and one in the Yukon. The wolverine has been a protected species in Idaho since 1965, but until recently little has been known about its distribution or population status in the state. In 1985, an Idaho survey of biologists and trappers was conducted via mail questionnaire to help determine the status and distribution of wolverines. Results from the survey indicated that wolverines appeared to survive in at least three major areas in the state—the Sawtooth/Smoky Mountains in central Idaho had the greatest concentration of sightings (Groves 1988). In 1989, winter field surveys in this area were initiated and these continued through 1990. In conjunction with these surveys, a wolverine poster seeking information on sightings was widely distributed and several newspaper stories on the surveys were released statewide.

After obtaining solid evidence that several wolverines were present in central Idaho, a wolverine interagency steering committee was formed in 1990 to guide and oversee a research project. A research and funding prospectus was written and submitted to the Forest Service, FWS, and IDFG. With help from local conservation-minded nongovernmental organizations (NGOs) and the Idaho congressional delegation, funding was received from the Forest Service and a study was initiated in winter 1991–1992. Through the summer of 1994, the research project captured and radio collared sixteen wolverines. Data were gathered on home range, activity patterns, behavior, population size, and habitat use. Additionally, wolverines were successfully photographed and individually identified at bait and trap stations (Copeland 1993; Copeland and Harris 1993). Results of the study have been reported to the interagency Western States Carnivore Group led by the Forest Service and will be incorporated in the conservation strategy for this species group.

Professional and Organizational Weaknesses

Anyone charged with managing an endangered species project may have learned a frustrating lesson—it is usually not the technical, biological problems that thwart conservation success. Instead, most problems can be attributed to professional and organizational responses to the task at hand. At the professional level, many biologists initially define their problems narrowly—often just in biological terms (species X habitat is being destroyed by factors Y and Z and the result is a decline in species X). At the organizational level, bureaucratic structure and agency values further confound the conservation challenge. Together these problems are often counterproductive to conservation.

Both sets of problems—professional and organizational—can become particularly troublesome at the recovery stage where political and socioeconomic factors further influence decisions. Fortunately, the candidate cases described in this chapter have not reached the stage where they are being conducted in a contentious, political atmosphere. As a result, there is still an opportunity to redress, to some extent, the professional and organizational difficulties before they become intractable impediments to successful conservation. The following sections detail the professional and organizational lessons gleaned from seven years' experience working on three candidate species. Although this experience is restricted geographically, the lessons apply to similar conservation problems and organizations.

Professional Weaknesses

Failure to adequately define a problem or to establish a criteria for measuring success in resolving it are two of the more commonly encountered professional weaknesses in many conservation programs for candidate and listed species. These weaknesses can plague a program from its inception and make it nearly impossible to reach an endpoint. Together with inexperienced and in some cases, unmotivated staff, these weaknesses—all encountered in the cases discussed in this chapter—can have substantial bearing on the ultimate success or failure of conservation programs.

Defining the Problem. The first inclination in developing conservation programs for these sensitive species was to define the problem in technical, biological terms (Schön 1983). Put simply, little biological information existed on any of these species, so the logical first step was to collect general biological information on where and in what habitats the species occurred, threats the species faced, and its population trend. In retrospect, conservation efforts for these species would have benefited from framing the problem more broadly (and more accurately) at the beginning. Here are some additional questions that might have been considered at the outset:

1. Where might political support or opposition to these projects come from within the IDFG, Forest Service, FWS, and the general public? How could programs be designed to capitalize on areas of actual and potential support and neutralize any opposition?

2. What policies and arrangements in the IDFG, FWS, and Forest Service would help or hinder these projects?

3. How could these projects best be conducted within the bureaucratic structure of the two principal agencies (IDFG and Forest Service)?

4. What sort of long-term funding would be needed to undertake inventory, monitoring, and research for these species, and where would the funds come from?

5. What sort of criteria could be used to measure success?

6. How could the findings be used to influence decisions on land management activities that would benefit the species?

Addressing such questions could have resulted in more effective and efficient conservation programs at all project phases, especially implementation (Brewer and deLeon 1983). For example, initial opposition by the IDFG Commission to its personnel conducting work on a "salamander" might have been foreseen and avoided. Fragmented and inconsistent survey efforts for the harlequin duck and Coeur d'Alene salamander could have been conducted more systematically with a better understanding of the Forest Service bureaucracy. Similarly, the agency's inertia in initiating a wolverine study might have been overcome several years earlier had NGO colleagues been enlisted at an earlier juncture.

A central constraint on conservation programs for all these species was the narrowly viewed conception of the biological problem itself. For example, motivated and well-trained biologists can sometimes fall prey to the "mythology of research facts." That is, previous research on a species may be taken as fact and additional or alternative explanations are not always sought. Such was the case with the California condor (*Gymnogyps californianus*; Chapter 8) recovery program, which originally relied heavily on earlier work by other researchers. Similarly, in studying harlequin ducks, previous research indicated that harlequins are ground nesters, yet we later found them to be cavity nesters as well (Cassirer et al. 1993). Another assumption—that the harlequin's stream use in Idaho would be similar to that observed in adjacent Wyoming (Wallen 1987)—also proved to be false: harlequins in Idaho use much smaller streams at greater distances upstream than originally anticipated (Cassirer and Groves 1992). The lesson here is to critically examine the quality and relevance of previous research. A more open-minded and less dogmatic approach can ultimately be more efficient in terms of time, money, and conservation success.

Education and Experience. Even with the best understanding of the conservation problem, lack of pertinent experience and educational training with regard to nongame species was an obstacle to successful species management. Traditional education and work experience often focused on habitat management and ecology of game species. As a result, agency professionals were generally ill prepared to address broad conservation problems related to nongame species. Often they were not only unfamiliar with the biology of species such as salamanders and forest owls, for example, but were also unaware of survey and research techniques for nongame species. This dichotomy between traditional wildlife management professionals and the emerging profession of conservation biology is a recognized, though controversial, issue. (See Clark 1983 and Edwards 1989.)

The limitations of traditional wildlife biology professionals appear to be particularly critical in the Forest Service, where the agency has NFMA policy requirements to maintain viable populations of sensitive species. Much of the work expected of a district wildlife biologist involves threatened, endangered, and sensitive species—either in conducting surveys or in "clearance work" related to environmental assessments for such projects as timber sales. More and more, this work requires knowledge of the life history of nongame species as well as the various aspects and tools of conservation biology: viable population analyses, metapopulation concepts, population genetics, geographical information systems, remote sensing, and landscape ecology/habitat fragmentation issues (see Chapter 12). Traditional wildlife management programs at many universities fail to provide sufficient educational training on these topics.

Problems stemming from a lack of appropriate training were compounded by the time demands of working within the Forest Service bureaucracy (attending numerous meetings, training sessions, writing environmental assessments and similar documents). As a result, the service's efforts to survey or monitor sensitive species like the Coeur d'Alene salamander or harlequin duck often fell by the wayside or were conducted hurriedly or inappropriately as other activities received a higher priority. A recent emphasis on in-service training, better internal program coordination, and development of conservation assessments for sensitive species has improved this situation somewhat in the service's Northern Region.

Although policy and regulatory pressures to address complex issues involving nongame species generally are not as severe in state agencies compared to federal agencies, the lack of such expertise has become more noticeable in IDFG in recent years. For example, the department's plans to harvest timber in a state wildlife management area created a public controversy with respect to impacts on neotropical migratory birds (Groves 1992). Regional biologists lacked the expertise to address the problems or to design and implement measures to monitor the impact of timber harvest activities on nongame birds. Such nongame expertise is generally deficient throughout the IDFG (Groves and Unsworth 1993).

Wildlife professionals expected to tackle complex issues in threatened, endangered, and sensitive species management need broad technical training and experience that encompasses facets of nongame and conservation biology. But just as important, this training needs to be broadened to include skills and knowledge from the organizational and policy sciences that will enable wildlife professionals to operate more effectively in the bureaucratic arena where decisions and policy are made.

Motivations and Values. A more difficult professional problem to remedy is the lack of motivation or interest. Although the IDFG has a legal mandate to conserve all wildlife species, there is, in practice, an overriding emphasis on economically important game species (Groves and Unsworth 1993). This emphasis prevails because game species provide the department's primary source of income (revenue from the sale of fishing and hunting licenses) and, relatedly, because the IDFG perceives a need to be especially responsive to the state's sportsmen. This perception is reflected in the department's budget and management practices: funding for nongame in the IDFG was only 1.5 percent of the department's budget in 1992. Similarly, in most national forests in Idaho the major concern of federal and state wildlife biologists is the welfare of elk (*Cervus elaphus*; Groves and Unsworth 1993).

Although the IDFG had numerous opportunities, requests, and the financial capacity (through Pittman-Robertson funds in particular) to provide matching funds for the Forest Service's sensitive species projects, only one project (the wolverine) in seven years received any state funding beyond a single staff person's salary. Not only was there little financial support, but state wildlife biologists were often actively discouraged from participating in nongame projects. This discouragement was manifested in several ways: nongame issues were routinely last (or nonexistent) on the agendas of wildlife staff meetings; thousands of dollars were spent annually on wildlife research (exclusive of Section 6 ESA funds) but rarely on a nongame species; and agency culture in the IDFG provided little incentive or reward for participating in such projects. As a result, there were few nongame biologists in the IDFG (two full-time staff funded by state funds in 1992 out of greater than sixty biologists statewide) and virtually no career ladder for recognition and advancement.

On the Forest Service side, the agency continues to be dominated by commodity resource values (primarily timber) despite federal laws calling for a more balanced use of commodity and noncommodity resources (1960 Multiple-Use Sustained Yield Act and 1976 NFMA). Much has been written about this imbalance (see Frissell et al. 1992 and Lawrence and Murphy 1992), and the spotted owl (*Strix occidentalis*; Chapter 3) and red-cockaded woodpecker (*Picoides borealis*; Chapter 7) cases indicate the problems. In the service's Northern and Intermountain regions, budget and staff continue to be stacked heavily in favor of commodity interests. Although biological staff have increased in recent years, many ranger districts in Idaho were without biologists during the time when the sensitive species programs discussed here were conducted.

Program Evaluation. An important component of every conservation program should be establishing criteria by which it can be evaluated. This step is referred to as the evaluation phase of the policy process where estimated performance is compared with what was expected and the differences are reconciled (Brewer and deLeon 1983; Chapter 17). Yet few recovery plans for endangered species, and probably fewer programs for candidate or sensitive species, establish specific criteria to evaluate program performance (GAO 1988; Tobin 1990). Although few recovery plans establish explicit population goals for a species (X number of nests with Y productivity within Z area over a certain period of time), even fewer programs appear to evaluate why these goals are or are not being attained. Furthermore, the criteria for evaluation are usually framed only in narrow biological terms, even though many other variables are major components of nearly every endangered or sensitive species program.

The sensitive species programs discussed in this chapter fell short of timely evaluation. Specific goals for programs were not developed at the beginning, nor were the criteria for success. In conducting work on the Coeur d'Alene salamander, for example, some general criteria to measure success might have been: inventory of a certain percentage of the range of distribution in a given time period; identification and measurement of critical habitat components; initiation of research on the impact of timber harvests; increased agency and public education on amphibians; and policy and program development within the IDFG for conservation of amphibians. In practice, all of these goals can be specified in measurable terms.

With evaluative criteria in place, good evaluation is possible. Success would have been measured not only by progress toward biological goals but also by goals focusing on educational and policy improvements. This is not to say that these projects have been unsuccessful. Clearly that was not the case. Progress has indeed been made toward conservation of these species as a result of these projects and programs. The important point is that a road map for guiding these programs—and evaluating if, when, and where we took wrong turns—would have increased their overall effectiveness and improved future programs. This evaluation is the basis of learning and improvement.

Adaptive management (Walters 1986; Walters and Holling 1990) offers a promising approach for evaluating sensitive species programs and contributing to greater conservation success. (Adaptive management is roughly equivalent to the "single-loop" learning described in Chapters 6 and 15 and distinct from "double-loop" learning.) Adaptive management uses data to construct a range of alternative management responses to a problem by which policy choices can be made—as opposed to selecting a single response and not knowing whether it is correct. In the harlequin duck studies, for example, we hypothesized that human disturbance of breeding streams, fluctuating water flows in unstable watersheds, and low productivity of aquatic invertebrates may be responsible for the observed lower densities of harlequins in Idaho compared to other sites (Cassirer and Groves 1991). Although attempts were made to test these hypotheses indirectly, future work might focus on experiments aimed at assessing the effects of limiting human access to breeding streams or deliberately disturbing nesting birds. Similarly, human disturbance may also be limiting wolverine

erations (see Harrison 1972). The IDFG has a considerable number of rules and expectations of its employees; many of these rules are formalized in a policy manual. Working at hunter "check stations," participating in the department's annual "free-fishing day," wearing uniforms in the field, and working through a rigid bureaucratic chain of command to receive authorization for actions are but a few of the rules.

Though many of these rules are well intended, the rigidity with which they are enforced can stifle individual initiative and productivity. This is particularly true for professionals whose orientation may not be toward the disciplines for which many of these policies were implemented—namely, traditional wildlife and fisheries management and game law enforcement. Because these disciplines continue to dominate the culture of the IDFG and many state game and fish agencies, biologists whose interests lie in the conservation of nongame species, rare plants, or ecosystems may continue to have trouble fitting into these agencies or being accepted.

Forest Service biologists, whether oriented toward game species or nonconsumptive biological resources, have had difficulty gaining acceptance into their organizational culture. A study of foresters, range managers, and wildlife biologists hired by the service in Regions 4 and 6 between 1978 and 1981 found considerable differences among these professions. Wildlife biologists had stronger professional commitments, were more idealistic in selecting their profession, and had greater difficulty fitting into the Forest Service than their forestry or range management peers (Kennedy and Mincolla 1982; Kennedy 1985). Experience in conducting sensitive species programs suggests that the stature of biologists in the Forest Service has risen in recent years with increased public and agency interest in conserving biodiversity. Nevertheless, these biologists appear to remain on the lower end of the bureaucratic pecking order as opposed to their more commodity-oriented colleagues. The continual strains of working at the lower end of a bureaucratic hierarchy can be costly, both personally and professionally. Conservation programs would benefit greatly from a more "generative" agency culture where rigidity is avoided, where all personnel are given incentive to think and work creatively and critically, and where biologists, particularly those working on the conservation of nonconsumptive resources, have a more equitable voice at the decision-making table.

Interagency Cooperation and Communication. Although there was considerable cooperation between the Forest Service and the IDFG on candidate and sensitive species programs, such was not the case with the FWS. Even though the FWS Idaho office was informed of all activities with these candidate species, it expressed little interest except when petitioned to list the Coeur d'Alene salamander as an endangered species. Although FWS staff did attend an interagency organizational meeting for the wolverine study, the FWS never provided any funding for the projects discussed in this chapter. Neither the FWS nor the IDFG showed much inclination to use ESA Section 6 funding for candidate species. All of this funding typically went to the state's high-profile endangered species (gray wolf, *Canis lupus;* woodland caribou, *Rangifer tarandus;* grizzly bear, *Ursus arctos horribilis;* peregrine falcon, *Falco peregrinus;* bald eagle, *Haliaeetus leucephalus*—something for which the FWS has been taken to task (GAO 1988).

The Forest Service expended considerable efforts to publicize its threatened, endangered, and sensitive species programs. Annual interagency meetings were held to review progress on these programs, and training sessions were conducted on sensitive species at either the forest or regional level. Although nearly always invited, FWS biologists were conspicuous by their absence. Noticeably missing too were IDFG personnel (such as regional wildlife biologists) beyond members of the Nongame and Endangered Wildlife Program. In addition to this lack of participation, some IDFG biologists criticized Forest Service biologists for spending too much time and resources on sensitive species instead of game species like elk (Groves and Unsworth 1993). This narrow approach to wildlife management by the IDFG was counterproductive to cooperation with the federal agency that is responsible for managing more habitat in the state than any other public or private entity.

Plans are under way, however, to improve cooperation between the FWS, Forest Service, and other agencies with regard to management of sensitive and candidate species (J. Gore, pers. comm.). A memorandum of understanding (1994) among the Forest Service, FWS, BLM, National Park Service, and National Marine Fisheries Service recognizes that a focused, cooperative effort in the management of candidate species could be beneficial and cost-effective to both agencies in preventing the need to list many candidate species as threatened or endangered. The memorandum calls for the development of conservation agreements for habitats or species that would preclude the need for listing as threatened or endangered. Using this federal memorandum as a template, agencies in Idaho (Forest Service, BLM, FWS, IDFG, and Idaho Department of Parks and Recreation) have recently signed a similar MOU (1994) designed to conserve candidate and sensitive species through coordinated, cooperative management and problem solving.

Lessons

Several key professional and organizational issues have been discussed in this chapter. The professional shortcomings include an excessively narrow problem definition, inadequate experience and education, motivation and values that run counter to conservation of candidate or sensitive species, and a lack of program evaluation. The organizational shortcomings include agency structure, agency culture, and interagency communication and cooperation. As these problems suggest several valuable lessons, steps for designing and conducting a model conservation program for a candidate or sensitive species are outlined here. These steps incorporate knowledge gained from the professional and organizational experiences discussed throughout the chapter.

Step 1: Define the Problem. The conservation problem and its context should be defined broadly to include all aspects—biological, political, policy, economic, and more. Think early and comprehensively about potentially controversial aspects of a project, such as dealing with private landowners or special interest groups. Funding for the wolverine program, for example, involved garnering support from wilderness advocates who were concerned about the impact of motorized recreation on wolverines and other wildlife.

Step 2: Funding, Funding, Funding. There is a saying in the real estate business that the three most important points in selling a home are location, location, and location. Much the same could be said about funding for a conservation program. Search widely for funding and do not give up if the agency that should fund the program cannot or will not. Look to outside organizations for funding (the National Fish and Wildlife Foundation, National Geographic, and others). If there is a special interest group that might put pressure on an agency to fund a program, do not hesitate to apply some leverage, but be aware of potential negative consequences to you and your program.

Step 3: Consult Experts. Seek multidisciplinary expertise in planning a program. Think about what factors (biological, professional, organizational) may limit or control the population under study or the program. For example, the assistance of a population ecologist/statistician might be needed if viability analyses are planned. Help might be sought from someone skilled in organizational management if a program becomes bogged down by bureaucratic inertia.

Step 4: Measure Success. Establish criteria and goals by which you can gauge the program and ultimately its success or failure. It is difficult to know where a program is going without a map and destination in mind. Plan some time for frequent, periodic evaluations. Do not confine the program to biological goals and criteria. Be sure to include professional and organizational goals. Set criteria to determine when a project is complete and the program should move on to another phase.

Step 5: Operate Effectively in a Bureaucracy. Do not expect to change bureaucracies. Instead, learn about the bureaucratic structure and culture of the agencies involved as early as possible, and figure out how to best work within this structure and culture. Remember that many agency policies are unwritten and learned only with time and experience. Consider putting together an advisory or steering committee consisting of species experts, appropriate agency biologists and line officers, and any other stakeholders (say, a key private landowner). Keep this committee and other interested parties well informed, a step that can help build consensus. Inform as many potentially interested parties as possible about the program and its projects. By doing so you will amass support for the program and mitigate any potential problems from detractors who might claim they were not kept informed. This tactic was used successfully in both the wolverine and harlequin duck conservation programs.

Step 6: Build Support. Encourage agency biologists to participate in the program from the beginning. If necessary, organize training sessions for biologists who want to help but lack the expertise. Such sessions with Forest Service biologists concerning projects discussed in this chapter not only helped them do a better job professionally but built enthusiasm and support for the Forest Service's sensitive species program as a whole. Try to obtain support for the program from the highest possible level in the bureaucracy. Remember that as personnel move from one job to another, commitment often moves too.

Step 7: Achieve Conservation. Do not forget the end goal of a program—halting a population decline, restoring or improving habitat, enacting sound management practices that will prevent future declines. In the midst of a research project, it is sometimes easy to forget why a program was initiated in the first place, and achieving actual conservation measures can be the most difficult part of a program. Try to ascertain the best avenues to achieve conservation within the political framework of the program. Develop a conservation plan, for example, but be sure to have separate plans for how it will be implemented.

These steps, of course, represent an ideal scenario. In the real world, most conservation biologists seem overwhelmed with work and finding the time to think broadly and deeply about a project can be difficult. But the payback of doing so could be well worth the effort. Although several professional and organizational problems have been highlighted in this chapter, the fact remains that these problems are but the tip of the iceberg compared to the problems that will ensue when a species or population needs listing as threatened or endangered.

Besides the greater magnitude of biological, organizational, and professional problems to deal with, the sheer financial costs of fixing problems skyrocket as a species becomes more endangered. A Forest Service official in the Northern Region remarked a few years ago that there were probably fewer harlequin ducks in the northern Rockies than grizzly bears. Although this is probably true, fewer dollars have been spent on all three of the candidate species discussed in this chapter—cumulatively over the life of those programs—than is spent in one year on grizzly bears in the northern Rockies. Thus there are strong biological, political, and financial incentives to tackle the conservation problems of candidate species before their plight worsens and achieving conservation becomes more difficult, more expensive, and less likely to succeed.

Acknowledgments

I am indebted to The Nature Conservancy and the Idaho Department of Fish and Game for funding during my tenure as coordinator and nongame biologist with the Idaho Conservation Data Center. Funding was provided by the Nongame and Endangered Wildlife Program and Pittman-Robertson funds. The Forest Service's Challenge Cost-Share Program, a contract with the service's Intermountain Research Station, and a grant from the National Geographic Society provided most of the funding for the three projects referred to in this chapter.

Special thanks are due to B. Ruediger, D. Atwood, and M. Rath for their work in launching regional Forest Service sensitive species programs. Although many people assisted with these projects, two deserve special credit for their considerable contributions: Frances Cassirer, wildlife researcher with the IDFG, was a lead biologist on the harlequin duck research project and also assisted in many ways with work on Coeur d'Alene salamander projects; Jeff Copeland, also a wildlife research biologist with the IDFG, is currently the principal investigator on the wolverine study. J. Gore (Forest Service, FWS), M. Maj (Forest Service), and V. Saab (Forest Service–Research) provided helpful reviews.

References

Brewer, G. D., and P. deLeon. 1983. *The Foundations of Policy Analysis.* Homewood, Ill.: Dorsey Press.

Cassirer, E. F., and C. R. Groves. 1991. *Harlequin Duck Ecology in Idaho: 1987–1990.* Boise: Idaho Department of Fish and Game.

———. 1992. *Ecology of Harlequin Ducks in Northern Idaho Progress Report 1991.* Report to the Intermountain Research Station, USFS. Boise: Idaho Department of Fish and Game.

Cassirer, E. F., C. R. Groves, and D. L. Genter. 1993. *Conservation Assessment for the Coeur d'Alene Salamander* (Plethodon idahoensis). Report prepared for USDA Forest Service Region 1. Boise: Idaho Department of Fish and Game.

Cassirer, E. F., C. R. Groves, and R. L. Wallen. 1991. Distribution and population status of harlequin ducks in Idaho. *Wilson Bulletin* 103:723–725.

Cassirer, E. F., G. Schirato, F. Sharpe, C. R. Groves, and R. N. Anderson. 1993. Cavity nesting by harlequin ducks in the Pacific Northwest. *Wilson Bulletin* 105:691–694.

Clark, T. W. 1983. Wildlife biology's need for a new philosophy. *Nongame Newsletter* 2:6–7.

———. 1988. The identity and images of wildlife professionals. *Renewable Resource Journal* (Winter):12–16.

Clark, T. W., R. Crete, and J. Cada. 1989. Designing and managing successful endangered species recovery programs. *Environmental Management* 13:159–170.

Clarke, J. N., and D. McCool. 1985. *Staking Out the Terrain: Power Differentials Among Natural Resource Management Agencies.* Albany: State University Press of New York.

Copeland, J. 1993. *Assessment of Snow-Tracking and Remote Camera Systems to Document Presence of Wolverines at Carrion Bait Stations.* Boise: Idaho Department of Fish and Game.

Copeland, J., and C. R. Groves. 1992. *Progress Report: Wolverine Ecology and Habitat Use in Central Idaho.* Boise: Idaho Department of Fish and Game.

Copeland, J., and C. Harris. 1993. *Progress Report: Wolverine Ecology and Habitat Use in Central Idaho.* Boise: Idaho Department of Fish and Game.

Edwards, T. C. 1989. The Wildlife Society and the Society for Conservation Biology: Strange but unwilling bedfellows. *Wildlife Society Bulletin* 17:340–343.

Federal Register. 1986. Endangered and threatened wildlife and plants; Findings on pending petitions and descriptions of progress on listing actions. USDI Fish and Wildlife Service, 50 *Code of Federal Regulations* pt. 17, 51(6):996–999.

———. 1991. Endangered and threatened wildlife and plants: Animal candidate review for listing as endangered or threatened species, proposed rule. pt. VIII. USDI Fish and Wildlife Service, 50 *Code of Federal Regulations* pt. 17, 56(225):58804–58836.

———. 1993. Plant taxa for listing as endangered or threatened: notice of review. pt. IV. USDI Fish and Wildlife Service, 50 *Code of Federal Regulations* pt. 17, 58:51144–51190.

Frissell, C. A., R. K. Nawa, and R. Noss. 1992. Is there any conservation biology in "New Perspectives?" A response to Salwasser. *Conservation Biology* 6:461–464.

General Accounting Office (GAO). 1988. *Endangered Species: Management Improvements Could Enhance Recovery Effort.* Report to the chairman, Subcommittee on Fisheries and Wildlife Conservation and the Environment, Committee on Merchant Marine and Fisheries, House of Representatives. GAP/RCED-89-5. Washington, DC: Government Printing Office.

Groves, C. R., and J. Unsworth. 1993. Wapiti and warblers: Integrating game and nongame management in Idaho. In *Proceedings of the Workshop on Management of Neotropical Migratory Birds,* D. Finch and P. Stangel (eds.). Fort Collins, Colo.: U.S. Forest Service Rocky Mountain Research Station.

Groves, C., R. Wallen, and F. Cassirer. 1990. Clown on the water. *Idaho Wildlife Magazine* 10:24–25.

Groves, C. R. 1988. Distribution of the wolverine in Idaho as determined by mail questionnaire. *Northwest Science* 62:181–185.

———. 1992. Appraisal of the Idaho Department of Fish and Game's proposal to establish a demonstration woodlot in Farragut State Park. Unpublished report. Boise: Idaho Department of Fish and Game.

Harrison, R. 1972. Understanding your organization's character. *Harvard Business Review* (May–June):119–128.

Idaho Department of Fish and Game (IDFG). 1992. *Proceedings of the Harlequin Duck Symposium,* 23–24 April 1992, Moscow, Idaho. Boise: Idaho Department of Fish and Game.

Kennedy, J. J. 1985. Viewing wildlife managers as a unique professional culture. *Wildlife Society Bulletin* 13:571–579.

Kennedy, J. J., and J. A. Mincolla. 1982. *Career Evolution of Young 400-Series U.S. Forest Service Professionals*. Project Report 1. Logan, Utah: College of Natural Resources.

Lawrence, N., and D. Murphy. 1992. New perspectives or old principles? *Conservation Biology* 6:465–468.

Meese, G. M. 1989. Saving endangered species: Implementing the Endangered Species Act. In *Preserving Communities and Corridors*, G. Mackintosh (ed.). Washington, DC: Defenders of Wildlife.

Moseley, R., and C. Groves. 1992. *Rare, Threatened, and Endangered Plants and Animals of Idaho*. Boise: Idaho Department of Fish and Game.

Schön, D. A. 1983. *The Reflective Practitioner: How Professionals Think in Action*. New York: Basic Books.

Scott, J. M., B. Csuti, K. Smith, J. E. Estes, and S. Caicco. 1991. Gap analysis of species richness and vegetation cover: An integrated biodiversity conservation strategy. In *Balancing on the Brink of Extinction: The Endangered Species Act and Lessons for the Future*, K. A. Kohm (ed). Washington, DC: Island Press.

Tobin, R. 1990. *The Expendable Future: U.S. Politics and the Protection of Biological Diversity*. Durham, N.C.: Duke University Press.

USFS Intermountain Region. 1989. *Action Plan for Threatened, Endangered, and Sensitive Species Activity Review*. Ogden, Utah: U.S. Forest Service.

USFS Northern Region. 1991. *Every Species Counts: Northern Region's TES Action Plan*. Missoula, Mont.: U.S. Forest Service.

Wallen, R. L. 1987. Habitat utilization by harlequin ducks in Grand Teton National Park. M.S. thesis, Montana State University.

Walters, C. J. 1986. *Adaptive Management of Renewable Resources*. New York: McGraw-Hill.

Walters, C. J., and C. S. Holling. 1990. Large-scale management experiments and learning by doing. *Ecology* 71:2060–2068.

Westrum, R. 1986. Management strategies and information failures. Paper presented at the NATO Advanced Research Workshop on Failure Analysis of Information Systems, Bad Winsheim, Germany, August 1986.

Wilcove, D. S. 1993. Getting ahead of the extinction curve. *Ecological Applications* 3:218–220.

Wilcove, D. S., M. McMillan, and K. C. Winston. 1993. What exactly is an endangered species? An analysis of the U.S. endangered species list: 1985–1991. *Conservation Biology* 7:87–93.

I I

The Australian Eastern Barred Bandicoot Recovery Program
Evaluation and Reorganization

**Gary N. Backhouse, Tim W. Clark,
and Richard P. Reading**

The eastern barred bandicoot, *Perameles gunnii* (Figure 11-1), is a rabbit-sized marsupial that once occurred in three states of south-eastern Australia: Victoria, South Australia, and Tasmania. The family Peramelidae suffered greatly following European settlement of Australia two hundred years ago. Of eleven species, three are now extinct. Of the extant species, four have experienced major declines (greater than 90 percent loss in range and abundance) and at least seven subspecies are extinct (Aitken 1979; Kennedy 1992).

Public concern about the eastern barred bandicoot's decline in Victoria was first noted in 1937 (Harper 1945) and again in the 1960s and 1970s (Wakefield 1971; Pizzey 1975). In 1960 the state government first investigated its status. Field surveys in the 1970s showed the species had substantially declined (greater than 99 percent loss in historic range and abundance; Seebeck 1979). Management guidelines were suggested in 1982, were later expanded, and have been more or less continuously implemented in varying ways and degrees since that time. Yet the species has declined further, and in November 1991 field surveys found the species near extinction in the wild. This crisis precipitated a detailed program evaluation from which it became obvious to all that the recovery program had failed (Reading et al. 1992). After exposing certain professional and organizational weaknesses of the program, participants set in place a new structure and process to rectify past problems in a renewed effort to recover the species. This new effort, as we shall see, has been more successful.

FIGURE 11-1.   The eastern barred bandicoot. (Photo by T. W. Clark.)

In this chapter we focus on the Victorian population of the eastern barred bandicoot. After describing the species and tracing its decline, we detail how the initial recovery effort failed to conserve the species, present some of the key lessons gleaned from this experience, and conclude with an update describing new organizational arrangements, management operations, and the improved status of the species.

From Abundance to Near Extinction

Historically the eastern barred bandicoot was distributed over three states. In South Australia, the last recorded specimen was taken in 1893 (Aitken 1983). In Tasmania, the species remains widely distributed but is declining due to habitat loss (Driessen and Hocking 1991; Robinson, Murray, and Sherwin 1991); recent DNA analysis has distinguished the Tasmanian population as a separate subspecies (Robinson et al. 1990). In Victoria, the species originally occupied about 2.8 million hectares, but the last wild population, in and near the city of Hamilton, had declined to virtual extinction by 1992. Some animals from the Hamilton population were taken into captivity beginning in 1988 and serve as stock for restoring the species.

Today the species is considered endangered in Victoria (DCNR 1990). Nationally, the species is considered vulnerable (ANZECC 1991). It has been listed as a threatened taxon under Victoria's Flora and Fauna Guarantee Act of 1988, and an action statement (a general management plan under the act) has been prepared (DCNR 1991).

Biology and Threats

Eastern barred bandicoots weigh up to 1 kilogram but average about 660 g (Clark and Seebeck 1990). Bandicoots rarely live more than two or three years in the wild and are highly fecund. Gestation is twelve or thirteen days, and maximum litters are five young (the mean is 2.4). Young remain in the pouch for about fifty-five days after which they become independent and disperse at three months after birth. Females may breed from three months of age and can give birth to another litter immediately after weaning the previous one. Although reproduction occurs throughout the year, it is depressed during late summer and ceases during drought. Under ideal conditions a single female can produce twenty young per year.

The ecology and population dynamics of the species are little known. Densities of 0.4 to 8.5 bandicoots per hectare have been reported (Heinsohn 1966; Brown 1985; Dufty 1988, 1991), but a more rigorous estimate of 1.5 per hectare has been calculated for the wild population in Victoria (Minta, Clark, and Goldstraw 1990). The species is omnivorous, feeding principally on invertebrates including earthworms, beetles, crickets, and caterpillars (Heinsohn 1966; Dufty 1988, 1991; Brown 1989). Some plant material, including fruit and bulbs of grasses, is also taken. Bandicoots are crepuscular and nocturnal, resting during the day in grass-lined nests.

The range of threats facing the eastern barred bandicoot is considerable and formidable (Brown 1989; Seebeck, Bennet, and Dufty 1990). Extensive habitat alteration and destruction have occurred throughout the bandicoot's range—through clearing, grazing by domestic stock and introduced herbivores, altered fire regimes, pasture "improvement," and human urban and suburban expansion. Over 99 percent of Victoria's native grasslands and grassy woodlands, in which the species formerly existed, have disappeared (Scarlett et al. 1992). There are also direct pressures by introduced carnivores—particularly domestic and feral cats (*Felis catus*) and red fox (*Vulpes vulpes*)—by motor vehicle collisions, by diseases, particularly toxoplasmosis spread by cats, and by the likely toxicity of pesticides on bandicoots and their invertebrate food supply (Brown 1989; Lenghaus, Obendorf, and Wright 1990; Seebeck, Bennet, and Dufty 1990). Random environmental events such as fire and drought have probably contributed to the decline as well (Lacy and Clark 1990).

Victoria's Last Wild Population

The decline of Victoria's last wild population, which survived in and near the city of Hamilton, accelerated in the 1980s. The species occupied about 3000 hectares in the 1970s. Range declined to about 1400 hectares by 1985 (Sherwin and Brown 1990), to about 600 hectares by 1988 (Dufty 1988), and to about only 80 hectares in 1992 (Reading, Fisher, and Goldstraw 1992). By late 1988 and early 1989, about 180 to 220 bandicoots remained (Dufty 1988; Minta, Clark, and Goldstraw 1990). Annual monitoring of population trends began in late 1989. In 1989, fifty-two individuals were captured; in 1990, thirty-six individuals were captured; in both 1991 and 1992, only three individuals were captured (Clark and Goldstraw 1991; Reading, Fisher, and Goldstraw 1992).

A population viability analysis (PVA), using a computer simulation model called Simpop, was conducted on Hamilton bandicoot population data by Lacy and Clark (1990). PVA results showed a decline of 25 percent a year and estimated the mean time to extinction at ten years given current management arrangements. Patrick and Myroniuk (1990) reanalyzed the data used by Lacy and Clark using the updated PVA program Vortex (Lacy 1993) and estimated the time to extinction at between 5.6 and 7.2 years. But the actual rate of decline was even greater than that modeled by either Lacy and Clark (1990) or Patrick and Myroniuk (1990), owing in part to the removal of some animals for a captive breeding program. Since then, Clark et al. (1992) have modeled the decline using the RAMAS/age simulation computer model and note that a skewed sex ratio and reduced litter sizes have significantly contributed to synergy between stochastic and deterministic forces, driving the small population to near extinction.

Management of the Wild Population

Considerable effort has gone into the bandicoot recovery effort in recent years. (For a summary see Clark and Seebeck 1990.) In 1982, the first interim management prescription was produced. Active management of wild bandicoots commenced on a modest scale in 1982 with some habitat enhancement. The state published a draft management plan in 1987 and released the final plan in 1989 (Brown 1989). State and Hamilton recovery teams were established in 1989 to aid and monitor the recovery effort. Subsequently, efforts to enhance and extend habitat, control predators, reduce road mortality, and educate the community were undertaken. Although these activities did not halt the bandicoot's decline, they may have slowed it slightly.

Beginning in 1988, some animals were taken into both semiwild and captive situations (Seebeck 1990). Animals were first placed in large captive breeding pens in Gellibrand Hill Regional Park near Melbourne. For the first three years of operation, these facilities contributed little to building the overall population, as animals were difficult to manage or even monitor. Many bandicoots died, disappeared from the pens, or were injured (Reading et al. 1991). Indeed, the population was sustained only by regular infusions of wild caught animals—which simply increased the rate of decline of the wild population while providing little benefit to overall species restoration. In addition to animals that escaped from breeding pens, beginning in 1989 others were purposely removed from the pens and released into a 400-hectare nature reserve within the park (Watson 1991). The reserve was enclosed by a somewhat predator-proof fence.

After much resistance, largely from government officials charged with recovering the species, a second 100-hectare site adjacent to the city of Hamilton was enclosed by a predator-proof fence in 1989 (Seebeck 1990). A few animals trapped during surveys of the Hamilton wild population were put inside this enclosure in an unplanned and unsystematic fashion. In 1989 a few bandicoots were placed in pens at Mooramong, a National Trust of Victoria sheep ranch near Skipton in southwestern Victoria. These animals were given minimal attention and were a minor part of the overall restoration effort.

Beginning in 1989, additional bandicoots were placed in an intensive captive breeding program managed by the Zoological Board of Victoria and held in its facilities at Melbourne Zoo, Healesville Sanctuary, and Werribee Zoological Park. As these animals ultimately bred well and increased to captive carrying capacity, many were released into Gellibrand Hill Nature Reserve and Mooramong.

At the end of 1991, the entire bandicoot "metapopulation" consisted of 109 animals scattered over these few sites. This number included the nearly extinct wild population and several semiwild and captive populations, none of which were self-sustaining. Time was clearly running out. The crisis ultimately resulted in an extensive in-house evaluation of the entire bandicoot conservation effort in 1991 (Reading et al. 1992).

Weaknesses in the Original Program

By late 1991, despite nearly a decade of research, intensified management, and the commitment of considerable resources, the conservation program had clearly failed. In view of the extreme urgency of the situation, a major program review was undertaken to examine why the program was unable to achieve species restoration, as well as to offer remedies for the program's shortcomings. A two-day evaluation involving seven principal participants resulted in a report that was widely circulated (Reading et al. 1992). The following discussion draws heavily on that report.

The evaluation cited several major weaknesses related to the program's organization and operation: professional shortcomings, lack of a strategic plan specifying targets for recovery, no timelines for essential actions, no assigned responsibilities, little attention to vital social or economic issues essential to bandicoot conservation (see Reading, Clark, and Arnold in press), crucial gaps in the information base required for species management, lack of effective leadership, poor communication and coordination among program participants, improper staffing, and ineffective operation of key activities. Many of these weaknesses stemmed from professional and organizational problems. Five are discussed in detail here.

Lack of Reliable Knowledge

The program's first major weakness was lack of reliable knowledge about the conservation problem. Prior to late 1989, there was little information about the wild or captive population segments. The program muddled along without knowing the true status of the bandicoot recovery challenge. This made it easy for many participants to believe that the program was working smoothly.

In reality, research conducted during the early and mid-1980s lacked rigorous experimental design and hypothesis testing. Not only were there meager data on long-term population mortality rates and causes or on habitat use patterns, but the existing information was largely anecdotal, descriptive, and based on unsystematic data collection, casual observation, or opinion. Radio telemetry was first tried in

1988, but field application did not occur until late 1992. In short, there was insufficient information to model the population and its habitat as a basis for understanding wild population trends and planning conservation measures.

Standardized yearly monitoring of the wild population was not initiated until late 1989—after the population was nearing extinction. Similarly, although semiwild populations were established in 1988 and 1989, standardized population monitoring did not commence until late 1991. Early research and data management of the semiwild and intensive breeding populations were inadequate also. Releases into both fenced enclosures and the wild occurred without experimental design. Until recently, the few records kept were not centralized for use by decision makers. Initial attempts at captive breeding occurred in large (0.5-hectare) pens and were based on the assumption that bandicoots would breed and progeny could be trapped for later release. Adequate management in such large pens proved impossible, however, and up to 50 percent of the pouch young simply "disappeared" (Watson 1991). Thus the bandicoot restoration program suffered from a chronic lack of rigorous biological science and reliable knowledge. Without a basis to understand the causes of the bandicoot's decline and offer realistic solutions, the restoration program was significantly handicapped.

Inadequate Definition of the Problem

The program's second major weakness was an inadequate definition of the conservation problem at hand. In the absence of reliable knowledge, many assumptions were made. These unverified assumptions led to an inadequate problem definition—which in turn led to delays in management action, an underappreciation of the situation's urgency, failure to explore the full array of causes of bandicoot decline, and a concentration on the biological dimensions of the problem to the exclusion of key sociological and political dimensions (see Reading, Clark, and Kellert 1991).

Recognition of the urgency of the bandicoot recovery challenge and the consequent need for a recovery plan took many years despite growing evidence of a vast decline over preceding decades. By the time a discernible program existed, the species had suffered a 99.9 percent decline and the single remaining wild population was nearing extinction. Rescuing a species from extinction under such conditions generally requires enormous resources and intensive management measures (IUCN 1987).

Because the bandicoot conservation problem was inadequately defined, an effective solution could not be formulated or implemented. Instead, early efforts to clarify the problem focused on biological and technical considerations and ignored a host of other essential concerns, such as power/authority questions, organizational problems, and socioeconomic issues directly affecting the bandicoot's fate (Reading, Clark, and Kellert 1991). Thus, the vague, problem definition came to dominate the early stages of the program, which focused on only a few aspects of the bandicoot's decline, such as habitat destruction and degradation, predation by feral and domestic cats, and road kills. Most management effort went into technical aspects of habitat protection and enhancement, and even then actions were essentially too little and too late. Other threats to the bandicoot's survival, including disease and pesticides, received virtually no attention.

Moreover, key figures in the recovery program generally displayed an excessively narrow focus on their area of specialty. This "bounded rationality" (Simon 1976) contributed to the consequent lack of attention to social, economic, organizational, and political issues at the heart of the bandicoot's endangerment. In short, program managers and key field personnel often lacked the expertise and breadth of view to recognize essential nonbiological dimensions of the problem. Furthermore, some lacked even the biological knowledge and skills necessary for bandicoot restoration, especially rigorous scientific experimentation and evaluation skills. Others were clearly overcommitted on other programs and could not allocate the necessary time and effort to the bandicoot program.

Overall, then, there was a general lack of individual accountability and little ability to work closely in teams or to solve even widely recognized shortcomings of the program. In conclusion, the bandicoot restoration program suffered from an inadequate definition of the conservation problem—a shortcoming that could be traced to a number of professional and organizational limitations.

Lack of Leadership

The program's third major weakness was a lack of strong, effective leadership. Leadership was largely absent from all levels, especially top-level senior managers, middle managers, and recovery teams. Over the years, the bandicoot program experienced a high turnover in officials and senior managers. During their short tenure, most of these people seemed little concerned with endangered species in general and eastern barred bandicoots in particular. Instead, they focused on other pressing issues. Even those who did show interest did so in limited, ephemeral ways. Although the eastern barred bandicoot is Victoria's rarest mammal, no official or senior manager considered the conservation effort to be a significant case deserving of special attention. In addition, lower-level managers rarely sought greater support from higher levels. As a result, the restoration effort never received high-level endorsement or benefited from high-level leadership.

Responsibility for leadership, therefore, was left to mid-level managers or recovery teams. Most members of the recovery teams were mid-level government managers or university faculty or students. Several of them were the same people who had failed to understand the urgency of the conservation task in the first place, who were satisfied with the sporadic, unreliable science being generated, and who carried a false sense of optimism about the entire program. As these people held much of the decision-making authority, there was little opportunity for others to assume leadership or challenge the assumptions, science, management, recovery team structure, or mode of decision making. It seemed that neither the disposition nor the skills necessary for successful leadership existed. The government agency's culture, in which most mid-level managers worked, did not offer incentives and rewards for strong and effective leadership of the bandicoot program. Nor did it rectify substandard performance. Moreover, the lead government agency underwent several major reorganizations and downsizings over the last several years of the bandicoot program. The organizational uncertainty had a disturbing effect on the attention and actions of many government participants, who were faced with early retirement, transfers, reassignments, or curtailment of current activities. This was hardly the kind of environment that promoted leadership.

In the absence of effective leadership, parts of the restoration program were dominated by parochial issues pushed by the Hamilton community and fueled by the local press at the expense of a strategic, statewide approach. As the frustration of working in a rigid, bureaucratic environment began to take its toll, there was a significant loss of enthusiasm among key members frustrated by poor program management and even some evidence of burnout. In conclusion, the program lacked leadership from top to bottom. Instead of a well-led campaign, the program was largely an ad hoc affair that was incrementally formed and incompletely implemented.

Absence of Comprehensive Management

The program's fourth major weakness was an absence of comprehensive management. This occurred largely because the program lacked a strategic master plan with clearly measurable goals. Although management guidelines were first developed in 1982 and a management plan was finalized in 1989, these documents simply suggested a number of approaches that could be taken for bandicoot conservation. They did not contain precise targets, timelines, or responsibilities and consequently were of little value in building, guiding, or evaluating the program.

Traditional bureaucratic structures were used to manage the program. Reporting relationships tended to develop along existing organizational channels. This system was rigid and thus not adaptable to rapidly changing issues. Traditional bureaucratic procedures were implemented and defended as the best way of operating. There was little recognition of the benefits of alternative organizational arrangements (Clark, Crete, and Cada 1989; Clark and Cragun 1991). Initially, even the changes resulting from the 1991 program review did little to alter the powerful program structure.

Poor communication hindered implementation of the recovery program. Record keeping was inadequate and neither centralized nor widely available as a basis for decision making and operations. Information failed to reach many key personnel or the public. As misinformation was common—sometimes based on rumor, misinterpretation, and poor reporting—this resulted in suspicions and strained relations among the agencies and people involved in the recovery effort. Constructive criticism of the program, particularly from outside the agency, was treated as unfounded and unfair. A defensive attitude resulted in the adoption of a siege mentality rather than effective consideration of the issues.

No contingency plans existed. So when extinction appeared imminent for the wild bandicoots at the end of 1991, several years earlier than predicted by PVA modeling, it came as a complete surprise to some program managers. There were simply no plans to deal with this situation when it inevitably arose. But the bandicoot crisis at the end of 1991 had other unforeseen consequences. Serious conflict arose during 1992 among top managers of the principal agencies responsible for implementing the recovery program. Some key senior managers had severe reservations about supporting the program because, despite the large expenditures for many years, bandicoots were still declining.

The program, as noted, developed in an ad hoc fashion. A poor understanding of the problem hindered the development of a larger program with clearer direction. Social, economic, and political considerations were largely ignored because they were poorly understood, considered too hard to solve, or not relevant—despite their documented importance in recovery programs elsewhere (Kellert 1985) and the obvious presence of social, economic, and political problems. There was little attention to public relations or education to build support for the program. As well, early inattention to powerful modeling tools—such as PVA modeling and decision analysis in planning and decision making—caused delays in decision making and led to several inferior decisions. Protocols and contingencies were inadequate for solving inevitable management problems.

In conclusion, the failure to develop a comprehensive, well-integrated, and goal-directed management response to the recovery challenge, including a plan to guide species restoration, significantly hindered efforts to conserve the bandicoot. And the largely ad hoc program that did emerge produced several suboptimal responses that in some cases only contributed to the species' decline.

Lack of Ongoing Evaluation

The program's fifth major weakness was the virtual absence of ongoing evaluation as a basis for improvement. Although there is always a reluctance to evaluate oneself, the process is absolutely essential to keep a program on course. When the 1989 management plan was finalized, it included virtually no measurable goals or objectives and few responsibilities or timelines. Subsequent guidelines and action plans developed to implement the management plan also lacked detail on objectives, responsibilities, and schedules. The consequences for the recovery program were profound. Evaluating past performance as a basis for setting future direction was impossible. The program appeared undirected, too costly, and excessively time consuming. Under such circumstances, it was impossible for the program to learn quickly or effectively from its own behavior.

The few intraagency reviews that did occur were subjective, informal, and lacked hard data to support conclusions. A combination of limited understanding of the issues facing bandicoot recovery and the lack of targets for the recovery program resulted in a vague idea of the expected outcomes of management actions. Poor program organization and record keeping made summarization and interpretation of data nearly impossible. Even the comparatively straightforward biological dimensions of the recovery effort were not monitored with sufficient rigor to permit annual assessments.

The program's termination or modification was not addressed, and program managers had little idea of when (or indeed if) successful recovery of the species would be achieved. Personnel working on bandicoot recovery were rarely encouraged to assess their own performance or the collective performance of the group, and little attempt was made to relate individual activities to the effectiveness of the overall recovery program. In conclusion, the critical role that evaluation plays in all complex programs was little appreciated. As a consequence, learning and fine-tuning were not possible—an omission that had severe ramifications for the restoration program.

Recent Reorganization and Progress

To their credit, many participants in the bandicoot recovery effort recognized the program's inability to recover the species and were therefore eager for critical self-analysis and ideas for improvement. Thus the atmosphere was conducive to the 1991 program review. Following that review, a major reorganization took place. In early 1992, a decision group was set up, a strategic planner was assigned, and four working groups were established. The new recovery program sought to establish reporting lines directly between senior decision makers and the field staff, regardless of agency allegiance. Although several problems remained, significant improvements resulted during 1992. The year began with about 109 bandicoots declining toward extinction; the year ended with more than 250 bandicoots in reintroduced wild, semiwild, and captive populations.

The missing leadership and remote decision-making authority in the original program were countered by creating a decision group composed of key senior managers from the two participating agencies—an arrangement that brought decision-making responsibility closer to the recovery effort. This group was supported by the four working groups responsible for implementing the program. These groups provided current information, made recommendations, and assisted evaluation.

Professionals in the four working groups were charged with developing specific targets. The captive breeding group, for example, used computer modeling to generate a numerical target of captive bandicoots—a goal for the group to achieve by the end of 1992. The decision group provided support. And in fact the captive breeding goal was exceeded. The 1992 program successes were due largely to the working groups, who were given the authority, guidance, and resources to develop and meet their own targets using their professional expertise.

The appointment of a strategic planner early in 1992—and the creation of the first recovery plan for the species (Backhouse 1992) to specify goals, targets, timelines, and responsibilities—helped put the program on a more objective footing. Other professionals with skills in economic and social analysis also assisted to improve program planning and implementation. These personnel helped turn the program around, and bandicoot numbers have increased, quite possibly for the first time since European settlement.

The ability to make changes to the organization, management, and staffing of the bandicoot program became possible during the early part of 1992 because of unforeseen reassignments in senior management and the resulting power vacuum. These changes permitted the adoption of a new organization, but further changes were not possible until after another fortuitous reassignment in senior management in early 1993. Once targets and responsibilities were set, several areas of the bandicoot program responded rapidly and positively.

Despite the marked improvement during 1992, the recovery program remained suboptimal. Some of the problems recognized in the 1991 review remain problematic, inadequately addressed, or unaddressed. Further improvements are required before the bandicoot program can be considered a model recovery program. The recovery plan for 1994 (Backhouse 1994) addresses some organizational and social problems by streamlining program management, further defining responsibility, and including community representation on the recovery team.

Lessons

If we are to improve our individual and collective performance in endangered species management, it is vital to point out the lessons. A number of important lessons are evident from this broad examination of the bandicoot recovery effort.

Lesson 1: Gain reliable knowledge early on about both the biological and nonbiological dimensions of the conservation problem. Gaining reliable knowledge about a small, endangered population is difficult under the best of circumstances. Not only is much of the required information absent and hard to obtain, but there are obviously time and resource limitations to obtaining knowledge. Nevertheless, the complexity, limitations, and scientific uncertainty cannot be permitted to paralyze the recovery program. If a species is allowed to decline to very small numbers before the conservation problem is investigated, the difficulty and costs of recovery increase significantly and the chances of success diminish accordingly. Moreover, such programs easily attract criticism from other programs, government, and the public.

Early acquisition of reliable knowledge is the best means of achieving conservation and carrying out a successful program. Although the new discipline of conservation biology was developed to address many of these problems, successful recovery also requires innovative biological and technical approaches to data collection, such as the use of closely related surrogate species in research. Obtaining rigorous, reliable biological and ecological knowledge is, in itself, a difficult and demanding job. And the need to use this information at the right time and in the right way compounds the species restoration challenge (Reading, Clark, and Kellert 1991).

Lesson 2: Recovery programs are most likely to succeed if they have a clear, comprehensive understanding of the conservation challenge. Comprehensive problem definition, as early as possible, is vital to recovery efforts. Narrow concentration on just a few dimensions inevitably leads to a poorly performing program. Recovery programs usually face a complex mix of biological, social, economic, political, and organizational challenges. As a result, the programs can be extremely complicated and may involve long-term commitments. A useful problem definition addresses all these elements.

Useful definitions reflect solutions to the problems they describe (Dery 1984). As Weiss (1989:97) states: "Problem definition . . . is not merely a label for a set of facts and perceptions. It is a package of ideas that includes at least implicitly an account of the causes and consequences of some circumstances that are deemed undesirable, and a theory about how a problem may be alleviated." Problems are rarely static; they change as the context changes and as knowledge is acquired. Problem definitions should be flexible and adaptable, as well, evolving as circumstances change and new information becomes available.

Lesson 3: Effective leadership is paramount in restoration programs.
Strong, effective leadership is needed at several levels—from the highest organizational levels to mid-level managers and recovery teams (Clark and Cragun 1991). Weak leaders and short tenure equal short time horizons and ineffective implementation (Fesler 1980; Straussman 1985). Leadership is required to project strategic vision, mobilize resources, make decisions, and inspire workers. Leaders must be simultaneously clear about what is needed to solve problems and open to new knowledge and opportunities. As Westrum (1988:14) notes: "[Leaders] may be the agents that bring about adequate cognitive performance, by providing the resources, the determination, the example, and the structures for adequate organizational thought. By encouraging free discussion they increase idea flow. By protecting essential channels they encourage early and successful reporting of problems and opportunities." Westrum discusses the traits of effective leaders, or "maestros," in more detail in Chapter 14. In sum, good leaders should possess human relations skills, administrative skills, and technical skills.

While some characteristics of leadership appear to be innate, other aspects can be taught and reinforced in either university or in-service agency programs (Westrum 1988). For working professionals, in-service training and regular updating of leadership skills are essential, especially in the face of the increasing complexity of recovery program management. Similarly, senior government leaders can be selectively recruited, cultivated, and supported.

Lesson 4: The organization and management of recovery programs must be well matched to the problems they face. Recovery programs can fail simply because of poor organization and management. While limited funding and lack of public support are frequently cited as a cause of failure, the root cause of a weak program is often inappropriate organization and management. This was certainly true in the bandicoot case. How a recovery program is organized dictates task assignments, resource allocation, information channels, control of communication, and more—all of which influence program effectiveness. Program managers must be aware of a range of organization and management options and employ the most appropriate ones.

To enhance performance, organizations should be well matched to their tasks (Clark and Westrum 1989; Clark, Crete, and Cada 1989). Traditional government bureaucracies are effective for routine activities but poorly matched to the uncertain, complex, and urgent tasks of recovering rare and endangered species. Recovery programs thus require organizations that remain task-oriented, encourage innovative and critical thought among all members, continually self-evaluate, reward new ideas, avoid overbureaucratization, create and maintain communication channels, and strive to improve performance by learning to learn. Given the organizational constraints inherent in government bureaucracies and the collaborative nature of many tasks, matching structure to task often requires the establishment of a parallel organization such as a special team.

Lesson 5: Regular and thorough program review is essential. Comprehensive program evaluation should be designed into all aspects and stages of recovery programs. Evaluations should be a formal part of the program and should lead to changes in the program based on the results. From the earliest stages of the recovery program, regular and comprehensive evaluations can lead to early problem identification and resolution. Evaluation should assess both people's and the program's performance. Regular evaluation ensures that the many constituencies of species restoration, especially senior managers, funders, and the local community, receive reports. This can be crucial to maintaining support for the program.

Evaluations should be included in the published record so that future recovery efforts can benefit from the lessons. The evaluations should be accurate and complete, however, and this is seldom the case. Evaluation should examine more than biological dimensions and should be undertaken by capable reviewers outside the organizations carrying out the work. As the bandicoot program illustrates so well, many problems are directly tied to the social, organizational, and political aspects of the program. Certainly the major program review in 1991 demonstrated the value of thorough, comprehensive evaluation (Reading et al. 1992).

This review of the eastern barred bandicoot recovery program has demonstrated that the program's weaknesses were primarily human and organizational, rather than biological. Confirmed improvement of the recovery program is essential, for the future survival of the species on mainland Australia depends on it.

Bandicoots declined for many years despite management aimed at restoration. Unlike many other endangered species, however, the eastern barred bandicoot has a very high reproductive capacity and the ability to survive and flourish in highly modified habitats. Indeed, the species' recoverability was demonstrated in 1992 when the number of bandicoots doubled following the establishment of a restructured goal-directed recovery program backed by professional expertise and resources. This augurs well for eventual recovery of the species. Yet simply increasing bandicoot numbers does not necessarily mean that the recovery program represents a model program to be emulated elsewhere. Given the right circumstances, even poorly organized and badly managed recovery programs can succeed at increasing species numbers.

While further increases in bandicoot numbers are envisaged, a substantial improvement in the status of the species will be more difficult to achieve. The last relict wild population is almost certainly beyond recovery and is now functionally extinct, if not actually extinct. Realistically, there is little hope of overcoming the problems associated with an expanding urban center, such as domestic pets, continuing decline in habitat, and road kills. Reintroductions will necessarily be focused within their former range, but probably away from Hamilton.

Captive breeding will continue as a major focus of the recovery program. A captive population of about a hundred bandicoots, including eighteen breeding pairs, will be maintained. This population should produce about seventy animals each year that can be used for reintroduction. Because of their small sizes, the captive and semiwild populations should be managed as one metapopulation, with an exchange of at least one reproductively successful individual between each population each generation. Commitment to a long-term strategy of reintroduction into unfenced habitat is essential. Among other things, this will require effective control of the red fox—the single biggest ecological obstacle to successful reintroductions of bandicoots.

Further organizational changes are essential to place responsibility for implementing the recovery program in the hands of those with a successful record in endangered species recovery. The 1993 recovery plan does just that (Backhouse 1993). Widening the range of expertise available to the recovery program is also essential for improved program performance. In particular, improved scientific methodology and design are required. There is also a need to obtain information on the biology and ecology of the species, initially concentrating on population dynamics and habitat requirements. There is an ongoing need, too, to improve the skills and training of participants in bandicoot recovery—especially in nonbiological areas such as social and economic research, organization, and management. Improved review and assessment of the program, at both individual and organizational levels, is required to enhance performance and promote individual and organizational learning.

The eastern barred bandicoot's future looks reasonably bright. For the foreseeable future we have a good chance of establishing and maintaining several populations throughout the species' former range in Victoria. Still, considerable challenges and risks lie ahead before the ultimate goal of bandicoot survival in the wild can be assured.

Acknowledgments

Our thanks to Peter Stroud (Werribee Zoo) for commenting on a draft of the chapter. A. Arnold, J. Fisher, and J. Seebeck of the Victorian Department of Conservation and Natural Resources and P. Myroniuk of the Melbourne Zoological Park provided invaluable insight during the 1991 program review.

References

Aitken, P. F. 1979. The status of endangered Australian wombats, bandicoots and the marsupial mole. In *The Status of Endangered Australasian Wildlife*, M. J. Tyler (ed.). Adelaide: Royal Zoological Society of South Australia.
————. 1983. Mammals. In *Natural History of the South-East*, M. J. Tyler, C. R. Twidale, J. K. Ling, and J. W. Holmes (eds.). Adelaide: Royal Zoological Society of South Australia.
Australian and New Zealand Environmental Conservation Council (ANZECC). 1991. *List of Australian Endangered Vertebrate Fauna, March 1991*. Canberra: CONCOM Ad Hoc Working Group on Endangered Fauna, Australian National Parks and Wildlife Service.
Backhouse, G. N. 1992. *Recovery Plan for the Eastern Barred Bandicoot Perameles gunnii*. Melbourne: Department of Conservation and Natural Resources.
————. 1994. *Recovery Plan for the Eastern Barred Bandicoot Perameles gunnii 1994–1996*. Melbourne: Department of Conservation and Natural Resources.
Brown, P. R. 1985. A preliminary report on investigations into the ecology and conservation of the mainland population of the eastern barred bandicoot *Perameles gunnii:* March 1983–June 1985. Unpublished report to World Wildlife Fund–Australia for Project 55.
————. 1989. Management plan for the conservation of the eastern barred

bandicoot, *Perameles gunnii,* in Victoria. *Arthur Rylah Institute Technical Report Series* 63:1–84.

Clark, T. W., and J. R. Cragun. 1991. Organization and management of endangered species programs. *Endangered Species Update* 8(8):1–4.

Clark, T. W., and P. Goldstraw. 1991. Summary of eastern barred bandicoot monitoring, Hamilton, Victoria, November 11–22, 1991. Unpublished report. Melbourne: Department of Conservation and Environment.

Clark, T. W., and J. H. Seebeck (eds.). 1990. *Management and Conservation of Small Populations.* Brookfield, Ill: Chicago Zoological Society.

Clark, T. W., and R. Westrum. 1989. High-performance teams in wildlife conservation: A species reintroduction and recovery example. *Environmental Management* 13:663–670.

Clark, T. W., R. Crete, and J. Cada. 1989. Designing and managing successful endangered species recovery programs. *Environmental Management* 13: 159–170.

Clark, T. W., J. P. Gibbs, and P. W. Goldstraw. 1992. Some demographics of eastern barred bandicoots (*Perameles gunnii*) extinction from the wild, 1988–1991, Victoria, Australia. Unpublished manuscript.

Department of Conservation and Natural Resources (DCNR). 1993. *Threatened Wildlife in Victoria.* Melbourne: DCNR.

———. 1991. *Eastern Barred Bandicoot Perameles gunnii Action Statement No. 4.* Melbourne: DCNR.

Dery, D. 1984. *Problem Definition in Policy Analysis.* Lawrence: University of Kansas Press.

Driessen, M. M., and G. J. Hocking. 1991. *The Eastern Barred Bandicoot Recovery Plan for Tasmania: Research Phase.* Hobart: Department of Parks, Wildlife, and Heritage.

Dufty, A. C. 1988. The distribution, population abundance, status, movement and activity of the eastern barred bandicoot, *Perameles gunnii,* at Hamilton. B.Sc. (honors) thesis, La Trobe University, Bundoora, Victoria.

———. 1991. Some population characteristics of *Perameles gunnii* in Victoria. *Wildlife Research* 18:355–366.

Fesler, J. W. 1980. *Public Administration: Theory and Practice.* Englewood Cliffs, N.J.: Prentice-Hall.

Harper, F. 1945. *Extinct and Vanishing Mammals of the Old World.* Baltimore: Lord Baltimore Press.

Heinsohn, G. E. 1966. Ecology and reproduction of the Tasmanian bandicoots (*Perameles gunnii* and *Isoodon obesulus*). *University of California Publications in Zoology* 80:1–107.

International Union for the Conservation of Nature and Natural Resources (IUCN). 1987. *The I.U.C.N. Policy Statement on Captive Breeding.* Gland, Switzerland: IUCN.

Kellert, S. R. 1985. Social and perceptual factors in endangered species management. *Journal of Wildlife Management* 49:528–536.

Kennedy, M. 1992. *Australasian Marsupials and Monotremes—An Action Plan for Their Conservation.* Gland, Switzerland: IUCN/Species Survival Commission.

Lacy, R. C. 1993. Vortex: A computation simulation model for population viability analysis. *Wildlife Research* 20:45–65.

Lacy, R. C., and T. W. Clark. 1990. Population viability assessment of the eastern barred bandicoot in Victoria. In *Management and Conservation of Small Populations,* T. W. Clark and J. H. Seebeck (eds.). Brookfield, Ill.: Chicago Zoological Society.

Lenghaus, C., D. L. Obendorf, and F. H. Wright. 1990. Veterinary aspects of *Perameles gunnii* biology with special reference to species conservation. In *Management and Conservation of Small Populations,* T. W. Clark and J. H. Seebeck (eds.). Brookfield, Ill.: Chicago Zoological Society.

Minta, S. C., T. W. Clark, and P. Goldstraw. 1990. Population estimates and characteristics of the eastern barred bandicoot in Victoria, with recommendations for population monitoring. In *Management and Conservation of Small Populations,* T. W. Clark and J. H. Seebeck (eds.). Brookfield, Ill.: Chicago Zoological Society.

Patrick, C., and P. Myroniuk. 1990. Preliminary population viability assessment of the eastern barred bandicoot: Selected wild and captive population scenarios. Unpublished report. Melbourne: Department of Conservation and Environment.

Pizzey, G. 1975. Have you seen this prowler? *Herald* (Melbourne), 13 Sept., p. 7.

Reading, R. P., T. W. Clark, and A. Arnold. In press. Values and attitudes of Hamilton, Victoria, residents toward the eastern barred bandicoot. *Anthrozoos.*

Reading, R. P., T. W. Clark, and S. R. Kellert. 1991. Towards an endangered species paradigm. *Endangered Species Update* 8(11):1–4.

Reading, R. P., J. Fisher, and P. Goldstraw. 1992. Results of eastern barred bandicoot monitoring in Hamilton, Victoria, November, 1992. Unpublished report. Melbourne: Department of Conservation and Natural Resources.

Reading, R. P., T. W. Clark, P. Goldstraw, A. Watson, and J. Seebeck. 1991. An overview of eastern barred bandicoot reintroduction programs in Victoria, Australia: With recommendations for future reintroductions. Unpublished report. Melbourne: Department of Conservation and Environment.

Reading, R. P., T. W. Clark, A. Arnold, J. Fisher, P. Myroniuk, and J. Seebeck. 1992. *Analysis of a Threatened Species Recovery Program: The Eastern Barred Bandicoot* (Perameles gunnii) *in Victoria, Australia.* Melbourne: Department of Conservation and Natural Resources.

Robinson, N. A., N. Murray, and W. Sherwin. 1991. A note on the status of the eastern barred bandicoot, *Perameles gunnii,* in Tasmania. *Wildlife Research* 18:451–457.

Robinson, N. A., W. B. Sherwin, N. D. Murray, and A. M. Graves. 1990. Application of conservation genetics to the eastern barred bandicoots, *Perameles gunnii*. In *Management and Conservation of Small Populations*, T. W. Clark and J. H. Seebeck (eds.). Brookfield, Ill.: Chicago Zoological Society.

Scarlett, N. H., S. J. Wallbrink, and K. McDougall. 1992. *Field Guide to Victoria's Native Grasslands*. Melbourne: Victoria Press.

Seebeck, J. H. 1979. Status of the barred bandicoot, *Perameles gunnii*, in Victoria: With a note on husbandry of a captive colony. *Wildlife Research* 6:255–264.

———. 1990. Recovery management of the eastern barred bandicoot in Victoria: Statewide strategy. In *Management and Conservation of Small Populations*, T. W. Clark and J. H. Seebeck (eds.). Brookfield, Ill.: Chicago Zoological Society.

Seebeck, J. H., A. F. Bennet, and A. C. Dufty. 1990. Status, distribution and biogeography of the eastern barred bandicoot, *Perameles gunnii*, in Victoria. In *Management and Conservation of Small Populations*, T. W. Clark and J. H. Seebeck (eds.). Brookfield, Ill.: Chicago Zoological Society.

Sherwin, W. B., and P. R. Brown. 1990. Problems in the estimation of the effective size of a population of the eastern barred bandicoot, *Perameles gunnii*, in Hamilton, Victoria. In *Bandicoots and Bilbies*, J. H. Seebeck, P. R. Brown, R. L. Wallis, and C. M. Kemper (eds.). Sydney: Surray Beatty & Sons.

Simon, H. 1976. *Administrative Behavior.* 3rd ed. New York: Free Press.

Straussman, J. D. 1985. *Public Administration.* New York: Holt, Rinehart.

Wakefield, N. A. 1971. Mammals of western Victoria. In *The Natural History of Western Victoria*, M. H. Douglas and L. O'Brien (eds.). Horsham: Australian Institute of Agricultural Science.

Watson, M. 1991. An overview of the captive breeding and re-introduction program for eastern barred bandicoots at Gellibrand Hill Park. Unpublished report. Melbourne: Department of Conservation and Environment.

Weiss, J. A. 1989. The powers of problem definition: The case of government paperwork. *Policy Sciences* 22:97–121.

Westrum, R. 1988. Organizational and interorganizational thought. Paper delivered at the World Bank Conference on Safety Control and Risk Management, October 1988, Washington, DC.

PART III

Theoretical Perspectives

Biologists should ask for the help of professionals
in other disciplines, including philosophers,
anthropologists, sociologists, and community
development workers. Otherwise, we risk espousing solutions that are theoretically robust, but
socially and politically naive.
MICHAEL E. SOULÉ

I believe that conservation biologists have a
responsibility to enter the policy arena and advocate both general principles and specific actions
needed to conserve biodiversity. . . . Scientists cannot simply hand over data to bureaucrats or politicians and expect them to make rational and prudent decisions about complex problems they know
and care little about.
REED NOSS

I 2

A Conservation Science Perspective
Conceptual and Experimental Improvements

Steven C. Minta and Peter M. Kareiva

Although ecology has always had connections with resource management and conservation, only recently has the science been asked to contribute to large-scale land use plans. The impetus for these new demands is, of course, the Endangered Species Act (ESA) and the National Environmental Protection Act. Several chapters in this volume, as well as our own observations, suggest that the connection between science and policy with respect to endangered species is disturbingly weak. For example, recovery plans for endangered species, which are supposed to use ecological data and insight to establish management guidelines, have been called "nothing more than a promise cloaked in prayer" (Soulé and Mills 1992:62). Indeed, a careful review of recovery plans for ninety-eight endangered plant species found a wholesale lack of the most basic of all ecological data—a species' capacity for population increase (Schemske et al. 1994). Part of the problem lies with the inertia of federal agencies and their resistance to change, as is well illustrated by the U.S. Forest Service's slowness in effectively dealing with spotted owls in the Northwest (see Chapter 3). And even when agencies do have the best of intentions, they often lack adequately trained personnel. But there are also fundamental problems with conservation science itself—weaknesses that hamper our effectiveness at protecting biodiversity and threatened species.

In this chapter we examine those aspects of conservation science that weaken its ability to contribute to policy decisions regarding endangered species. We begin by reviewing key ideas from ecological theory that are at the heart of conservation science. The unsettled status of ecological theory, we suggest, provides unsteady ground indeed for endangered species management. After proposing solid foundations upon which to base a conservation science for the future, we end with a précis of recent technical advances and theoretical ideas that are being promoted as a better basis for predicting and managing change.

The "New Ecology" and Its Implications

When the ESA was enacted, ecology was enamored of niche overlap theory and Lotka-Volterra models and filled with a vision of some soon-to-be-realized grand theory encoded in the form of rigorous mathematical statements. No longer is there much faith that ecology has a general theory. Some have even charged that the science of ecology is fatally ill (Peters 1991). Although there might be debate among ecologists about the prospects for theory, there is a widespread consensus that every environmental problem or endangered species must be treated as a special case (Slobodkin 1988; Hairston 1989; Botkin 1990). At a practical level this means that the management of endangered species must be based on detailed ecological information rather than platitudes concerning habitat fragmentation, life history theory, or disturbance theory (Simberloff 1988).

Apart from a growing skepticism about all-encompassing models, the most important change in ecological thinking pertinent to conservation has been the abandonment of "the balance of nature" as a supreme principle. It is worth noting, however, that the notion of some idealized balance of nature remains an important component of the lay public's view of environmental issues—thus worries about environmental degradation are often discussed in terms of this critical (almost spiritual) balance being upset (see Gore 1992). But contrary to the public's perception, "the classical paradigm in ecology, with its emphasis on the stable state, its suggestion of natural systems as closed and self-regulating, and its resonance with the nonscientific idea of the balance of nature, can no longer serve as an adequate foundation for conservation" (Pickett et al. 1992:84; see also Holling 1987 and Botkin 1990). Instead of a "balance of nature," ecologists now think in terms of "the flux of nature," which focuses on the processes that generate or maintain the species in a broad landscape. The abandonment of an equilibrium-dominated (or "balance") view is not simply an academic issue. It has far-reaching consequences for practical decisions—such as how we view forest fires. If one seeks to maintain equilibrium, for example, then forest fires should be prevented as much as possible because they perturb the steady state. But if one views ecosystems in terms of perpetual change and inexorable turnover of species generated by mosaics of disturbance, then forest fires might be favored because they are considered part of a natural disturbance regime that maintains diversity in the system.

Hand-in-hand with a deemphasis of equilibrium-based, closed populations has come greater attention to landscape-level phenomena. If the existence of local populations depends on dispersal from elsewhere rather than on a local self-maintaining system of checks and balances, then obviously we must consider broad regions in order to understand each species' persistence. Within the last decade, a major new journal and scientific society have arisen to deal with landscape ecology, and much of this field is geared toward conservation issues. The realization that species must be studied at a landscape level has also spurred greater dialogue between population biologists and ecosystem scientists. Ecosystem scientists have always been concerned with large-scale processes, such as watersheds and whole lakes, and have drawn their inspiration from studies that dwarf by several orders of magnitude the classic experiments of population and community ecology (Kareiva and Andersen 1988). By being forced to think of population dynamics at a landscape level, population ecologists have suddenly found themselves discussing issues that were once the exclusive domain of ecosystem science.

Yet the overturning of the classic "equilibrium" paradigm, the shift to landscape perspectives, and the tighter connections between population biology and ecosystem science have not coalesced into a well-defined theory. Indeed, these new ideas are barely mentioned in general ecology textbooks. Nonetheless, in the practical world these ideas are already having profound effects. It is now accepted that species-based conservation alone is incomplete and ineffectual and that the emphasis should broaden to habitat, landscape, or ecosystem approaches. Although the ESA was not intended to be a land management act (Rohlf 1989), we have gotten into land use planning through the back door because delisting many "big" species requires a more holistic, comprehensive, and proactive conservation strategy.

There remains tension between the population and ecosystem perspectives, however, as though the two viewpoints were mutually exclusive. But as Soulé and Mills (1992:65) point out, the conflict of ecosystem versus population viewpoints is absurd: "The dichotomy is fatally flawed conceptually and operationally. A pure ecosystem approach is just as contradictory as a pure species approach." Ecosystem management must attend to the viability of certain critical species even if they are not yet endangered—system viability cannot be dissociated from species viability (Soulé and Mills 1992; Franklin 1993). We turn now to the basic tenets of a unified species-based and ecosystem-based approach to maintaining biodiversity.

The Joining of Ecosystem Science and Population Biology

A Top-Down Approach to Ecosystems: Levels and Units of Study

Ecosystems have been defined somewhat arbitrarily as any part of the universe chosen as an area of interest. The line around that area is the ecosystem boundary, and anything crossing the boundary is input or output (Agee and Johnson 1988). Population and community ecologists tend to view ecosystems as networks of interacting populations. The biota constitute the ecosystem, and abiotic components such as soil or sediments are external influences. The biota may interact with the abiotic environment, but the environment is largely treated as the backdrop or context within which biotic interactions occur. The process-functional approach, by contrast, embodied by such classic experiments as the Hubbard Brook watershed manipulations, tends to view ecosystems as including both organisms and physical components (Likens 1985; O'Neill et al. 1986; Higashi and Burns 1991). Energetics is often emphasized as the central focus of ecosystem inquiry. This approach is able to deal with cyclic pathways that are often unobservable in population-community problems but nevertheless form the feedback essential to ecosystem maintenance. In its extreme form, the functional approach implies that energy flow and nutrient cycling are somehow more important than the biotic entities performing the function. The ecosystem, in fact, is a dual organization of these two dimensions (O'Neill et al. 1986).

Although The Ecosystem is often promoted as one of the most compelling concepts in ecology (Cherrett 1989; Higashi and Burns 1991; Burns 1992), it is not easy to identify the fundamental units of ecosystem study or the common metrics that should be applied to ecosystem analysis. In population biology, by way of contrast, fitness and finite capacity for population growth emerge as fundamental attributes. Moreover, the evolutionary mechanism of natural selection provides a clear rule for change and genes are discrete units of analysis within distinct taxonomic units (species). At the community level, populations are the obvious unit of analysis and interactions within the food web (predation, competition, and so forth) are the obvious focus of our inquiries. The rule that governs change within communities has always been coevolution in changing environments. For most population and community-level analyses, we arbitrarily delineate a study area that is feasible and effective to test our hypotheses—we control and confine our "experiment" as much as possible. We measure along three dimensions: space, time, and the relation or interaction among entities (individuals compete at one scale, for example, populations at another). What, then, are the key units of study for ecosystem analysis?

Landscapes must become the fundamental units for ecosystem studies (Figure 12-1). Landscapes are large enough to contain the spectrum of multiscaled disturbance processes that are the source of changing structure and function within ecosystems—from fallen trees to forest fragmentation. Studies of landscapes will allow us to examine cycles and feedback in response to disturbance and to address issues of sustainability, resistance, resilience, health, and stress (Costanza et al. 1992). Even though the notion of landscape is arbitrary in size, in practice it will generally be chosen on the basis of human management goals. By admitting the artificial bias of our decisions about what constitutes a landscape, we accept an unavoidable reality of the modern world: humans define ecosystems specifically for the purposes of studying *and* managing change. For conservation science, human values are implicitly central because the field is goal-directed and based on the tacit recognition of two things: first, that recent human disturbance is pervasive and falls outside earth's historic evolutionary and ecological time scales; second, that present land management is grossly mismatched to the long-term control of undesired disturbances and to the scientific experimentation necessary for adaptive learning (management). We end up with land management at a grand scale—one that most often matches the scale of the prominent endangered species or land areas with those regions in which humans have become a dominant force. For practical purposes, then, the ecosystem concept in conservation must be applied to areas at or below the size of biogeographic provinces (Chesapeake, Yellowstone, Everglades). For experimental purposes, conservation science should focus on landscapes.

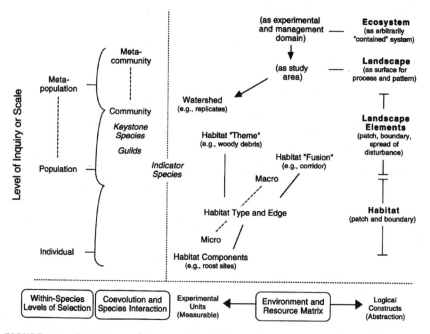

FIGURE 12-1. Dimensions of the experimental basis of conservation science. Assuming a human-centered approach to the "large" species, ecosystems are on the order of 10^3 or 10^4 square kilometers. The focus of inquiry is often set by matching the within-species or among-species level, which is fixed logically and experimentally, with some environmental scale (for which there is a gradient from concept to measurable units). The habitat and its resource components should be quantified to provide good experimental units for sampling and linkages to the individual and population, which are units of selection and analysis. Alternatively, community interactions may be studied in their own right with little reference to measurable environmental components (food webs, keystone species)—although guilds and indicator species are often used to represent the emergent properties of habitat types and associations.

Landscapes: Ecosystem Study Areas

The organism can generally be regarded as the basic unit in ecology (Lomnicki 1988; Cohen et al. 1993) and the species as the fundamental unit of biodiversity (Wilson 1992). Mechanistic ecology ultimately reduces to the individual level, and individual properties provide the experimental basis of interspecific interactions (Schoener 1986; DeAngelis and Gross 1992; Werner 1992). The individual as a whole is the target of selection (Mayr 1988), evolutionary forces act upon individual populations (Roughgarden 1989), and evolutionary changes take place within the context of ecosystems (Levin 1992a)—these are central tenets of the evolutionary synthesis that has made a decisive contribution to the unification of biology (Smocovitis 1992;

Mayr 1993). These considerations should go a long way toward unifying conservation genetics (promoting the fitness of targeted populations: Soulé and Mills 1992) with conservation ecology (Lynch and Lande 1993). Levin (1992a) says that conservation biology manages to span the middle ground between population biology and ecosystems science. The emerging idea of landscape may be a strong catalyst for improving experimental conservation science (Hansson and Angelstam 1991; Pickett et al. 1992).

Landscape ecology has focused primarily on the spatial patterns of landscape mosaics and interaction among their elements; it offers a way to consider environmental heterogeneity or patchiness in spatially explicit terms (Kolasa and Pickett 1991). The problem of ecological pattern is inseparable from the problem of the generation and maintenance of diversity (Levin 1992a). Natural and human-induced disturbances configure the heterogeneity that we observe and, in turn, influence the nature and spread of future disturbances (Turner and Dale 1991; Hobbs and Huenneke 1992). Fahrig (1990) claims that the population biologist's use of the terms "spatiotemporal variability" or simply "variability" is often synonymous with usage of the term "disturbance" by community and ecosystem ecologists (see also Turner and Dale 1991). The description of pattern is the description of variation, and the quantification of variations requires the determination of scales (Levin 1992a).

Since populations exhibit patchiness and variability on a range of spatial and temporal scales, the definition of commonness or rarity depends on the scale (Schoener 1987). Ecosystems exhibit analogous properties, including organizational scales (Levin 1992a). Landscapes have generally been viewed at a kilometers-wide scale, however, and human interactions with landscapes have been emphasized (Forman 1990; Wiens 1992). Prominent terrestrial ecosystems conform well to a surface mosaic where environmental change is at the scales set by the most pressing problems of agriculture, forestry, and ecotoxicology. Thus, in practice, landscapes are experimental domains peculiar to human-centered terrestrial ecosystems. (But see Wiens and Milne 1989 and Milne 1992.) Otherwise we would probably have analogs in marine ecosystems that are studied at a wide range of scales (Mann and Lazier 1991; Nicol and de la Mare 1993). Experiments in ecology, however, typically collect data in small study areas for only a few seasons (Kareiva and Andersen 1988; Risser 1991). To what extent does the understanding we obtain from these experiments extrapolate to large-scale phenomena (Kareiva 1989; Kareiva et al. 1993)? At this stage we have only models to rely on when answering this question, since multiscale experiments are nonexistent.

The view outlined here lends credence to Turner's (1989) conclusion that unlike many other scientific "disciplines," landscape ecology does not yet have a central body of concepts that give it a clear identity and direction. Similarly, the literature is predominantly descriptive, and "landscape ecology is not a particularly quantitative discipline, nor is it much concerned with theory or the formalities of hypothesis-testing" (Wiens 1992:149; but see Turner and Gardner 1991). Wiens et al. (1993:370) continue: "The emphasis of landscape ecology on holism has made it difficult to unravel the mechanisms underlying landscape dynamics, and the focus on landscapes at the scale of human perception has constrained statistical hypothesis testing and precluded the use of replicated experiments, limiting the scope of research questions." These writers believe that landscape ecology must integrate individual and population approaches if it is to be mechanistically based. Mechanistic models will yield simplified abstractions that can cover a wide variety of ecological circumstances (Schoener 1986; Murdoch 1993). Dunn et al. (1991) believe that research in landscape ecology should focus on how landscape dynamics interacts with species' tolerances in time and space (see also Ehleringer and Field 1993).

Landscape ecology has many superficial similarities with the study of metapopulations (Hanski and Gilpin 1991). Metapopulations are systems of local populations connected by dispersing individuals. Landscape ecology uses much of the same language, and it is interested in many of the same questions as metapopulation studies: survival of species, communities, and ecosystems in fragmented habitats; how to distinguish the matrix (the distinction between habitat patches and their surroundings); the origin, size, and shape of habitat patches. Metapopulation models form a bridge between the traditionally separate domains of population ecology (local abundance) and biogeography (regional occurrence; Gotelli 1991 and Opdam 1991). Despite the parallels, however, there are profound differences between metapopulation studies and the present practice of landscape ecology. The exploration of metapopulations is based on the well-studied behavior of local populations (Hanski and Gilpin 1991). Metapopulation studies have developed deductively through extensive use of mathematical models and extensive checking of these advances against observational data and experimental systems. Landscape ecology, on the other hand, is holistic in approach and takes as its object of study the entire landscape. Its methodology at this time is mainly descriptive as opposed to deductive.

It seems clear that the persistence of many species depends on their being constituted as a metapopulation distributed among many patches (May and Southwood 1990). But translating this notion into anything useful is quite difficult—the major hurdle is determining the scale of the relevant patchiness and then quantifying the processes that govern local extinction and colonization. We believe these goals could be advanced by more studies that systematically measure patterns of movement among patches, both within and between generations. Such studies are needed to underpin a quantitative understanding of metapopulation persistence (Doak and Mills 1994).

Watersheds: Replicate Units Within Landscapes

Many geographers, hydrologists, and biologists would agree that the watershed is a natural experimental and planning unit, and it has emerged as a practical unit of study at the ecosystem level (Hornbeck and Swank 1992; Naiman 1992). A watershed is defined as the geographic region within which water drains into a particular river, stream, or body of water. Watershed boundaries are defined by the ridges separating watersheds. Every activity on the surface of the land within a watershed sends its effects into the water (Brinson 1988). In theory, the term "watershed" can refer to any scale of water flow in any sized ecosystem.

The watershed-ecosystem concept recognizes the role of geomorphology in ecosystem functioning and the importance of the terrestrial-aquatic linkage to an extraordinary diversity and density of species (Naiman and Decamps 1990; Holland et al. 1991). The river continuum concept represents a synthesis of these ideas and proposes that watersheds provide a continuous gradient of physical conditions that affect the storage and transport of organic matter and the response of biotic communities (Minshall 1988). The link between upland use and downstream effects constitutes the central tenet of watershed management (Brooks et al. 1991). When Kratz et al. (1991) examined the relation between the spatial positioning of different locations in a landscape and the annual variability exhibited at each location, they found that water movement across the landscape was the key factor determining variability patterns.

The watershed concept is appealing from an experimental perspective because structures and functions have long been subject to rigorous measurement and modeling. The watershed conforms to the

cartographer's model of the earth's surface (planimetry) and is susceptible to tractable hydrological models based on flow, transport, perco-
lation, diffusion, filtration, seepage, evaporation, compartments, and
networks (Higashi and Burns 1991; Beven and Moore 1993). Watershed elements are easily measurable (precipitation, evapotranspira-
tion, water flow, erosion). Our best knowledge concerns sediments
and pollutants, which determine water quality, and wetland habitat
analysis. These factors influence species in many ways, leading to numerous forms of assessment. (See Williamson and Hamilton 1989, for
example, and Karr 1991.)

With this perspective of a watershed, we should be able to plan and
develop long-term, sustainable solutions to many natural resource
problems—and, at the same time, avoid many kinds of environmental
degradation (Naiman 1992). Watershed management offers a unifying theme to help us understand the effects of land use on wildlife
and habitat (Payne 1992; Satterlund and Adams 1992). As naturally
scalable, spatially distinct units, watersheds are ideal candidates for
replicates within landscapes (as in Likens 1985). Despite the real
world of disaggregated political and economic actions, water and its
constituents flow downhill and ignore political boundaries.

Habitat: An Old But Essential Concept

Long before "landscape ecology" or "metapopulations" became fash-
ionable, resource managers emphasized the importance of habitat
analysis. Certainly it would be silly to attempt to manage a species
without knowing its habitat requirements. After all, the main point of
much conservation is the recognition that different habitats provide
different food, water, cover, and other life requisites. But unlike watersheds, which have a natural metric, habitat is completely defined
by the organism and perceived and segregated by humans. Probably
no other ecological construct has proved more exasperating. For example: "I don't think we have a clue what habitat is and what it
isn't. . . . I can tell you that I think there's a hell of a lot of sandbagging
going on for several reasons. Can you define water? No. Can you define clear water? I doubt it. Can you define dirty water? The same can
be asked for soil, air, and most of our resources. It's silly to listen to
these repetitive harangues about how you can't do anything about
something because you can't define it" (L. D. Harris in Hudson
1991:75). In fact, we use the idea of habitat quite effectively, and
knowing how habitats are used has yielded the best results for
restoring and managing species. Topography and soils interacting with

historical disturbances conveniently results in natural habitat types that can be mapped to the surface of a landscape. Certainly gradients between types are problematic, as are the definitions of apparent types, but we have a good toehold in most applications.

The ESA focuses legal attention on the notion of critical habitat. Interestingly, critical habitat may correspond to specific resources or habitat components, such as nest trees or rookeries. Habitat components are resources that are easily measured and quantified. Sometimes they are discrete objects that may be found in many habitat types, but most often they are correlated with a particular habitat type. Habitat "fusions," such as migration routes or calving grounds, are likely to depend on a particular distribution and arrangement of multiple habitat types and topography. Changes at ecotones have been hypothesized as early indicators of environmental change (Yahner 1988; Naiman and Decamps 1990). In addition, we can think of habitat "themes" as common elements that span habitat types—for example, coarse woody debris in forested systems (Harmon and Hua 1991). Much of the landscape literature is devoted to corridors, barriers, and boundaries. (See, for example, Holland et al. 1991; Hudson 1991; Saunders and Hobbs 1991; Hansen and di Castri 1992.) The most clearly defined and measurable landscape elements are reducible or explicitly tied to habitat components, types, and boundaries (Figure 12-1).

These diverse perspectives of "habitat" drive home several points. Habitats are not mutually exclusive mapping units, even when viewed from the standpoint of a single species. Habitat fusions and themes are likely indicators of critical processes in the population, community, or ecosystem. We should continue searching for these key links between structure and function because habitat is one of the most measurable features of the environment. In searching we would do well to give greater emphasis to organism-oriented analysis and modeling (Rosenzweig 1991). We also might find more meaning in "habitat fragmentation" and its multitude of measurements (Harris and Silva-Lopez 1992; Groom and Schumaker 1993).

Detecting, Predicting, and Managing Change

Conservation science has spawned numerous tools for dealing with the problems of scarcity and diversity (Soulé 1986, 1987). Indeed, many conservation biologists view their main task as providing the intellectual and technological tools that will anticipate, prevent, minimize, and repair ecological damage (Soulé and Kohm 1989). Some techniques are primarily advances in biological methods (habitat restoration, DNA sequencing, captive breeding, reintroduction), others are conceptual, and still others emphasize direct applications of quantitative ecology.

Statistical ecology is a diverse set of quantitative techniques that deal with troublesome features of ecological data: high variances, temporal and spatial variation, unusual distributions, nonlinearity, and problems of replication. As well, conservation science has drawn attention to spatial pattern analysis, dispersion and movement, population dynamics (key factors), estimation of population parameters, resource selection, risk assessment, effect/impact analysis, and species abundance and diversity. Special problems concern the design of observational and experimental studies that involve inherently small sample sizes and the need for minimal disturbance. Pleas in the late 1950s for making ecology more of an experimental science were not heeded for another fifteen years (Hairston 1989). Despite the increasing emphasis, not all experimenters paid attention to the requirements of experimental design and analysis (Hurlbert 1984; Hairston 1989). Simulation has aided analysis and even substituted for our inability to come up with logistically feasible designs at the large scales required by many conservation issues.

Traditionally, mathematical ecology has been associated with analytic models and making sense of observations. The computer explosion has produced a merging of mathematical and statistical biology into the integrative field of computational biology. Its grand challenge is to elucidate the central problems of genetics, organismic structure/function relationships, complex hierarchical biological systems, and global change (Levin 1992b). Conservation biology has already reaped many rewards from models based on simulation and computer-intensive data analysis. (See the examples in Crowley 1992 and Levin 1992b.) Analytic and simulation models are perfect partners for simplifying the complex. We can make better use of data and better refine relationships. For the sake of brevity, we concentrate here on specific improvements. Other topics are reviewed elsewhere (Simberloff 1988; Bartell et al. 1992; Moore et al. 1992; Doak and Mills 1994).

Population Monitoring

Ultimately, the evaluation of an endangered species program requires accurate monitoring of target populations. Detecting change in numbers and vital rates is essential for timely listing and delisting of species. Recent work pertinent to conservation science is concerned with overcoming problems with small samples, model bias (assumptions), error estimation (confidence intervals), trend analysis, and intrusive methods (disruptive or harmful to animal). Advances in specific models and methods are addressed by a vast literature. (See the reviews in Seber 1986 and 1993.) The rapid progress in model selection and analysis, however, has outstripped improvements in experimental design and on-the-ground techniques for gathering data.

The work of Skalski and Robson (1992; see also Thompson 1992) is directed at experimental designs to devise the field study that most efficiently allocates sampling efforts. By carefully planning from the outset, we gain accuracy, reliability, and repeatability in monitoring techniques and protocols. Moreover, design features are matched to costs and human effort as a function of predetermined levels of accuracy. Adaptive designs adjust sampling procedures during surveys and can be very effective for sampling rare, clustered populations (Thompson 1992). Many cases—among them the grizzly bear (*Ursus arctos*), spotted owl (*Strix occidentalis*), and black-footed ferret (*Mustela nigripes*)—are painful monuments to the lack of careful preparation. The need for planning cannot be overemphasized. The inertia of a poor design is difficult to overcome once it has become entrenched in the system.

Of course, successful monitoring is not just about statistics and sampling design. Creative approaches for collecting data are testimonial to the value of experienced fieldworkers, who not only inspire or devise most of them but form the front line in attaining quality data. Among these creative approaches are the use of remote animal-triggered photography, animal signs (tracks and scats), recorded playbacks or vocalizations, lures or baits, hair snagging, and telemetry data.

Population Viability Analysis
Population viability anaysis (PVA) is one of the keystone ideas of conservation biology. (For a lucid review see Boyce 1992 and 1993.) It is the process of estimating the probability of persistence of a population for some arbitrary time into the future (Soulé 1987)—and, frequently, the probability of extinction or expected time to extinction (Dennis et al. 1991). Performing a PVA entails using biological data as the basis for a simulation model for the population.

Lebreton and Clobert (1991) and Boyce, however, conclude that such applications are premature because we cannot reliably estimate the extinction probability for any species. Given extreme environmental variance, maintaining minimum viable populations may not be as reasonable a target as managing for spatial configuration and location of habitats. A more appropriate approach for many species may be to model the habitat for the species, various strategies for managing this habitat, and the consequences of density dependence and interspecific interactions. PVAs are often compromised because of inadequate basic natural history and a lack of connection to management actions. Yet PVAs can be a valuable tool for formalizing our current understanding of the ecology of a population or species. When done properly this involves working closely with natural resource managers to develop a long-term iterative process of modeling and research that can reveal a great deal about how to manage a species. Thus PVA can be viewed as a variation on the notion of adaptive management (Holling 1978; Walters 1986). Although management agencies may resist programs that might pose a risk to threatened species, creative manipulations may be allowed if they promise to enhance conditions for the species of concern (Boyce 1993). At the very least, agency/academic coordination for conservation programs would be stimulated. The most celebrated cases are represented in this volume. In the spotted owl case, for instance, PVA successively added realism—culminating in a dynamic model interfaced with landscapes using a geographic information system (Lamberson et al. 1992; McKelvey et al. 1993). Resistance to adaptive management will probably lead to increasing reliance on simulating management options and spatial structure using increasingly sophisticated models (Stacey and Taper 1992; Wootton and Bell 1992).

When the evolution of PVA is viewed along these lines, one can easily envision species-specific systems that integrate demography, habitat, and spatial structure. This approach would subsume the many species-habitat models or techniques that are categorized as single-species models and habitat-analysis models. (See the reviews in Verner et al. 1986 and Morrison et al. 1992.) Such models are currently used by many key agencies in endangered species programs. Single-species models include simple correlation or presence/absence models, statistical models, Habitat Suitability Index models of the FWS, Habitat Capability models of the USFS, and Pattern Recognition models. Multispecies or community models include the Integrated Habitat Inventory and Classification System of the BLM, various habitat capability models, the lifeform system of the USFS, and community guild models. The latter two models cluster species with similar habitat requirements for feeding and reproduction. Habitat-analysis models overlap with single-species models and include the Wildlife and Fish Habitat Relationships program of the USFS, the Habitat Evaluation Procedures of the FWS, simulation models (such as Dynast), optimization models (such as Forplan), and economic analysis models.

Surprisingly, species-habitat models have rarely been validated. When models have been tested, they have fared poorly but have frequently done better than "personal opinion." Clearly, assumptions and the quality and resolution of the input data have played a major role. Formal models are only tentative, however, and simply serve as a starting point from which one can constantly improve and refine. In constructing a PVA it is critical that investigators carefully match three design elements for a model: assumptions, simplicity (variable choices, numbers, and relations), and data quality/availability (sampling and measurement error).

In summary, large-scale PVA-based modeling systems of the scope described here will require enormous databases (Norton and Mitchel 1993). To be realistic about long-range population projections, the landscape should be allowed to change over time—dramatically increasing the size of databases and computational requirements. For example, Pulliam et al. (1992) present a model designed to elucidate the effects of landscape-level variation in habitat dispersion on the size and extinction probability of avian populations in a region managed for timber production. In the model, habitat suitability and availability within the landscape change annually as a function of timber harvest and management strategies. The best platform for dealing with this kind of situation is a geographic information system (GIS).

Community-Habitat Themes: Keystones, Indicators, Guilds

Experience suggests that a narrow focus on local diversity, short time frames, and low levels of biological organization has unfortunate consequences for biodiversity. Thus it is essential to consider each phenomenon in a broad, ultimately global, context—that is, to think big. Long before landscape ecology became popular, community ecologists were attempting to reduce complexity and find order: "What do we understand of the assembly of communities? Obviously we know that there is a large amount of organization in the connections among species. But how much? The answer is unknown for any kind of community. . . . We know some keystone species, some assembly rules, some processes of competition and symbiosis that serve as a weak gravitational force" (Wilson 1992:180). Whether a species in fact occurs in a suitable habitat is due largely to chance, but for most organisms the chance is strongly affected by the identity of the species already present.

Ecologists define keystone species as those whose removal is expected to result in the disappearance of much of the assemblage. Mills et al. (1993) make a convincing argument for dropping the label "keystone species;" instead, they advocate the study of interaction strengths because it recognizes the complexity, as well as the temporal and spatial variability, of interactions. (See also Cohen et al. 1993.) Certainly the concept has drawn our attention to differing interaction strengths in food webs and to the pattern in which only a few species have strong interactions that affect community composition. The prevalence of cascades subsequent to single-species perturbations suggests that strong interactions are relatively common in well-studied systems (Paine 1992).

Similar arguments have been made against the concept of indicator species (Landres et al. 1988; see also Conner 1988). First, no single criterion can be used. Effective indicator species are stenotopic or sensitive to environmental changes, have low levels of variability in response to the environmental conditions of interest, and are year-round residents in the target area. Large body size and large area requirements are sometimes considered important criteria. More practical criteria include facility of monitoring and the indicator's relatively rapid response to changes in conditions. Second, no single species can be used (Landres et al. 1988; Noss 1990). Third, we should be more objective in the way we choose species. Kremen (1992) presents a simple protocol, based on ordination techniques, for establishing the indicator properties of a group of organisms and for selecting an indicator species subset for more intensive monitoring. Ryti (1992) describes an algorithm for selecting species in a focal taxon that can act as a biodiversity umbrella; by selecting reserves that include all species in the focal taxon, we also include species in other taxa. The most intensive and experimentally sound use of indicators has been in aquatic systems (Karr 1991; Moyle and Leidy 1992). Carpenter et al. (1993) report whole-lake experiments suggesting that key indicator species give the best warning signal for major ecosystem deterioration, long before chemical or physical changes are evident.

Simberloff and Dayan (1991) describe how the guild concept became inextricably linked with the idea of indicator species, except that indicators were transformed to "guild indicators" rather than indicators of entire communities. Verner (1984) has defined a "management guild" as exactly those species that respond similarly to changes in their environment. Guilds are normally assigned on the basis of habitat and feeding grounds, however, in the absence of empirical research into how they respond to environmental changes.

The meaning and value of keystone species, indicator species, and guilds of species will always be questioned. Scientists will continue using these ideas—if only as a shorthand for communicating with policymakers, managers, and the public—but labeling species as members of a guild or as keystones should be based more on experimental data in the future and less on "common sense." Predators, especially top carnivores, are likely to be especially useful to resource managers as indicator species because of their wide-ranging habits and integration of lower levels of prey.

Remote Sensing, GIS, and Gap Analysis

A geographic information system (GIS) is a computer system designed to allow users to collect, manage, and analyze large volumes of spatially (and temporally) referenced and associated attribute data (Maguire et al. 1991; Laurini and Thompson 1992). The exponential increase in sensor technology and digital image data-processing capabilities within the last decade has created a revolution in our ability to sense and measure from remote platforms (Hobbs and Mooney 1990; Vane and Goetz 1993). The range of spectral bands, the detailed resolution, the global coverage—all have encouraged applications to species and resource issues (Heit and Shortreid 1991; Stoms et al. 1992; Kovar and Nachtnebel 1993). More important, remote sensing and GIS have vastly expanded our capability for analysis, modeling, and management at large scales from landscape to global. (See Saunders and Hobbs 1991; Turner and Gardner 1991; Young and Green 1993.)

Despite the undeniable promise of GIS, there is also a long list of worries. One of the biggest problems is that training has not kept pace with the proliferation of GIS and remote-sensing technology, and users may be intimidated by the tremendous mass and variety of data. Consequently, many biologists and resource managers cannot take advantage of its potential for analysis and modeling; instead they gain most of their insights from visual displays and simple manipulations (such as generating overlays). A far deeper problem involves questionable data quality and the errors associated with deriving or combining raw data. Awareness of such issues has brought to the forefront the need for cross-disciplinary collaboration in GIS development and application. GIS is rapidly advancing from a "solution in search of a problem" to a more problem-oriented approach in resource sciences. Integrated PVAs, discussed earlier, represent an example of a species approach. A biodiversity approach has enormous potential, and the most renowned example is gap analysis.

We have learned that merely setting aside habitat often leads to unsatisfactory results at best and to irreversible simplification at worst. A crash program is needed to carry out extensive surveys and mapping to identify areas that are critical for the protection of natural and genetic resources (Soulé and Kohm 1989; Wilson 1992). Gap analysis is a "quick-assay" methodology that identifies the gaps in representation of biodiversity in areas managed exclusively or primarily for the long-term maintenance of populations of native species and natural ecosystems. Gap analysis uses vegetation types and vertebrate and butterfly species (as well as other taxa, such as vascular plants, if adequate distributional data are available) as indicators of biodiversity. The correlation of topographic relief and elevational diversity with vegetative diversity is acknowledged as integral to the nature and choice of landscape-sized samples. The goal is to ensure that all ecosystems and areas rich in species diversity are represented adequately in biodiversity management areas. (See also Margules and Austin 1991 and Ryti 1992.) The originators (Scott et al. 1993) believe that this proactive strategy will eliminate the need to list many species as threatened or endangered in the future. Gap analysis deserves critical examination because it is likely to play a key role in the National Biological Survey.

As the methods and models of gap analysis are still evolving, they are not yet fully described. (See Scott et al. 1993 and the related publications they cite.) Gap analysis generally results in maps and tables summarizing the predicted distribution and conservation status of vegetative types and species as well as an evaluation that can be used to develop a conservation strategy. With assistance from gap analysis experts, decision makers can gauge the value of the results for large-scale planning and environmental assessment.

Because gap analysis is still in its early stages, no mapping project has been completed and scrutinized with respect to its accuracy. The potential sources of error are enormous—ranging from misinterpretation of remote-sensing data and flawed range maps to crude vertebrate habitat models and fundamental biases arising as a result of lumping together quantitative data into a limited number of discrete states. (See, for example, Goodchild et al. 1992 and Rastetter et al. 1992.) Moreover, recent studies of biodiversity patterns elsewhere suggest that certain premises of gap analysis may not hold true. An analysis of diversity hotspots for five different taxa in England, for example, revealed so little overlap in hotspots among taxa that each group of organisms would probably need its own separate conservation strategy or gap project (Prendergast et al. 1993).

The most important considerations in evaluating a gap approach are error estimation, interaction, and accumulation. (See, for example, Goodchild and Gopal 1989; Heuvelink et al. 1989; Lanter and Veregin 1992; Stoms et al. 1992.) Data and model validation procedures also need to be standardized, and examination of minimal mapping units for likely scales of analysis is necessary (Stoms 1992). Variable and data weighting is of considerable value, too, since some data sources are more reliable than others, some variables are more accurate or more important, and many databases are highly nonrandom in their areal coverage and intensity. All of these considerations will vary in importance depending on the investment and spatial scope of analysis.

Gap analysis is so ambitious that, despite continuous caveats from its originators, extraordinary expectations have been built-up. Although the far-thinking developers have promoted a glowing image of gap analysis, they never intended it to be embraced as the Grail that will provide exact answers for precise species and land management. Yet that turn-key format is eagerly sought by politicians, resource managers, and conservation advocates—gap analysis translates to the nonspecialist in a most appealing and intuitive way. Biologists, however, realize that constructing and implementing such a strategy will have to be simple and sketchy at first, with increasing realism and complexity built in as both refinement and acceptance are gained over the years.

Lessons

The division between academic science and management science are breaking down as more sophisticated technology, computational methods, and theoretical complexity are applied to species and ecological systems (see Chapters 17 and 18). Increasing specialization of knowledge and training should eventually reach a threshold where separation and isolation become absurd and unjustifiable. Resource managers often complain that researchers do not supply results that are applicable, practical, or understandable (Chapter 15). An immediate incentive toward eliminating this schism would be to regulate government funding on the basis of representative participation in, and peer review of, research projects. The National Biological Survey is the beginning of such a trend. Levin (1992b, 1992c) discusses how efforts are being made to promote interdisciplinary collaboration and explains how this interaction can be strengthened and encouraged through attention to modes and levels of support. (See also Wilson 1992 and Kingsolver et al. 1993.)

Computational biology is presently the key driving force for most advances in the field. It will provide simulated experimentation based on available data until large-scale or long-term experiments can be undertaken. Unification and standardization of data are more likely when study areas are large landscapes that transcend administrational boundaries. Currently, GIS-centered platforms offer the best architecture for handling massive data structures, intensive modeling, and statistical analysis. This approach will facilitate experimental design, particularly using principles of adaptive management. As researchers and managers from different sectors of society begin working together, secrecy and ownership (of both land and data) may weaken. We may have to wait years or decades for the best long-term data, but in many cases we already have historical data that are not being used or are being sequestered. With the right incentives, some of the data backlog for key bioregions and endangered species can be analyzed and interpreted. If we cannot establish minimal cooperation in the form of data sharing, there is little hope for bigger efforts such as regional centers for collaborative research and training (Soulé and Kohm 1989; Wilson 1992).

Perhaps the most promising prospect for institutionalizing these changes is represented by the formation of the National Biological Survey. In the spring of 1993, Secretary of the Interior Bruce Babbitt announced his proposal to establish a National Biological Survey within the Department of the Interior (Babbitt 1993). Through an internal reorganization within the department,

> the Secretary proposes to combine substantial portions of the biological research and survey activities of eight Departmental bureaus—the Fish and Wildlife Service; National Park Service; Bureau of Land Management; Bureau of Reclamation; Minerals Management Service; Office of Surface

Mining; Geological Survey; and Bureau of Mines—into a new National Biological Survey (NBS). . . . The mission of the NBS is to gather, analyze and disseminate the biological information necessary for the wise stewardship of our Nation's natural resources, and to foster understanding of biological systems and the benefits they provide to society (p. 1).

The objectives are to reduce overlap, duplication, and costs of research efforts while establishing a renewed leadership and focus and allowing effective priority setting by departmental managers (Chapter 17). In addition, the department's credibility would be enhanced and land managers would have greater incentive to rely upon more timely, objective scientific information. The most specific objective is "to develop an anticipatory, proactive biological science program that will enable land and resource managers to develop comprehensive ecosystem management strategies, thus avoiding costs and conflicts such as those involved in several past Endangered Species Act crises" (p. 1).

Thus the NBS will focus on national, regional, ecosystem, and landscape-level needs. It will undertake a coordinated inventory and monitoring program and "identify, in a proactive fashion, chronic declines of species and natural habitats," so that "research results from this program will enable land and resource managers to adopt ecosystem-based management strategies to protect potentially imperiled species at reduced levels of cost and conflict" (p. 2). Seven "activities" are listed to accomplish the NBS mission; four of them involve "ecosystem" as a key construct and a fifth involves global biological resources. The NBS will assume responsibility for the national and regional inventory and monitoring of biological diversity, for resources of national significance (such as certain endangered species), and for the development of standards and protocols for conducting inventories and monitoring programs.

We think the general plan for the NBS will catalyze improvements in conservation science. The first and most difficult step will be to overcome the reluctance to actively and adaptively manage and experiment. Orians (1993) musters a cogent argument for strong action despite the uncertainties and suggests we use environmental impact statements as hypotheses that can be tested by setting up ecological monitoring following any large-scale action. Quick sampling methods must be devised to take inventories, plan use patterns, and monitor the results (Ehrlich 1992; Wilson 1992). Since we cannot make accurate, long-range predictions about ecological systems, the closest proxy for designed management is that of species and processes, such as strongly interacting species (predators, for example). Decision making should be formalized so that it can be repeated and audited, as with risk analysis, decision analysis, and the Delphi method (Maguire 1986; Bartell et al. 1992).

No matter how strong our science, no matter how sound our policy, we cannot seem to compensate for an expanding human population. Very few people realistically envision a world of increasing humans with decreasing rates of environmental degradation. Therefore, "hope for biodiversity rests on the blip theory of human population . . . environmentalists pray that the breeding binge will be a transient blip rather than a surge toward permanent planetary obesity" (M. Soulé in Grumbine 1992:xi). Inevitably, an ever-expanding human population will require increasingly severe forms of triage (e.g., Hardin 1980, McIntyre et al. 1992), as we select and administer ever fewer species. An improved conservation science is our best shot at fully informed economic and ethical decisions.

Acknowledgments

Additional funding was generously provided by the National Council for Air and Stream Improvement. Discussions with W. Brewster, D. Casey, T. Clark, L. Irwin, C. Larson, R. Mace, R. Reading, T. Schoener, W. Settle, and M. Soulé were most helpful.

References

Agee, J. K., and D. R. Johnson (eds.). 1988. *Ecosystem Management for Parks and Wilderness*. Seattle: University of Washington Press.

Babbitt, B. 1993. Regarding: a proposal to establish a National Biological Survey within the Department of the Interior. Memo to all Cabinet Secretaries. U.S. Department of the Interior, Washington, D.C. Dated March 17, 1993.

Bartell, S. M., R. H. Gardner, and R. V. O'Neill. 1992. *Ecological Risk Estimation*. Boca Raton, Fla.: Lewis Publishers.

Beven, K. J., and I. D. Moore (eds.). 1993. *Terrain Analysis and Distributed Modelling in Hydrology*. New York: Wiley.

Botkin, D. B. 1990. *Discordant Harmonies: A New Ecology for the Twenty-First Century*. New York: Oxford University Press.

Boyce, M. S. 1992. Population viability analysis. *Annual Review of Ecology and Systematics* 23:481–506.

————. 1993. Population viability analysis: Adaptive management for threatened and endangered species. *Transactions of the North American Wildlife and Natural Resources Conference*. Washington, D.C.: Wildlife Management Institute.

Brinson, M. M. 1988. Strategies for assessing the cumulative effects of wetland alternation on water quality. *Environmental Management* 12:655–662.

Brooks, K. N., P. F. Ffolliott, H. M. Gregersen, and J. L. Thames. 1991. *Hydrology and Management of Watersheds*. Ames: Iowa State University.

Burns, T. P. 1992. Ecosystem: A powerful concept and paradigm for ecology. *ESA Bulletin* 73:39–43.

Carpenter, S. R., T. M. Frost, J. F. Kitchell, and T. K. Kratz. 1993. Species dynamics and global environmental change: A perspective from ecosystem experiments. In *Biotic Interactions and Global Change*, P. M. Kareiva, J. G. Kingsolver, and R. B. Huey (eds.). Sunderland, Mass.: Sinauer Associates.

Cherrett, J. M. 1989. Key concepts: The results of a survey of our members' opinions. In *Ecological Concepts: The Contribution of Ecology to an Understanding of the Natural World*, J. M. Cherrett (ed.). Cambridge, Mass.: Blackwell Scientific.

Cohen, J. E., R. A. Beaver, S. H. Cousins, et al. 1993. Improving food webs. *Ecology* 74:252–258.·

Conner, R. N. 1988. Wildlife populations: Minimally viable or ecologically functional? *Wildlife Society Bulletin* 16:80–84.

Costanza, R., B. G. Norton, and B. D. Haskell (eds.). 1992. *Ecosystem Health: New Goals for Environmental Management*. Washington, DC: Island Press.

Crowley, P. H. 1992. Resampling methods for computation-intensive data analysis in ecology and evolution. *Annual Review of Ecology and Systematics* 23:405–447.

DeAngelis, D. L., and L. J. Gross (eds.). 1992. *Individual-Based Models and Approaches in Ecology: Populations, Communities and Ecosystems*. New York: Chapman & Hall.

Dennis, B., P. L. Munholland, and J. M. Scott. 1991. Estimation of growth and extinction parameters for endangered species. *Ecological Monographs* 61: 115–143.

Doak, D. F., and L. S. Mills. 1994. A useful role for theory in conservation. *Ecology* 75:615–626.

Dunn, C. P., D. M. Sharpe, G. R. Guntenspergen, F. Stearns, and Z. Yang. 1991. Methods for analyzing temporal changes in landscape pattern. In *Quantitative Methods in Landscape Ecology*, M. G. Turner and R. H. Gardner (eds.). New York: Springer-Verlag.

Ehleringer, J. R., and C. B. Field (eds.). 1993. *Scaling Physiological Processes: Leaf to Globe*. New York: Academic Press.

Ehrlich, P. R. 1992. Population biology of checkerspot butterflies and the preservation of global biodiversity. *Oikos* 63:6–12.

Fahrig, L. 1990. Interacting effects of disturbance and dispersal on individual selection and population stability. *Comments on Theoretical Biology* 1:275–297.

Forman, R.T.T. 1990. The beginnings of landscape ecology in America. In *Changing Landscapes: An Ecological Perspective*, I. S. Zonneveld and R.T.T. Forman (eds.). New York: Springer-Verlag.

Franklin, J. F. 1993. Preserving biodiversity: Species, ecosystems, or landscapes? *Ecological Applications* 3:202–205.

Goodchild, M. F., and S. Gopal (eds.). 1989. *Accuracy of Spatial Databases*. Bristol, Pa.: Taylor & Francis.

Goodchild, M. F., S. Guoqing, and Y. Shiren. 1992. Development and test of an error model for categorical data. *International Journal of Geographic Information Systems* 6:87–104.

Gore, A. 1992. *Earth in the Balance*. New York: Penguin Books.

Gotelli, N. J. 1991. Metapopulation models: The rescue effect, the propagule rain, and the core-satellite hypothesis. *American Naturalist* 138: 768–776.

Groom, M. J., and N. Schumaker. 1993. Evaluating landscape change: Patterns of worldwide deforestation and local fragmentation. In *Biotic Interactions and Global Change*, P. M. Kareiva, J. G. Kingsolver, and R. B. Huey (eds.). Sunderland, Mass.: Sinauer Associates.

Grumbine, R. E. 1992. *Ghost Bears: Exploring the Biodiversity Crisis*. Washington, DC: Island Press.

Hairston, N. G. 1989. *Ecological Experiments: Purpose, Design, and Execution*. New York: Cambridge University Press.

Hansen, A. J., and F. di Castri (eds.). 1992. *Landscape Boundaries: Consequences for Biotic Diversity and Ecological Flows*. SCOPE Book Series. Ecological Studies, vol. 92. New York: Springer-Verlag.

Hanski, I., and M. Gilpin. 1991. Metapopulation dynamics: Brief history and conceptual domain. In *Metapopulation Dynamics: Empirical and Theoretical Investigations*, M. Gilpin and I. Hanski (eds.). New York: Academic Press.

Hansson, L., and P. Angelstam. 1991. Landscape ecology as a theoretical basis for nature conservation. *Landscape Ecology* 5:191–201.

Hardin, G. 1980. *Promethean Ethics: Living with Death, Competition, and Triage*. Seattle: University of Washington Press.

Harmon, M. E., and C. Hua. 1991. Coarse woody debris dynamics in two old-growth ecosystems. *BioScience* 41:604–610.

Harris, L. D., and G. Silva-Lopez. 1992. Forest fragmentation and the conservation of biological diversity. In *Conservation Biology*, P. L. Fiedler and S. K. Jain (eds.). New York: Chapman & Hall.

Heit, M., and A. Shortreid (eds.). 1991. *GIS Applications in Natural Resources*. Fort Collins, Colo.: GIS World.

Heuvelink, G. B., M.P.A. Burrough, and A. Stein. 1989. Propagation of errors in spatial modeling with GIS. *International Journal of Geographic Information Systems* 3:303–322.

Higashi, M., and T. P. Burns (eds.). 1991. *Theoretical Studies of Ecosystems: The Network Perspective*. New York: Cambridge University Press.

Hobbs, R. J., and L. F. Huenneke. 1992. Disturbance, diversity, and invasion: Implications for conservation. *Conservation Biology* 6:324–337.

Hobbs, R. J., and H. A. Mooney (eds.). 1990. *Remote Sensing of Biosphere Functioning*. Ecological Studies, vol. 79. New York: Springer-Verlag.

Holland, M. M., P. G. Risser, and R. J. Naiman (eds.). 1991. *Ecotones: The Role of Landscape Boundaries in the Management and Restoration of Changing Environments*. New York: Chapman & Hall.

Holling, C. S. (ed.). 1978. *Adaptive Environmental Assessment and Management*. New York: Wiley.

———. 1987. Simplifying the complex: The paradigms of ecological function and structure. *European Journal of Operational Research* 30:139–146.

Hornbeck, J. W., and W. T. Swank. 1992. Watershed ecosystem analysis as a basis for multiple-use management of eastern forests. *Ecological Applications* 2:238–247.

Hudson, W. E. (ed.). 1991. *Landscape Linkages and Biodiversity*. Washington, DC: Island Press.

Hurlbert, S. H. 1984. Pseudoreplication and the design of ecological field experiments. *Ecological Monographs* 54:187–211.

Kareiva, P. M. 1989. Renewing the dialogue between theory and experiments in population ecology. In *Perspectives in Ecological Theory*, J. Roughgarden, R. M. May, and S. A. Levin (eds.). Princeton, N.J.: Princeton University Press.

Kareiva, P. M., and M. Andersen. 1988. Spatial aspects of species interactions: The wedding of models and experiments. In *Community Ecology*, A. Hastings (ed.). Lecture Notes in Biomathematics 77. New York: Springer-Verlag.

Kareiva, P. M., J. G. Kingsolver, and R. B. Huey. 1993. Introduction. In *Biotic Interactions and Global Change*, P. M. Kareiva, J. G. Kingsolver, and R. B. Huey (eds.). Sunderland, Mass.: Sinauer Associates.

Karr, J. R. 1991. Biological integrity: A long-neglected aspect of water resource management. *Ecological Applications* 1:66–84.

Kingsolver, J. G., R. B. Huey, and P. M. Kareiva. 1993. An agenda for population and community research on global change. In *Biotic Interactions and Global Change*, P. M. Kareiva, J. G. Kingsolver, and R. B. Huey (eds.). Sunderland, Mass.: Sinauer Associates.

Kolasa, J., and S.T.A. Pickett (eds.). 1991. *Ecological Heterogeneity*. New York: Springer-Verlag.

Kovar, K., and H. P. Nachtnebel (eds.). 1993. *Application of Geographic Information Systems in Hydrology and Water Resources Management*. IAHS Publication no. 211. Wallingford, U.K.: IAHS Press.

Kratz, T. K., B. J. Benson, E. R. Blood, G. L. Cunningham, and R. A. Dahlgren. 1991. The influence of landscape position on temporal variability in four North American ecosystems. *American Naturalist* 138:355–378.

Kremen, C. 1992. Assessing the indicator properties of species assemblages for natural areas monitoring. *Ecological Applications* 2:203–217.

Lamberson, R. H., R. McKelvey, B. R. Noon, and C. Voss. 1992. A dynamic analysis of northern spotted owl viability in a fragmented forest landscape. *Conservation Biology* 6:505–512.

Landres, P. B., J. Verner, and J. W. Thomas. 1988. Ecological uses of vertebrate indicator species: A critique. *Conservation Biology* 2:316–328.

Lanter, D. P., and H. Veregin. 1992. A research paradigm for propagating error in layer-based GIS. *Photogrammetric Engineering and Remote Sensing* 58: 825–833.

Laurini, R., and D. Thompson. 1992. *Fundamentals of Spatial Information Systems*. APIC Series, no. 37. New York: Academic Press.

Lebreton, J. D., and J. Clobert. 1991. Bird population dynamics, management, and conservation: The role of mathematical modelling. In *Bird Population Studies: Relevance to Conservation and Management*, C. M. Perrins, J. D. Lebreton, and G.J.M. Hirons (eds.). Oxford: Oxford University Press.

Levin, S. A. 1992a. The problem of pattern and scale in ecology. *Ecology* 73:1943–1967.

————. (ed.). 1992b. *Mathematics and Biology: The Interface—Challenges and Opportunities*. NSF and U.S. Department of Energy Workshop, PUB-710. Berkeley: Lawrence Berkeley Laboratory, University of California.

————. 1992c. Orchestrating environmental research and assessment. *Ecological Applications* 2:103–106.

Likens, G. E. 1985. An experimental approach for the study of ecosystems. *Journal of Ecology* 73:381–396.

Lomnicki, A. 1988. *Population Ecology of Individuals*. Princeton, N.J.: Princeton University Press.

Lynch, M.,and R. Lande. 1993. Evolution and extinction in response to environmental change. In *Biotic Interactions and Global Change*, P. M. Kareiva, J. G. Kingsolver, and R. B. Huey (eds.). Sunderland, Mass.: Sinauer Associates.

Maguire, D. J., M. F. Goodchild, and D. W. Rhind (eds.). 1991. *Geographical Information Systems: Principles and Applications*. 2 vols. New York: Wiley.

Maguire, L. A. 1986. Using decision analysis to manage endangered species populations. *Journal of Environmental Management* 22:245–260.

Mann, K. H., and J.R.N. Lazier. 1991. *Dynamics of Marine Ecosystems: Biological-Physical Interactions in the Oceans*. New York: Blackwell Scientific.

Margules, C. R., and M. P. Austin (eds.). 1991. *Nature Conservation: Cost Effective Biological Surveys and Data Analysis*. Melbourne: CSIRO. Distributed by International Specialized Services, Portland, Oregon.

May, R. M., and T.R.E. Southwood. 1990. Introduction. In *Living in a Patchy Environment*, B. Shorrocks and I. R. Swingland (eds.). New York: Oxford University Press.

Mayr, E. 1988. *Toward a New Philosophy of Biology: Observations of an Evolutionist*. Cambridge: The Belknap Press, Harvard University Press.

————. 1993. What was the evolutionary synthesis? *TREE* 8:31–34.

McIntyre, S., G. W. Barrett, R. L. Kitching, and H. F. Recher. 1992. Species triage—seeing beyond wounded rhinos. *Conservation Biology* 6:604–606.

McKelvey, K., B. R. Noon, and R. H. Lamberson. 1993. Conservation planning for species occupying fragmented landscapes: The case of the northern spotted owl. In *Biotic Interactions and Global Change*, P. M. Kareiva, J. G. Kingsolver, and R. B. Huey (eds.). Sunderland, Mass.: Sinauer Associates.

Mills, L. S., M. E. Soulé, and D. F. Doak. 1993. The keystone-species concept in ecology and conservation. *BioScience* 43:219–224.

Milne, B. T. 1992. Spatial aggregation and neutral models in fractal landscapes. *American Naturalist* 139:32–57.

Minshall, G. W. 1988. Stream ecosystem theory: A global perspective. *Journal of the North American Benthological Society* 7:263–288.

Moore, H.S.M., W. V. Holt, and G. M. Mace. 1992. *Biotechnology and the Conservation of Genetic Diversity*. New York: Oxford University Press.

Morrison, M. L., B. G. Marcot, and R. W. Mannan. 1992. *Wildlife-Habitat Relationships: Concepts and Applications*. Madison: University of Wisconsin Press.

Moyle, P. B., and R. A. Leidy. 1992. Biodiversity in aquatic ecosystems: Evidence from fish faunas. In *Conservation Biology*, P. L. Fiedler and S. K. Jain (eds.). New York: Chapman & Hall.

Murdoch, W. W. 1993. Individual-based models for predicting effects of global change. In *Biotic Interactions and Global Change*, P. M. Kareiva, J. G. Kingsolver, and R. B. Huey (eds.). Sunderland, Mass.: Sinauer Associates.

Naiman, R. J. (ed.). 1992. *Watershed Management: Balancing Sustainability and Environmental Change*. New York: Springer-Verlag.

Naiman, R. J., and H. Decamps (eds.). 1990. *The Ecology and Management of Aquatic-Terrestrial Ecotones*. Man and the Biosphere Series. Carnforth, U.K.: UNESCO/Parthenon.

Nicol, S., and W. de la Mare. 1993. Ecosystem management and the Antarctic krill. *American Scientist* 81:36–47.

Norton, T. W., and N. D. Mitchel. 1993. Application of generalized additive models to wildlife modelling and population viability analysis. In *GIS '93 Symposium*. Vancouver, B.C.: GIS '93.

Noss, R. F. 1990. Indicators for monitoring biodiversity: A hierarchical approach. *Conservation Biology* 4:355–364.

———. 1992. Issues of scale in conservation biology. In *Conservation Biology*, P. L. Fiedler and S. K. Jain (eds.). New York: Chapman & Hall.

O'Neill, R. V., D. L. DeAngelis, J. B. Waide, and T.F.H. Allen. 1986. *A Hierarchical Concept of Ecosystems*. Princeton, N.J.: Princeton University Press.

Opdam, P. 1991. Metapopulation theory and habitat fragmentation: A review of holarctic breeding bird studies. *Landscape Ecology* 5:93–106.

Orians, G. H. 1993. Policy Implications of global climate change. In *Biotic Interactions and Global Change*, P. M. Kareiva, J. G. Kingsolver, and R. B. Huey (eds.). Sunderland, Mass.: Sinauer Associates.

Paine, R. T. 1992. Food-web analysis through field measurement of per capita interaction strength. *Nature* 355:73–75.

Payne, N. F. 1992. *Techniques for Wildlife Habitat Management of Wetlands*. New York: McGraw-Hill.

Peters, R. H. 1991. *A Critique for Ecology*. New York: Cambridge University Press.

Pickett, S.T.A., V. T. Parker, and P. L. Fiedler. 1992. The new paradigm in ecology: Implications for conservation biology above the species level. In *Conservation Biology*, P. L. Fiedler and S. K. Jain (eds.). New York: Chapman & Hall.

Prendergast, J., R. Quinn, J. Lawton, B. Eversham, and D. Gibbons. 1993. Rare species, the coincidence of diversity hotspots, and conservation strategies. *Nature* 365:335–337.

Pulliam, H. R., J. B. Dunning, and J. Liu. 1992. Population dynamics in complex landscapes: A case study. *Ecological Applications* 2:165–177.

Rastetter, E. B., A. W. King, B. J. Cosby, G. M. Hornberger, R. V. O'Neill, and J. E. Hobbie. 1992. Aggregating fine-scale ecological knowledge to model coarser-scale attributes of ecosystems. *Ecological Applications* 2:55–70.

Risser, P. G. (ed.). 1991. *Long-Term Ecological Research: An International Perspective*. Scope 47. New York: Wiley.

Rohlf, D. J. 1989. *The Endangered Species Act: A Guide to Its Protection and Implementation*. Stanford, Calif.: Stanford Environmental Law Society.

Rosenzweig, M. L. 1991. Habitat selection and population interactions: The search for mechanism. *American Naturalist* 137:S5–S28.

Roughgarden, J. 1989. The structure and assembly of communities. In *Perspectives in Ecological Theory*, J. Roughgarden, R. M. May, and S. A. Levin (eds.). Princeton, N.J.: Princeton University Press.

Ryti, R. T. 1992. Effect of the focal taxon on the selection of nature reserves. *Ecological Applications* 2:404–410.

Satterlund, D. R., and P. W. Adams. 1992. *Wildland Watershed Management*. 2nd ed. New York: Wiley.

Saunders, D. A., and R. J. Hobbs (eds.). 1991. *Nature Conservation 2: The Role of Corridors*. New South Wales, Australia: Surrey Beatty & Sons.

Schemske, D. W., B. C. Husband, M. H. Ruckelshaus, C. Goodwillie, I. M. Parker, and J. G. Bishop. 1994. Evaluating approaches to the conservation of rare and endangered plants. *Ecology* 75:584–606.

Schoener, T. W. 1986. Mechanistic approaches to community ecology: A new reductionism? *American Zoologist* 26:81–106.

——. 1987. The geographical distribution of rarity. *Oecologia* 74:161–173.

Scott, J. M., F. Davis, B. Csuti, R. Noss, B. Butterfield, C. Groves, H. Anderson, S. Caicco, F. D'Erchia, T. C. Edwards, Jr., J. Ulliman, and R. G. Wright. 1993. Gap analysis: A geographic approach to protection of biological diversity. *Wildlife Monographs* 123:1–41.

Seber, G.A.F. 1986. A review of estimating animal abundance. *Biometrics* 42:267–292.

——. 1992. A review of estimating animal abundance II. *International Statistical Review* 60:129–166.

Simberloff, D. 1988. The contribution of population and community biology to conservation science. *Annual Review of Ecology and Systematics* 19:473–511.

Simberloff, D., and T. Dayan. 1991. The guild concept and the structure of ecological communities. *Annual Review of Ecology and Systematics* 22:115–143.

Skalski, J. R., and D. S. Robson. 1992. *Techniques for Wildlife Investigations: Design and Analysis of Capture Data.* New York: Academic Press.

Slobodkin, L. B. 1988. Intellectual problems of applied ecology. *BioScience* 38: 337–342.

Smocovitis, V. B. 1992. Unifying biology—the evolutionary synthesis and evolutionary biology. *Journal of the History of Biology* 25:1–65.

Soulé, M. E. (ed.). 1986. *Conservation Biology: The Science of Scarcity and Diversity.* Sunderland, Mass.: Sinauer Associates.

———. 1987. *Viable Populations for Conservation.* New York: Cambridge University Press.

Soulé, M. E., and K. A. Kohm (eds.). 1989. *Research Priorities for Conservation Biology.* Published in cooperation with the Society for Conservation Biology. Island Press Critical Issues Series. Washington, DC: Island Press.

Soulé, M. E., and L. S. Mills. 1992. Conservation genetics and conservation biology: A troubled marriage. In *Conservation of Biodiversity for Sustainable Development.* O.T. Sandlund, K. Hindar, and A.H.D. Brown (eds). pp 55–69. Oslo, Norway: Scandinavian University Press.

Stacey, P. B., and M. Taper. 1992. Environmental variation and the persistence of small populations. *Ecological Applications* 2:18–29.

Stoms, D. M. 1992. Effects of habitat map generalization in biodiversity assessment. *Photogrammetric Engineering and Remote Sensing* 58:1587–1591.

Stoms, D. M., F. W. Davis, and C. B. Cogan. 1992. Sensitivity of wildlife habitat models to uncertainties in GIS data. *Photogrammetric Engineering and Remote Sensing* 58:843–850.

Thompson, S. K. 1992. *Sampling.* New York: Wiley.

Turner, M. G. 1989. Landscape ecology: The effect of pattern and process. *Annual Review of Ecology and Systematics* 20:171–197.

Turner, M. G., and V. H. Dale. 1991. Modeling landscape disturbance. In *Quantitative Methods in Landscape Ecology*, M. G. Turner and R. H. Gardner (eds.). New York: Springer-Verlag.

Turner, M. G., and R. H. Gardner (eds.). 1991. *Quantitative Methods in Landscape Ecology.* New York: Springer-Verlag.

Vane, G., and A.F.H. Goetz. 1993. Terrestrial imaging spectrometry: Current status, future trends. *Remote Sensing of the Environment* 44:117–126.

Verner, J. 1984. The guild concept applied to management of bird populations. *Environmental Management* 8:1–14.

Verner, J., M. L. Morrison, and C. J. Ralph (eds.). 1986. *Wildlife 2000: Modeling Habitat Relationships of Terrestrial Vertebrates.* Madison: University Wisconsin Press.

Wagner, F. H. 1989. American wildlife management at the crossroads. *Wildlife Society Bulletin* 17:354–360.

Walters, C. J. 1986. *Adaptive Management of Renewable Resources*. New York: Mc-Graw-Hill.

Werner, E. E. 1992. Individual behavior and higher-order species interactions. *American Naturalist* 140:S5–S32.

Wiens, J. A. 1992. What is landscape ecology, really? *Landscape Ecology* 7:149–150.

Wiens, J. A., and B. T. Milne. 1989. Scaling of "landscapes" in landscape ecology, or, landscape ecology from a beetle's perspective. *Landscape Ecology* 3:87–96.

Wiens, J. A., N. C. Stenseth, B. Van Horne, and R. A. Ims. 1993. Ecological mechanisms and landscape ecology. *Oikos* 66:369–380.

Williamson, S. C., and K. Hamilton. 1989. *Annotated Bibliography of Ecological Cumulative Impacts Assessment*. U.S. Fish and Wildlife Service Biological Report 89(11). Washington, DC: U.S. Fish and Wildlife Service.

Wilson, E. O. 1992. *The Diversity of Life*. Cambridge: Belknap Press of Harvard University Press.

Wootton, J. T., and D. A. Bell. 1992. A metapopulation model of the peregrine falcon in California: Viability and management strategies. *Ecological Applications* 2:307–321.

Yahner, R. H. 1988. Changes in wildlife communities near edges. *Conservation Biology* 2:333–339.

Young, R. H., and D. Green (eds.). 1993. *Landscape Ecology and Geographic Information Systems*. Bristol, Pa: Taylor & Francis.

13

A Conflict Management Perspective
Applying the Principles of
Alternative Dispute Resolution

**Julia M. Wondolleck, Steven L. Yaffee,
and James E. Crowfoot**

The scientific, social, and political aspects of endangered species conservation are inherently complex and often conflicting. Rarely are there simple ways to reconcile them. Machlis (1992:161) has observed that "biologists, ecologists and conservationists have increasingly grasped a harsh disciplinary reality: solutions to biological problems lie in social, cultural, and economic systems." The cases presented in this book provide dramatic testimony to this reality. They depict situations that many would argue require immediate action, yet despite the efforts and evidence of scientists and others, action is not forthcoming or is, at best, very slow. The scientists and resource managers involved in these cases express frustration at foot-dragging by agency officials, lack of funding, indifference by the public and other scientists, and a political system that appears unresponsive to their pleas.

This chapter examines the conceptual basis for alternative dispute resolution (ADR) processes and considers their relevance for making endangered species decisions. We describe a number of endangered species situations where ADR has been employed and offer a series of lessons for managers and decision makers. Clearly there are both benefits and liabilities to the use of these approaches. In most circumstances, good process makes good sense. Yet those responsible for creating and implementing decision-making processes are often ignorant of all the human behavioral variables and social system dynamics that go into good process. Whether a formalized negotiation approach makes sense or not depends on one's objectives, the alternative actions possible, and the technical and political characteristics of the issue. Keep in mind, however, that traditional approaches have not yielded great success in saving endangered species. Trying out alternatives may be worth it simply because this historical baseline is not very encouraging.

The Problem: Decision-Making Processes

Part of the frustration encountered by resource managers comes from an incomplete understanding of the social and political aspects of endangered species conservation. Many of the programs detailed in this book represent important public choices that must be made and implemented by a diverse set of people and groups. Some of the cases—the spotted owl (*Strix occidentalis caurina*) and grizzly bear (*Ursus arctos horribilis*), for example—are major public disputes with implications for human activity over millions of acres. Even some of the more localized cases, such as the California condor (*Gymnogyps californianus*) and Florida panther (*Felis concolor coryi*), require the tacit or active support of numerous scientists, politicians, agencies, and private groups. Without such support, even the best-intentioned conservation strategies will fail: dying a quiet death on a recovery team's shelf, languishing in administrative appeals, political delays, or court challenges, or simply failing to deal with the full set of issues that contribute to the endangered species problem at hand.

Many traditional decision-making processes fail at mobilizing the understanding, trust, and capabilities needed for effective action. Conventional administrative processes often fail to produce a thorough understanding of the full range of issues, concerns, and values to prescribe an effective solution. Nor do they generally build stakeholders' support for any preferred direction. Legislative processes involve limited technical understanding and are oriented toward compromises that are often ineffective on the ground. Judicial processes are adversarial, preempting the creativity and cooperation needed to craft effective courses of action. The result of many of these traditional processes is either an impasse or else a decision that cannot be implemented.

The characteristics of endangered species conservation disputes tend to exacerbate the problems of traditional decision-making processes. Often there is considerable uncertainty, which allows disagreements over facts, methods, and theories to fester—and with sensitive species, it is often difficult to resolve the uncertainty because of constraints on sampling and experimenting. Due to this uncertainty, a considerable amount of professional judgment must be exercised. Yet a lack of trust in the agencies formally charged with decision making means that their judgment will be treated with skepticism. And often those with a stake in the outcome do not feel their interests are being accurately heard or adequately acted upon. The fact that endangered species conservation yields fairly intangible long-term benefits, while imposing tangible short-term costs, leads parties to different conclusions about whether or not action is warranted. The rapid changes occurring in many habitats and the shifts in public support for environmental protection cause existing practices to be questioned and challenged. Such a situation is ripe for prolonged and antagonistic conflict. A traditional decision-making process that tries to make a decision unilaterally will almost invariably be met with dissatisfaction.

There may be a better way. Decision-making approaches that seek to institute a process of ongoing collaborative problem solving among diverse stakeholders can at times build the understanding, support, and carefully crafted decisions needed to achieve effective on-the-ground action. For more than twenty years, collaborative negotiations between representatives of government agencies, business interests, communities, and environmentalists have resolved disputes ranging from development of a local shopping mall or townhouse complex to construction of a dam threatening an endangered species. The content of federal rules governing such factors as wood stove emissions and pesticide use has been determined through consensus-based negotiations. Landfills have been sited, national forest plans have been developed, and the terms by which offshore oil exploration may proceed have all been achieved using multiparty collaborative-negotiation processes. (See Wondolleck 1979, 1988; Scott and Hirsh 1983; Susskind and McMahon 1985; Bingham 1986.)

A Conceptual Look at ADR

Formal dispute resolution efforts are known by many different names: environmental mediation, environmental negotiation, alternative dispute resolution, and environmental dispute settlement. (See Bacow and Wheeler 1984; Moore 1986; Amy 1987; Susskind and Cruikshank 1987; Carpenter and Kennedy 1988; Crowfoot and Wondolleck 1990.) Such approaches strive to help divergent interests deal with their differences in a productive, collaborative manner, rather than relying on traditional administrative or judicial mechanisms that tend to focus on small pieces of a larger problem and result in adversarial "winner-take-all" outcomes. In the discussion that follows, we will refer to this broad class of approaches as alternative dispute resolution.

One of the underlying premises of dispute resolution is that *how* one goes about making decisions—the process—has a direct bearing on the ultimate viability of the decisions reached. In struggling with complex environmental and natural resource issues, we focus too often on the question of what should be done—What are some alternative outcomes? Which is the best outcome?—and too little time addressing the question of *how* the decision should be made. How is a process question that raises other questions: What information is needed, and how is it going to be acquired? Who will be affected and should therefore be involved? Whose actions are necessary to effective decision making and implementation? Or, conversely, who can block these efforts and thus should be involved? What criteria should be used in divising alternatives? What criteria should be used in judging these alternatives? How can groups interact to arrive at a mutually acceptable decision? How will implementation and monitoring be ensured? Process considerations frequently are dismissed in environmental decisions in the belief that scientific analysis alone can inform decision making. In fact, the process is seldom considered at all.

Paying attention to process is necessary in order to build good solutions to the problems presented by endangered species conservation. Conservation strategies will succeed on the ground only to the extent that they deal with all the issues of concern. A strategy for owl conservation that does not consider ways to offset the impacts of the strategy is unlikely to succeed. An approach to wolf reintroduction that does not consider the concerns of implementing agencies and adjacent landowners is unlikely to solve the problem. A resolution to fishing rights controversies that does not involve and educate affected groups will rarely work. Technically valid strategies that do not consider the realities of implementation resources are unlikely to yield much success on the ground. Better process can yield better conservation strategies.

Paying attention to process is necessary, too, in order to build consensus around a set of actions. In the eyes of many, scientific evidence is not enough justification for action. Science alone is not adequately convincing to those whose support is needed. If those with a stake in an issue are not part of the decision making, they may not accept, support, and carry out the decisions. Often an inadequate process creates a lack of trust and insufficient information, as those who are not part of the process do not understand legal requirements, administrative constraints, or technical possibilities. Because we have a tendency to caricature those we mistrust, inadequate process allows those who are excluded from decision making to vilify the decision makers and vice versa.

Finally, inadequate process results in a lack of ownership among those not involved in making a decision, thereby limiting their support for this decision. Yet without some measure of broad-based support for action, it is very difficult to bring about change. There must exist an incentive to take action: a personal, organizational, or societal sense of urgency and desire for change. Experience has shown that this desire can be fostered through a group's substantive involvement in decision making through trusted and accountable representatives.

Not all disputes are amenable to resolution through a collaborative decision-making process, of course. Clearly the interests of all parties must not be entirely incompatible; no party should be violating a fundamental belief or premise of their own organization (Moore 1986:11). Gerald Cormick (1980:28), a professional mediator of environmental disputes, suggests four additional prerequisites that should be satisfied before effective dispute resolution can be brought about:

1. All parties must recognize "the necessity of other parties participating in the decision-making process as coequals." This does not mean that everyone must agree with and support the interests of the other parties but, rather, that each realizes that all parties need to work together to address the problem successfully.

2. Each party must have "sufficient power or influence to exercise some sanction over the ability of other parties to take unilateral action." In other words, no individual party can impose its will on the other. Each party must have the power to prevent a decision from taking effect—by filing a lawsuit or an administrative appeal, encouraging public protest, fostering political intervention, or some other means.

3. Representatives of stakeholders must "be able to commit themselves and their constituents to the implementation and support of any agreement reached." If a party is unable to guarantee its support of a collaboratively developed decision, the other participants will question the usefulness of working with that party.

4. All parties must share a sense of urgency to resolve the dispute. Endangered species advocates may clearly view action as needed.

For a collaborative effort to be productive, however, the other parties must similarly perceive a need to take action in order to meet their interests.

Many endangered species disputes do not meet these criteria. Fundamental values may separate disputants, and it is hard for them to find ways to accommodate their differences in negotiated solutions. In some situations, it is difficult for a diverse set of stakeholders, such as a coalition of environmental groups, to agree on a scheme of representation so that all their concerns are brought into the negotiation accurately and their representative has the power to bind them all to a settlement. In other situations, certain parties are benefited by the status quo; in this case, delaying settlement is their best strategy for defending their interests.

For some endangered species situations, however, ADR processes can improve decision making. Sometimes groups can agree to disagree on value differences and find ways to proceed. In other situations, mediators find ways to assist large sets of stakeholders to organize themselves and represent their interests in negotiations. Parties who are favored by inaction can have the incentives facing them changed by the strategic behavior of their opponents. In the spotted owl case, for example, timber interests were favored by delays through much of the history of the case because older forest plans allowed more cutting. But lawsuits filed by environmental groups resulted in injunctions on timber harvest and changed the nature of the incentives facing timber interests considerably.

Even when ADR is appropriate, the literature makes it clear that bad process can doom such efforts. More than twenty years of experience in attempting to resolve environmental disputes has led to an understanding of the key elements of an effective process. It is essential that appropriate and adequate representation be achieved. Parties having a stake in the outcome as well as an ability to influence its implementation, either positively or negatively, should be represented. (See Cormick 1980; Harter 1982; Moore 1986; Susskind and Cruikshank 1987; Carpenter and Kennedy 1988.) This does not mean that every person with a stake must be present but, rather, that they must either be adequately represented by someone else or at least they must formally sanction the effort. Those having responsibility for making a final decision must either be present or formally sanction the effort, thereby lending legitimacy to the process and its outcome (Cormick 1980).

The process should be structured in a manner which ensures that everyone agrees on the nature of the problem to be addressed and the scope of the issues to be discussed. (See Harter 1982; Susskind and Cruikshank 1987; Carpenter and Kennedy 1988.) In other words, everyone should be there to solve the same problem and they should all agree on its definition. A common objective for the collaborative effort should be determined at the outset. Moreover, the process should provide for joint fact-finding when there is uncertainty or when people have different judgments about the implications of the data. Furthermore, the process should encourage the joint development and then evaluation of various alternatives (Moore 1986; Susskind and Cruikshank 1987; Wondolleck 1988). If the parties have never participated in a collaborative negotiation process or are having trouble communicating because of past tensions or other reasons, a third-party mediator or facilitator might be called upon to assist in structuring and running the process (Moore 1986).

Finally, "there should be reasonable assurances that affected government agencies will implement the agreement if it is reached" (Cormick 1980:29). In other words, the formal decision makers should be at the table themselves, appropriately represented at the table, or publicly sanctioning the process and its outcome. This does not mean that the agencies must consent to be bound by an agreement; indeed, agencies like the U.S. Fish and Wildlife Service (FWS) or the U.S. Forest Service are constrained by law from making such a commitment. But it does mean that if the process is carried out effectively, there is a good chance that any agreement will be the core of a proposed agency decision that would then move through the agency's normal decision-making procedures, including public review.

Good process is not easy. Communication between diverse individuals and groups is often problematic. Defusing emotional situations in a way that leads to long-term problem solving is not a skill of many resource professionals. Creating a situation where groups can set aside their positions and focus on their interests is a difficult task. Finding creative ways to deal with uncertainty or differences of perspective requires hard work that is often constrained by organizational or professional norms. Nevertheless, there are ways to do these things. And since decision making is unavoidable, finding ways to design and implement better choice processes is important.

While some situations can benefit from well-crafted, ADR-style processes, conflict need not always be viewed as something to be managed for purposes of resolution. Conflict plays many important social roles. It serves an essential function in raising public awareness about a situation, influencing public opinion, and, in turn, stimulating an administrative or political response that leads to action on behalf of a species. Often conflict changes the strategic power of different interests in ways that might otherwise take a long time to have an effect. If the conflict is dissipated early, its long-term effects may be dampened. Conflict is also a driving force for social change. At times, small-scale responses are incapable of dealing with the magnitude of the underlying problem. Sometimes the only way conservative institutions are forced to deal with big problems is through protracted, messy conflict. Because protection of biological diversity is such a fundamental element of environmental protection but is at odds with many deeply-entrenched societal practices, protracted conflict can be expected on this issue.

Regardless of these benefits of conflict, there are times when neither the stakeholders nor society at large are benefited by a pervasive state of impasse. And in endangered species conservation, impasses that delay action are not in the best interests of the species. For some groups in some situations, there may be strategic advantage in avoiding negotiated settlements and building conflict. For those groups, whether such an approach is appropriate depends on what is likely to occur without negotiation. Experience with endangered species tells us that business as usual is usually bad for conservation. At times, multiparty collaborative negotiations can yield better outcomes. The challenge is to find a means to structure opportunities for problem solving in ways that encourage all parties to take each interest seriously and persevere at finding ways to work out their differences, arrive at decisions, and support their implementation.

Negotiations Involving Endangered Species

The idea of encouraging collaboration among those with a stake in endangered species conservation is not a new one.[1] Consultation and coordination on fisheries and wildlife issues have been an aspect of federal law for more than fifty years (Bean 1977). Since 1973, the FWS and the National Marine Fisheries Service (NMFS) have consulted with other agencies and developers through the Section 7 interagency consultation requirement of the Endangered Species Act (ESA) to devise specific mitigation measures that would protect listed species. These negotiations have resulted in protective measures in over 15,000 specific consultations, at times significantly modifying a project (Yaffee 1991).

In the past fifteen years, a number of endangered species versus development disputes have employed elements of multiparty collaborative decision making. One early case that provides a textbook example of the structure and promise of a formal dispute resolution effort is the Grayrocks Dam case. As early as 1978, representatives of regional and national environmental organizations, farmers, developers, power companies, and the governors of the states of Wyoming and Nebraska pursued collaborative negotiations to settle their differences about construction of the Grayrocks Dam near Wheatland, Wyoming (Wondolleck 1979). Construction of the dam would affect water flow in the North Platte River, in turn affecting the critical habitat of the endangered whooping crane (*Grus americana*) nearly 240 miles downstream.

After several years of litigation and congressional battles, the parties themselves were able to address their interests in a creative way that satisfied all. The final settlement allowed the project to move forward in exchange for commitments to minimum water flows, a series of mitigation measures, and the establishment of a $7.5 million trust fund for research and the purchase of water rights for the whooping crane. While some individuals not party to these negotiations might argue that the negotiation was inappropriate, those involved still argue today that their chosen course of action was superior to the others available at the time and that the outcome was better for the crane than what probably would have occurred without the negotiations.

Apart from this early case, endangered species conservationists have participated in collaborative negotiation at all levels of implementation of the ESA. Their experience lends insight into when and how ADR can be useful as well as the cautions that must accompany its use. Here we describe their experiences in two cases: habitat conservation planning and ecosystem-based negotiations.

Habitat Conservation Plans

There have been many attempts to address the needs of endangered species collaboratively at the habitat or ecosystem level. The development of habitat conservation plans (HCPs) under Section 10(a) of the ESA is one example of a holistic attempt to ensure that an endangered species habitat is not slowly whittled away by incremental development. Section 10(a) allows the FWS or NMFS to grant a permit for the incidental taking of a listed species—if a HCP is prepared and approved. Collaborative negotiations among developers, local, state, and federal agencies, and endangered species conservationists have led to plans governing where and how development may occur within the critical habitat of the mission blue butterfly (*Plebejus icarioides missionensis*) on San Bruno Mountain, south of San Francisco, the Coachella Valley fringe-toed lizard (*Uma inornata*) in southern California's desert region, as well as in a number of other situations to date in California, Texas, and Florida. Bean et al. (1991) and Beatley (1994) present a valuable description and analysis of collaborative processes applied to habitat conservation planning.

Few HCP processes have been without their detractors. Some still argue, in principle, that it is wrong to compromise what little remains of a species' habitat in providing for development activity. Those who participated in the development of these plans, however, maintain that the species are much better protected, even with the compromises, than they might be by the uncertain outcome of a protracted political or legal fight. Supporters continue to advocate the collaborative development of HCPs in ways that directly incorporate the knowledge and expertise of endangered species conservationists and build upon our common understanding and efforts on behalf of specific species.

The San Bruno Mountain HCP.   San Bruno Mountain is the last large open space left in the northern San Francisco peninsula. Not surprisingly, the mountain had been the source of a bitter battle between local environmentalists and developers well before negotiations on the San Bruno Mountain HCP began. The mountain is also home to the San Bruno elfin butterfly (*Callophrys mossii bayensis*) and mission blue butterfly—both listed as federally endangered—as well as the Callippe silverspot butterfly (*Speyeria callippe callippe*), which was a candidate for listing at the time of this dispute.

An early agreement between San Mateo County and a local developer left one-third of the mountain available for development and the remainder preserved as parkland. When the FWS proposed to list and designate critical habitat for the Callippe silverspot butterfly, however, negotiations were undertaken to develop a HCP that was to become the model for subsequent HCPs. Participants in this process included the Committee to Save San Bruno (a coalition consisting of the local Committee for Green Foothills, a local chapter of the Sierra Club, and the local chapter of the National Audubon Society), Visitacion Associates (a developer), several local cities, the FWS, and the California Departments of Fish and Game and Forestry. The San Bruno Mountain HCP was finalized in November 1982. It allowed for development of 10 percent of the butterflies' habitat with the remainder of the area becoming state, county, or city-owned parkland. It also provided for "enhancement" of poor-quality habitat and additional research on the butterfly populations and included restrictions on grading and construction activities. Finally, the HCP set up a three-tiered amendment process to allow some flexibility in carrying out the plan.

Not all environmental group participants signed the agreement, and those who ultimately supported the plan did so warily. Implementation has not been a quiet affair. In fact, it is still not clear whether or not the plan will help the butterflies. There have been amendments to the HCP and there have been lawsuits challenging the ecological soundness of the agreement. At the time, the parties thought the process was cumbersome enough to be a disincentive for people who might want to alter the HCP. Their sanguine outlook was unfounded. One participant noted that "wars broke out continually for the next four years." As an example, he cites an episode in which the developer proposed a major boundary change only two months after the agreement was signed. The Committee to Save San Bruno felt "betrayed." The developer was angry because the committee was "unreasonable." Since 1983, the HCP has been modified several times. Some amendments have had a significant impact on butterfly habitat, though mitigation measures were added as a condition of approval. One participant noted: "We didn't really think there would be so many amendments. The agreement turned out to be more ephemeral than we had expected." Despite efforts to anticipate problems, according to one participant, "it's nearly impossible to do."

Environmentalists, in general, are not satisfied with the HCP, though representatives of groups that endorsed it say the plan was the best they could do given the hostile political climate that prevailed in the early 1980s. Some explain that they were extremely concerned about protecting the ESA, which was under attack in Congress at the time. They were also concerned about the developer's threat to sabotage the Callippe silverspot butterfly listing and delist the endangered butterflies. One participant commented: "Reagan and Watt had taken over. Trade associations in Washington, D.C., were demanding repeal of the ESA. The Tellico Dam case was fresh in the public's mind. We thought that if we created a conflict between a housing development in San Francisco and an insect, albeit an attractive one, we might threaten the act itself. The case had the potential to generate a lot of publicity. So we decided to negotiate a principled compromise."

Coachella Valley Fringe-Toed Lizard HCP. For several years, environmental groups in the Coachella Valley, a 300-square-mile area of southern California desert, had been working to preserve a portion of the desert ecosystem. One of their aims was to protect the Coachella Valley fringe-toed lizard, a federally listed endangered species. Because the area was one of the fastest-growing regions in southern California, they faced tremendous local political opposition to their efforts. It had been a lonely and decidedly uphill battle until 1982—the year Congress amended the ESA to allow incidental taking permits and the Sunrise Development Company decided to build the Palm Valley Country Club on more than 400 acres of lizard habitat. The requirement that an HCP be prepared before any development permits could be issued led to a multiparty negotiation involving the FWS, Bureau of Land Management (BLM), California Department of Fish and Game, California Nature Conservancy, Coachella Valley Ecological Reserve Foundation, Agua Caliente Indian Tribe, several local governments, and the Sunrise Development Company.

In 1985, the groups reached agreement on an HCP. It set aside close to 17,000 acres in three permanent reserves. These protected lands contain 7838 acres of habitat potentially occupiable by lizards, roughly 10 percent of what then existed. Additional habitat was protected through informal agreements with the BLM and Southern California Edison Power Company to manage adjacent lands in a way that would not harm the lizard. From the perspective of endangered species conservationists, implementation of the HCP shows mixed results. Enforcement of the plan's provisions, as well as restoration and enhancement of lizard habitat, have both been successful. The lizard's fate, however, is still uncertain. Monitoring shows that biological assumptions about the lizards were incorrect. Funding too has been problematic. The pool of money created from development fees—to be used for land acquisition and maintenance—has not filled as quickly as expected.

Alan Muth of the Coachella Valley Ecological Reserve Foundation, a leader in early efforts to preserve the desert ecosystem and a participant in the HCP process, argues that "the process worked here. We did the best we could using the best information at the time. Now we have to wait and see what happens." He is reluctant, though, to advocate amending the HCP to remedy some of its shortcomings. "Opening up the process for amendments," he says, "is like opening a can of worms for both strengthening and weakening amendments." While the plan has been criticized for not protecting enough lizard habitat, Muth does not think the criticism can be justified: "If you examine what was left [undeveloped] in the valley that could potentially serve as good lizard habitat, we got almost everything protected in the HCP. Admittedly, that was a small percentage of the lizard's historical range, but we had to have lands that could be maintained ecologically as good habitat over the long haul."

Ecosystem-Based Negotiations

At yet another level of ESA implementation, there have been at least two attempts to apply ADR principles and procedures in developing comprehensive management plans for ecosystems supporting a number of endangered species. The focus of these two processes—recovery planning for the endangered fishes of the Upper Colorado River and the Platte River Management Joint Study—has been to involve all affected parties in collaboratively assessing the status and needs of the species and, with this understanding, to then create management plans governing how future development should be evaluated. Neither process has been without its difficulties. And despite many years of effort, both processes have yet to be completed. At one level, both have encountered procedural problems—in particular, poorly facilitated sessions—that have led to frustration and unproductive periods of time. Lack of resources to fully support the processes, as well as wavering political interest, have hindered their progress too. At another level, the complexity of the issue, the lack of data, and the degree of uncertainty have led to frustration and conflicting opinions about the proper course of action.

As in the HCP processes described earlier, endangered species conservationists have participated in these larger efforts but with a healthy dose of caution. They want to ensure the long-term recovery and survival of the species; but they do not want to sanction, by virtue of their participation in these processes, continued exploitation of the resources upon which the species depend. They fear that uncertainties will be overlooked and that development projects will cause the irretrievable demise of the species. At the same time, however, these conservationists grasp the necessity of building a common understanding among key parties and accumulating knowledge about the species and the ecosystem upon which they depend in order to make wise decisions. And, too, they believe they have greater influence as part of this effort than as outside critics.

Recovery Planning for the Endangered Fishes of the Upper Colorado River Basin. The recovery program for several endangered fish species in the Upper Colorado River Basin is a collaborative effort on the part of several environmental organizations, water development groups, and state and federal agencies. The Colorado squawfish (*Ptychocheilus lucius*), humpback chub (*Gila cypha*), and bonytail (*Gila elegans*) are federally listed endangered species in the river while the razorback sucker (*Xyrauchen texanus*) is extremely rare and a candidate for federal listing. These four species are referred to collectively as the endangered fishes of the Upper Colorado River Basin.

The clash between the endangered fish species of the Upper Colorado and development interests along the river came to a head when federal agencies attempted to fulfill their requirements under Section 7 of the ESA. In 1978, faced with the Bureau of Reclamation's need for opinions and given the limited biological data on the endangered Colorado River fish, the FWS agreed to cooperate with the bureau in an intensive three-year study to determine the flows necessary to preclude jeopardy to the fish. This study was expanded when the 1982 amendments to the ESA directed federal agencies to coordinate with state and local agencies to resolve water resource issues in concert with the conservation of endangered species. This amendment was added to the ESA specifically to address conflicts between water development and conservation of the endangered species in the Upper Colorado and Platte River basins.

The FWS took the lead in organizing a coordinating committee—with representatives from state and federal agencies, water interests, and environmentalists—to evaluate alternatives and develop a program for recovering the fishes in a manner consistent with state water rights systems. The Recovery Implementation Program for the Endangered Fish Species in the Upper Colorado River Basin was finalized in September 1987 (U.S. Department of Interior 1987). It includes a range of conservation activities that have been determined necessary to protect and recover the Upper Colorado's rare fish species while allowing for new water development. The ultimate goal of the program is the recovery and delisting of the three endangered species and the management of the razorback sucker so that it does not need the protection of the ESA. The implementation plan requires a cooperative federal/state/private effort and includes programs to acquire instream flows for the fish pursuant to state law. The plan calls for a change in operation of Bureau of Reclamation dams to benefit the fish. Other programs, such as stocking native fish, controlling stocking of sport fish, and habitat development, are also key elements of the proposal. The programs are designed to mitigate the impacts of water projects and allow the FWS to issue nonjeopardy biological opinions.

Most participants both praise and criticize the effort. While many appreciate elements of the process used to reach decisions, they also worry that the outcome is less than satisfactory. Several aspects of the program have been controversial—particularly the recovery elements related to habitat improvement, management of predators or competitors, and the operation of federal reservoirs to meet the habitat needs of the fish. Nevertheless, most groups are glad they participated. One environmental participant believes that some subsequent FWS opinions have been "stronger than we ever would have gotten without the agreement. It's all made possible by the education of the developers and the water users—and the environmentalists too—so that there is a much better understanding of the issues and that's because of the 'hand's on' nature of the program." A FWS participant commented that "the process was successful in that it brought people together, and . . . once the plan was approved, it brought money to the issue." A water development participant agreed: "The process is working. There is an allocation of water for the fish in place, and several other possibilities for habitat are being explored. We get along quite well, and all of us are trying to make it work."

Another environmental group representative reflects that "the best way to improve habitat is to change the way the big reservoirs are operated. This program makes it possible for that to occur. [But] the program is moving at a glacial pace, and I don't know if the species will be saved." Others cite the success of the program in terms of the value of the research that has been initiated and funded as a result. While the environmentalists appreciate the improved understanding and attention to the species' needs, they are still uncertain about the long-term effect of the effort. As one participant noted: "One of the reasons the Colorado process is a 'success' is because it hasn't been confronted with any real issues yet."

Platte River Management Joint Study.   The Platte River Management Joint Study is a collaborative effort, modeled after that used on the

Upper Colorado River, to develop a viable management plan for a
225-mile stretch of the Platte. The conflict at issue in this case is a
common theme in the arid West: an ongoing struggle between a conservation community that wants water left in the river to benefit
wildlife versus a development community that wishes to divert water
to meet growing agricultural and municipal needs. The 80-mile Big
Bend segment of the river is especially valued as habitat for a variety
of species dependent upon the natural river system with its open waters, exposed sandbars, and adjacent wetlands. These species include
the endangered whooping crane, least tern (*Sterna antillarum*), and
bald eagle (*Haliaeetus leucocephalus*), the threatened piping plover
(*Charadrius melodus*), four-fifths of the world population of sandhill
cranes (*Grus canadensis*), over 7 million ducks and geese, and hundreds
of other bird species.

Participants in this process have the goal of developing a detailed,
technical plan for managing the Platte that would allow additional
water development to proceed in compliance with the protective requirements of the ESA. Without conflicting with existing water laws,
the plan would assure adequate habitat for the endangered, threatened, and other migratory bird species that depend on the Platte.
The Bureau of Reclamation and FWS are lead agencies in the effort.
The states of Wyoming and Colorado, the Army Corps of Engineers
and several other federal agencies, water development interests, and
the environmental community are all participants in the plan. The
State of Nebraska withdrew from the process after a few years of involvement, choosing to pursue its interests through litigation.

Despite their fears and discomfort with the collaborative process,
neither side of the conflict was willing to accept a continuing impasse
that prevented further water development in the basin and failed to
protect the essential habitats of the Central Platte. Ken Strom, manager of Audubon's Lillian Annette Rowe Wildlife Sanctuary and a participant in the process, describes the dilemma faced by his organization in deciding whether and how to participate in this process:

> We are constantly asking ourselves the question: Is it worth
> our time and money to participate? We want an outcome
> that is a legitimate scientific outcome and one that will not
> be challenged in court. Much of our early internal debate fo-
> cused on the fact that simply by participating we were adding
> legitimacy to the process and we weren't sure if we wanted
> to do this. We thought about letting them go ahead and then

attacking the outcome legally, but realized we can't afford to fight this dam by dam. There was also great potential damage by not participating in terms of what would be developed if we were not there.

Water development interests made a similar assessment. According to one participant:

> Water interests perceived that this was not the first time they'd had difficulty surmounting Section 7. They realized a mechanism had to be found to "accommodate" the ESA. So water development interests petitioned the Secretary of Interior, who told the FWS and the BR to work out a consensus resolution.

Some progress has been made, but it has not been easy. At this point, ten years into the process, the immediate goals have been modified. The target now is completion of a conceptual plan for protecting the endangered species of the Big Bend area, leaving development of a detailed implementation plan to a later effort. Most participants in this process feel that, despite the difficulties encountered thus far, it is important to persevere—maintaining a dialogue will at least allow them to understand each other's concerns. A conservation group participant is pleased that the process has conveyed to others a "greater knowledge of what we're trying to do and what we're trying to protect." A FWS participant commented: "I'm encouraged by [the process]. I would be more encouraged if we had reached the point we're at earlier and more encouraged if I could tell you there were no points of contention left and if Nebraska was a full partner." But the bottom line, he says, is that the process should continue. A water development participant concurs: "This isn't a winner-take-all deal; if we don't find a basis of understanding then we all lose."

Lessons

The collaborative dispute settlement processes described in this chapter depict certain dilemmas for the conservationists. In each case, frustration, disappointment, and shortcomings are readily expressed by those involved. In almost the same breath, however, they say that their participation was appropriate and that, while far from an easy and fully successful route, it was the most appropriate approach—given their options. There are two key lessons to take from these experiences.

First, multiparty collaboration designed to address the conflicting needs of endangered species and development interests is never an easy process. Success is not guaranteed. Collaborative processes cannot magically convert a complicated situation into a simple problem. Conservation of endangered species is complex; collaborative negotiations are bound to reflect that complexity.

Second, these processes do hold some promise for endangered species advocates. While ADR processes are not a godsend that will resolve all the difficulties inherent in endangered species conservation, they can help in certain key ways. Collaboration can spur efforts that promote conservation measures on behalf of a species. It can assist groups in building a better understanding of a species and its needs and, in so doing, lead to better decisions on behalf of the species. It establishes formal and informal networks of relationships among participants that can then be enlisted to bring about future action to protect the species.

But to realize the promise of dispute resolution requires that endangered species advocates maximize their influence in these processes. This means they must capitalize on the potential benefits of the conflict to endangered species by increasing public awareness, encouraging active public involvement, and stimulating political pressure in support of the species. It also means they must be skilled in the negotiation process itself. And, finally, it means that an appropriately structured and well-managed process be employed, one that is grounded in the experience of the ADR field.

In some of the cases presented elsewhere in this book—in particular the California condor and the red-cockaded woodpecker (*Picoides borealis*)—efforts were made to bring together all the affected groups and agencies with authority to take action on behalf of the species. These efforts, however, were not always successful. At times they were poorly managed. At times their intent was less than genuine, as some groups participated in hopes of delaying action or co-opting the efforts of the endangered species advocates. At times the collaborations were not useful because the wrong people or organizations were present or key issues were not on the agenda.

Clearly there have been poorly executed ADR processes that tarnish the image of how helpful such collaborations might be. Many are frustrated by misdirected and purposeless meetings. Some fear the potential for co-optation. The solution, however, is not to avoid meetings and interactions with other parties. The solution is to make the overall process, along with its meetings, work—by ensuring that a common objective is established at the outset and the key issues of concern are on the agenda, by taking care that the appropriate interests and organizations are represented (and, moreover, represented by the right people), by providing for professional facilitation when this assistance would be useful, and by ensuring that the information necessary to the dialogue is not only available but usable by all. Applying ADR experience with other environmental and natural resource issues to endangered species can help conservationists in planning an appropriate process and deciding whether or not to participate. (See Harter 1982; Susskind and Cruikshank 1987; Crowfoot and Wondolleck 1990.)

The question of whether ADR promotes endangered species conservation is not a simple either/or question. Any decision-making or regulatory process can benefit from the lessons of the dispute resolution field. Even if the process consists of an FWS official consulting with a development agency staff member on the potential impact of a proposed project, good process skills can assist in sorting through the full range of concerns and finding ways to meet them creatively. Good process can yield benefits even if decision making moves forward without a complex, multiparty negotiation. As more parties enter the discussions and more issues are laid on the table, it becomes even more important that processes be designed and implemented to address the concerns of involved groups, gather essential information, and facilitate dialogue and problem solving.

The collaborative processes discussed in this chapter clearly have the potential for advancing endangered species conservation efforts. If used appropriately and managed carefully, dispute resolution processes can be helpful in coordinating and stimulating efforts among the divergent groups with a stake in the conservation of endangered species. Even if comprehensive settlements are not achieved on every issue, creating a forum for information exchange and communication about group concerns can promote the effectiveness of the ultimate resolution. Indeed, one opportunity for incorporating ADR-style processes in endangered species cases may lie in data negotiation—a process in which the parties can focus on the data and theory underlying a dispute and thereby potentially reduce the uncertainty facing decision makers (Ozawa 1991).

While dispute resolution processes have considerable potential for building better decisions, they require the strategic involvement of those advocating the interests of an endangered species. The benefits of these processes are not automatic; the difficulties must be confronted head on. Endangered species conservationists who enter these collaborative processes must do so knowledgeably, understanding their interests and their options and strategically pursuing them. In the words of Rowland (1992:504): "As an environmental advocate, I tend to share my colleagues' skepticism of mediation as a means to protect the myriad animal and plant species that compete for living space with a swelling human population. I also believe, however, that in the right circumstances, mediation can be a useful adjunct to other established processes."

We offer three key recommendations to decision makers and other participants at both the policy and the organizational levels:

1. Provide opportunities for collaboration. The single most influential reason why collaborative efforts to develop workable plans for protecting species are not pursued is simply because there is no opportunity.

2. Educate individuals and organizations about process options so that they can knowledgeably take advantage of the opportunities or strategically bypass them.

3. Furnish the necessary support for these collaborative efforts. Provide the resources that enable groups to participate and that allow the process to be planned and managed properly. Additionally, provide the experience and skills—particularly through training and using the results of environmental dispute settlement research—for individuals and organizations to participate effectively.

For organizations and individuals considering joining a dispute resolution process, these recommendations should prove useful:

1. Understand the context in which the endangered species situation exists, and analyze the options afforded by this context. In other words: clarify your organization's alternatives in the situation. Only then can you decide on the appropriate course of action.

2. Participate in structuring and guiding a collaborative process. The structure and management of such processes are not carved in stone; they must be shaped to fit the immediate situation.

3. Make sure your issues are on the agenda, stay on the agenda, and get addressed to your satisfaction.

4. Educate and train your professionals to be effective in advancing the endangered species effort through whatever course of action is chosen.

5. Make sure the right people are representing your organization and remain accountable to it. Collaboration and advocacy require different skills than do scientific assessment and analysis. Inform your members about the organization's objectives in collaborative processes, the progress that has been made, and the decisions that may result.

Note

1. The quotations and conclusions presented in this section are based on research in progress conducted by Julia Wondolleck and Steven Yaffee for the Administrative Conference of the United States, Washington, D.C.

References

Amy, D. J. 1987. *The Politics of Environmental Mediation.* New York: Columbia University Press.

Bacow, L. S., and M. Wheeler. 1984. *Environmental Dispute Resolution.* New York: Plenum.

Bean, M. J. 1977. *The Evolution of National Wildlife Law.* Washington, DC: Council on Environmental Quality.

Bean, M. J., S. G. Fitzgerald, and M. A. O'Connell. 1991. *Reconciling Conflicts Under the Endangered Species Act: The Habitat Conservation Planning Experience.* Washington, DC: World Wildlife Fund.

Beatley, T. 1994. *Habitat Conservation Planning: Endangered Species and Urban Growth.* Austin: University of Texas Press.

Bingham, G. A. 1986. *Resolving Environmental Disputes: A Decade of Experience.* Washington, DC: Conservation Foundation.

Carpenter, S., and W.J.D. Kennedy. 1988. *Managing Public Disputes: A Practical Guide to Handling Conflict and Reaching Agreements.* San Francisco: Jossey-Bass.

Cormick, G. 1980. The theory and practice of environmental mediation. *Environmental Professional* 2(1):24–33.

Crowfoot, J., and J. M. Wondolleck. 1990. *Environmental Disputes: Community Involvement in Conflict Resolution.* Washington, DC: Island Press.

Harter, P. J. 1982. Negotiating regulations: A cure for the malaise? *Georgetown Law Journal* 71(1):1–118.

Machlis, G. E. 1992. The contribution of sociology to biodiversity research and management. *Biological Conservation* 62:161–170.

Moore, C. 1986. *The Mediation Process: Practical Strategies for Managing Conflict.* San Francisco: Jossey-Bass.

Ozawa, C. 1991. *Recasting Science: Consensual Procedures in Public Policy Making.* Boulder: Westview Press.

Rowland, M. 1992. Bargaining for life: Protecting biodiversity through mediated agreements. *Environmental Law* 22(2):502–527.

Scott, T. J., and L. Hirsh. 1983. Negotiating conflict on Georges Bank. *Environmental Impact Assessment Review* 1(3):561–576.

Susskind, L., and J. Cruikshank. 1987. *Breaking the Impasse: Consensual Approaches to Resolving Public Disputes.* New York: Basic Books.

Susskind, L., and G. McMahon. 1985. The theory and practice of negotiated rulemaking. *Yale Journal of Regulation* 3(1):133–165.

U.S. Department of the Interior. 1987. *Recovery Implementation Program for Endangered Fish Species in the Upper Colorado River Basin.* Final Report, 29 September 1987. Washington, DC: U.S. Fish and Wildlife Service.

Wondolleck, J. M. 1979. Bargaining for the environment. Master's thesis, Massachusetts Institute of Technology.

———. 1985. The importance of process in resolving environmental disputes. *Environmental Impact Assessment Review* 5(4):341–356.

———. 1988. *Public Lands Conflict and Resolution: Managing National Forest Disputes.* New York: Plenum.

Yaffee, S. L. 1991. Avoiding endangered species/development conflicts through interagency consultation. In *Balancing on the Brink of Extinction: The Endangered Species Act and Lessons for the Future,* K. Kohm (ed.). Washington, DC: Island Press.

14

An Organizational Perspective
Designing Recovery Teams
from the Inside Out

Ron Westrum

Recovery teams are at the heart of endangered species conservation. Teams of various sizes are used to plan restoration efforts and carry out needed actions. Even though the structure and activities of these teams vary considerably, they are likely to share common properties and face common problems. In this chapter I want to focus on these common concerns and sketch a strategy for designing successful teams (Clark and Westrum 1989). The term "recovery team" as used here includes not only the formal U.S. Fish and Wildlife Service teams that plan for species recovery and produce recovery plans but also the research and management teams and working groups that form around the endangered species conservation task. The membership of these various teams—recovery teams, field teams, management teams, and working groups—may be constant or ever-changing. Teams may have a common core membership while people on the periphery come and go. The team itself may be relatively transient—a "task force"—or more permanent. All these teams are grouped together here under the term "recovery team." Also, to simplify, I will use the generic term "wildlife manager" to cover all the members of these teams even though in reality not all team members are wildlife managers in the strict sense of the word.

Since nearly all the work of species research and management is carried out by teams, successful endangered species restoration depends on making these teams effective (Clark et al. 1989; Clark and Cragun in press). One of the key tasks of management at all levels should be to promote effective teams. The question is how to do this. In what follows I will examine teams in a general way and then relate the discussion specifically to the task of conserving endangered species. First, I will consider reasons why teams fail and thus identify ways in which teams should be strengthened. Second, I will describe the underlying theory for designing successful recovery teams. Third, I will discuss various approaches for getting real recovery teams to approximate this ideal. My own background is in the study of creativity and technology management, so I will draw upon examples from these areas to illustrate points about endangered species recovery teams and the overall task of endangered species conservation.

Endangered Species: The Human Envelope

The world of endangered species is constantly in motion. Not only are a species' behavior and habitat in constant motion, but the human behavior of wildlife managers and the human community at large are also in motion. This complicates the team's task: the weaving of a complex and ever-changing cocoon around the endangered species. Keeping this cocoon strong in the winds of change requires both an ability to manage a recovery team effectively and also to do effective networking. By "networking" I mean developing cooperative arrangements between organizations that can accomplish what no single organization or agency can accomplish. No endangered species is likely to thrive without interorganizational efforts. Yet neither good team management nor networking skills may thrive in a bureaucratic environment—which often characterizes endangered species recovery efforts. Systematically teaching these skills to newcomers may be even less likely.

The wildlife manager should have more than a passing knowledge of organization and management theory. Let me suggest a useful metaphor that expresses why this knowledge is so important. When I give my basic undergraduate course on technology, I always teach my students to think about technologies as protected by a human envelope. This envelope is made up of designers, managers, operators, and maintenance people. Seeing humans as the protectors of technology is the reverse of the way we usually think about the matter. We usually see technology as protecting *us*. But if the human envelope fails, then the technology will fail too—and with it the technology's ability to do what it is supposed to do in feeding us, protecting us, transporting us, and fulfilling our other purposes.

The endangered species has a human envelope too. That human envelope is the recovery team and associated structures. The leaders of the recovery team need to understand organizational and interorganizational dynamics—for if the organization is not handled well, the species will not be handled well. To succeed in recovery requires an effective human envelope. An effective organization in turn requires skilled leadership. Thus the wildlife manager must understand a lot about managing teams of human beings. To aid the wildlife manager, I have added a short list of books at the end of the chapter.

The wildlife manager also has to learn from previous and ongoing efforts of other teams. Some teams are more effective than others. Why? What do successful teams do? Species recovery must be understood as a process. It is both a process on the part of the recovering species and also a process on the part of the recovery team. So the wildlife manager must be an expert on process improvement, both in terms of biological knowledge and in terms of organizational skills. Just as in sports, knowing how to run a team comes from studying those who are the top practitioners: the maestros. The expert biologist is not necessarily the expert manager (and vice versa). But the wise wildlife manager learns from both.

Why Do Recovery Efforts Fail?

Recovery efforts fail for four basic reasons: intention, incompetence, ignorance, and ill fortune. Intentional failures come from contrary or inadequate motivation among those in control—an agency might have more pressing goals, for instance, or might actually want the species to go extinct. Incompetence, on the other hand, is not a matter of motivation but of skill—species recovery efforts can fail (resulting in extinction) because a team is not properly set up, staffed, or directed to be effective. And then there is ignorance. Ignorance can still frustrate teams that are well motivated and intelligent but simply do not have enough knowledge to save the species. Finally there is bad luck. In principle every risk can be foreseen, but this is often hard to accomplish in real life. Hurricanes, chemical spills, and outbreaks of diseases are difficult to predict and control. Nonetheless, crises are handled very differently by effective and ineffective teams, so even here good management may prevent a setback from becoming a catastrophe. We now look at each of these factors in more detail.

Intention: "Round Up the Usual Suspects"

Many wildlife managers, some of whom entered the profession because they liked hunting and fishing, see managing game species as their primary responsibility. Some wildlife managers may not be interested in saving endangered species because they do not see it as part of their role. In addition, endangered species issues almost always involve legal and interorganizational hassles, whether from other government departments, nongovernmental organizations, or private interests. Thus survival of a threatened or endangered species may be seen by many managers as more of a problem than a desired outcome. And saving an endangered species is always difficult. For all these reasons, a manager's mind perhaps naturally focuses on more routine or tractable problems. When endangered species are seen as a problem for the wildlife manager, the net (and occasionally even the favored) result may well be extermination through "benign neglect." Such neglect may not be conscious and could never be publicly admitted, but it may be real enough—and the consequences could be fatal for the species.

There is a famous scene in the film *Casablanca* when the police inspector, having just seen the character Rick, played by Humphrey Bogart, shoot a man, orders his men to "round up the usual suspects." Thus the inspector, well aware of who committed the crime, knows that his orders cannot possibly result in arresting the perpetrator. His intention, then, is not to solve the crime but to maintain the appearance of wanting to apprehend the criminal. This kind of cynical action—giving orders knowing that they cannot possibly result in the outcome apparently sought—may be found in officials in wildlife bureaucracies. The real goal of these managers may be to aid constituencies whose activities can harm endangered wildlife. But the managers have to look as if they are complying with the Endangered Species Act, because it is the law. In fact, managers are often in the position of abiding by conflicting mandates as they carry out their duties. Endangered species, however, may perish in the process.

Some contributors to this book have described officials who displayed neutral or even antagonistic sentiments toward endangered species. David Mattson and John Craighead (Chapter 5) discuss at length the various ways that species endangerment tends to sink lower on the priority list in land management bureaucracies, which on the whole seem to be more interested in their timber, mining, oil, and tourist constituencies. It is not too surprising, then, that allocation of funds, reward systems, and recruitment all seem to work against preserving the species in favor of these other interests. Similarly Jerome Jackson's description of the U.S. Army's attitude toward the red-cockaded woodpecker in Chapter 7 could only be described as neutral at best and at worst downright hostile. Army commanders complied with the letter of the law in ways that destroyed woodpecker habitat. Even more serious actions are cited by Steven Yaffee (Chapter 3) in the Northwest, where efforts to protect old-growth forests through protecting the northern spotted owl have led to death threats against "Woodsy the Owl" paraders and attempts to have the environmental film *The Lorax* removed from school libraries. Many of these sentiments are shared inside as well as outside of land management bureaucracies.

Consider, too, that in many cases endangered species teams are formed through interagency networking. Such teams may include members with very different motivations. On the one hand, there may be highly cooperative people; on the other, turf protection or even revenge may be the dominant motives for some members. My colleague Gordon Moss, a student of cooperation, thinks that a common vision is the most important feature of true cooperation. The extent to which a group develops a common vision is shaped by past experience, personal ties, and, above all, leadership. A recovery effort without a common vision is bound to fail. Yet common vision is difficult to achieve.

The common vision may not be focused on species preservation, however. I found particularly interesting Steven Yaffee's observation that the leadership elite of the Forest Service is dominated by World War II veterans with a "can do" spirit. This timber-extraction-oriented culture will not be easy to change. Nor will its short-term rationality. But rather than seeing the Forest Service as the villain, it may be more valuable to view it as being efficient—at the wrong task. Again, the problem is intention. To what ultimate ends are the system's energies devoted?

Incompetence: "No Cause for Alarm"

Intention is not sufficient to ensure good performance. Wildlife managers may fail because they are not properly equipped intellectually for effective endangered species management. What previous training in endangered species conservation has the manager had? What habits of thought and action? The wildlife manager may bring to the task of species management a set of methods that work very well on a large population of a game species. When the deer population gets too low, for instance, reduce hunting and enhance habitat quality. When it gets too high, issue more hunting licenses.

But paradigms that work well with large populations may fail with smaller, endangered ones. Endangered species problems may involve unexpected responses on the part of the animal population to environmental changes. "Business as usual" may work with a large, stable population. It may fail utterly with a smaller, endangered one. When expected population responses fail to materialize, the manager may not know what to do next. In this case, incompetence results from a failure to match the manager's skills to the conservation task. The mismatch will be even greater when managing endangered species is only one task among many handled by the same person. The incompatibility of the mental habits needed for routine population management and those demanded by the endangered species task will become more pronounced.

One example of the "business as usual" approach might be seen in the problems of the Wyoming ferret population (Chapter 4). At a time when field teams were reporting that the black-footed ferret (*Mustela nigripes*) population was rapidly declining, the supervisor of the Wyoming Game and Fish Department's Biological Services was reported as having said: "It was like the Chicken Little Syndrome. . . . Everyone started to panic. But there is no cause for alarm. No one talked to [the U.S.] Fish and Wildlife Service or the [Wyoming] Game and Fish. We still cannot see these ferrets as being in serious danger" (Ontiverroz 1985:A1). And indeed Wyoming Game and Fish did not believe ferrets at that time were endangered. Yet they most certainly were. How could a bureaucracy possibly ignore data important to one of its key tasks? The answer is probably that its habits of thought were inappropriate for the task.

Part of the reason for bureaucratic incompetence may be that those who seek work in wildlife agencies are ill fitted to the task of endangered species recovery. We can speculate that some managers who are inclined toward readily visible indicators of success will be frustrated by the lack of such signs. In fact, those inclined toward theory or intuition may be better at endangered species management, since much species "behavior" is hard to measure precisely and filling in data "gaps" may be necessary. Decision in the face of uncertainty is characteristic of endangered species management (Maguire 1986). Wildlife managers who are bad at handling ambiguous data will not be happy with endangered species work. Endangered species, rare and difficult to find, will not be easily captured in the mental nets of such people.

Then there is politics. All wildlife management involves politics, of course, but endangered species survival has special problems. It may require recovery teams to forge and manage a fragile coalition of contending or even fractious groups. It is no exaggeration to say that the health of the protected species may be linked to the health of the coalition. Maintaining such coalitions may require networking skills more typical of agricultural extension agents or entrepreneurs. How well are wildlife managers trained in these skills? Not very well! They are not given enough training. Certainly they can learn these skills; but if they do not, the species is likely to suffer. Competence in endangered species protection includes interpersonal skills as well as biological knowledge.

It is obvious that wildlife personnel may face quite different problems, depending on the organizations where they work. It may be helpful to think of the wildlife manager's work environment as falling into three broad categories—the pathological, the bureaucratic, and the generative—according to the way it handles information. In Table 14-1, I suggest some dominant features of these different organizational environments. There are certainly other features to consider. One implication of Table 14-1 is that successful organizations have a

TABLE 14-1

How Organizations Treat Information

Pathological Organization	Bureaucratic Organization	Generative Organization
Don't want to know	May not find out	Actively seek information
Messengers are shot	Messengers are listened to—if they arrive	Messengers are trained
Responsibility is shirked	Responsibility is compartmentalized	Responsibility is shared
Bridging is discouraged	Bridging is allowed but neglected	Bridging is rewarded
Failure is punished or covered up	Organization is just and merciful	Inquiry and redirection
New ideas are crushed	New ideas present problems	New ideas are welcomed

free flow of information. The key to successful information flow in organizations is this criterion:

> The organization is able to make use of information, observations, or ideas wherever they exist within the system, without regard for the location or status of the person or group having such information, observations, or ideas.

While other characteristics may also be indicative of organizational culture, I have found information flow to be a key measure of performance. Organizations that are generative are very good at making use of the information they have, but such organizations are rare. Most wildlife organizations probably have a bureaucratic culture. A bureaucratic culture may work reasonably well for routine species management, but it is definitely inappropriate for the endangered species recovery task (Clark 1985). And wildlife managers who find themselves in pathological organizations may have to spend more time worrying about their own survival than that of the species they wish to protect.

The first task of a wildlife manager whose job concerns threatened or endangered species is to create a generative environment. To track a threatened species requires effective use of all the available information. The manager must ensure that information held by some people in the network can be shared so it can be used by the team as a whole. Thus "weak signals" from the species are likely to be noted and acted upon, rather than being ignored or glossed over. An endangered species needs a recovery team whose members feel free to think, inquire, and speak their minds. Getting a team to perform in this way is no small accomplishment.

Ignorance: Intellectual Blunders

"To catch a rabbit, first think like a rabbit."
OLD PROVERB

Managers of endangered species may fail because they do not know the right answer. They may be willing, smart, and able, but the variables that make a difference may still elude them. This was very much the historical situation with the California condor described by Noel Snyder in Chapter 8. The condor team was doing its best, but its efforts were still not working. Similarly, frog populations all over the world seem to be declining or disappearing. What is happening? In spite of intensive research, the answer remains elusive (Yoffe 1992).

The problem of ignorance is particularly serious because of the speed with which an endangered species' situation can change. Ordinarily delay is perfectly acceptable in finding answers to biological problems. Formulation of hypotheses can take place in a leisurely time frame. With an endangered species, however, this time frame is short. Analyses, tests, and interventions must all take place fast. And if the intervention does not work, there must be a contingency plan to go rapidly into effect. Chapter 8 indicates that recovery plans are often irrelevant because they can never be prepared fast enough to provide guidance. By the time the plan is finished, the situation has usually changed. Interim documents, prepared to communicate information rather than to promulgate policy, would seem to be far more useful. Yet legal and bureaucratic requirements may mandate focusing many critical hours on developing a formal recovery plan.

Despite the need for rapid thinking and problem solving, a perfectly good team may find itself hamstrung by its own preconceptions. Consider what happened with the California condor recovery effort. A very useful Ph.D. thesis on the condor was done by Carl Koford (1953). This thesis became the intellectual model for subsequent decades, and its findings shaped thinking about how condors might be saved. For decades it was assumed that the key to preserving condors was maintaining habitat. Condors were also supposed to be extraordinarily sensitive to human presence. Both these assumptions were later proved wrong. Noel Snyder describes the "intellectual stasis" caused by these erroneous assumptions. It was not even Koford's own findings, but rather the interpretation given them by his professor, Miller, that appears to have caused the difficulties. Both concepts and data deriving from Koford's field notes were heavily processed by Koford and Miller working in concert. The resulting doctrine involved several assumptions unsupported by the original data. Miller's interpretations dominated discussion about the condor for decades. The ideas fit together in an amalgam which insisted that habitat preservation was the key, prevented the use of telemetry, and prevented the kind of studies that eventually pointed to lead poisoning (condors eating carrion that had been shot) as the key mortality factor. These ideas in turn were promoted by interest groups that found the implications supportive of their own goals. Surprisingly, the basic ideas were never given a thorough examination but were treated "like a religion." This is a fine example of the dangers of groupthink (see Janis 1982).

The team's thinking must proceed rapidly so that it can respond quickly to changes in the endangered species' situation. And it must consider a broad range of possibilities to avoid fixing prematurely on a conclusion. Both Francis Bacon and Sherlock Holmes warn us against premature conclusions. But human beings like conclusions, so this is a tall order. Maintaining intellectual flexibility requires sharp analytical skills and presupposes an adequate fund of knowledge. Yet recovery teams may have many more implementers than they have analysts. Reflection on practice is valuable but rare (Schön 1983). Moreover, teams need to become aware of their assumptions. The condor case shows what can happen when a predecessor's ideas are accepted without evaluating their validity. "Off-the-shelf" standard techniques may not work on all species. The team may have to improvise a custom-tailored solution. The key is this: the approach must reflect the dynamics of the problems facing the species.

This level of mental flexibility is unusual among bureaucrats. Many officials will be oriented toward the application of known principles, not the invention of new ones. Yet in many cases the successful recovery approach represents an invention. So what do we know about creativity in organizations? We might turn for inspiration to what we know about technological inventors. Based on my own interviews with Jacob Rabinow, prolific inventor, and James Hillier, former head of RCA Laboratories and a key inventor of the electron microscope, I speculate that successful inventors need extensive knowledge of devices and techniques. This might suggest that the successful wildlife biologist would be an experienced fieldworker with a vast mental fund of examples. The key to solving problems may well lie with this fund of examples. Problems in turn may be worked on through sharing these examples by "telling stories." I will have more to say about this later.

Ill Fortune

Things often do not go right for recovery programs. Weather, predators, and hostile humans may not cooperate with otherwise well-scripted plans. The Wyoming black-footed ferret recovery effort in the 1980s suddenly found a rapidly dwindling population due to canine distemper (Clark and Westrum 1987; Chapter 4). The response of government organizations was slow and nearly fatal to the ferret population. Yet contagious disease is only one item on a long list of things that can go wrong. It may happen, for instance, that two endangered species have a predator/prey relationship—and increased success of the predator species may inhibit recovery of the prey species.

Two points need to be made about ill luck. The first is that the team may broaden its perspectives to bring under control (or at least influence) aspects of the environment that might be seen as beyond the brief of the recovery team. When a species is introduced that may represent a threat to livestock or human life—wolves or grizzly bears, for instance—the team may consider the surrounding community's attitudes to be part of the system that needs influencing. Or maybe not. At some point, the team's ability to extend its influence will reach its limits.

The second point is that ill luck can be bettered or worsened by what the team does. Whatever the origin of the problem, whether caused by the team's mistake or the environment's vagaries, a well-led team will react more rapidly and appropriately to the unfortunate event than one less ably directed.

Keys to Success

The key features of a successful recovery are simple in principle but difficult to achieve in practice. These features include: solid biological understanding of the dynamics of the species and its habitat; a competent team for implementation; and control of the political process shaping the species recovery efforts. Each of these items deserves detailed discussion.

Understanding

It is obvious that understanding is a key factor in successful recovery, but exactly *what* understanding is necessary is not so clear. What is really needed?

To begin with, the relevant knowledge is not necessarily easy to express. It may be closer to craft knowledge, based on skill, rather than theoretical knowledge, based on deep understanding (Rasmussen 1986). Much of it may be intuitive or metaphorical and therefore hard to verbalize. This may also mean that the articulateness of the reports may have little to do with the coherence of the recovery effort. And it is concrete knowledge—knowledge of what to do next rather than abstract propositions. In the immediate present, endangered species need practical tinkerers, not abstract thinkers. The key thing is to know what to do about *this* species in *this* environment, not species generally. Nonetheless, recovery team leaders need to be thinkers, not just doers. They must be able to interpret the outcomes of their tinkering.

Many people have suggested that what Morris Stein (1983) has called "physionomic metaphors" are important in understanding complex things and processes. A physionomic metaphor provides a bridge between the person trying to understand a situation and the situation being understood. The metaphor allows you to "feel" yourself into the place of the thing analyzed. Several inventors and technologists have cited this as the key to understanding the behavior of devices. William B. McLean, for instance, inventor of the Sidewinder missile, used this style of thinking. It was also mentioned to me by Jacob Rabinow, Howard Wilcox, Paul MacCready, and Raymond Damadian, all highly creative inventors. Damadian insisted that it was how all creative scientists thought. I imagine that intuition about biological systems operates in much the same way.

One consequence of the intuitive nature of recovery-oriented knowledge is that it may be less valued than declarative knowledge. It may be seen as less scientifically useful than knowledge leading to biological theory.[1] Therefore it may remain in the professional's head rather than being shared through publication. Because of its "applied" character, it may even be undervalued by the person who has it. In contrast to textbooks and data banks of formal biological knowledge, there are apparently few data banks of information relevant to species recovery. Probably much of it is contained in the form of stories told by successful (and unsuccessful) practitioners.

Bureaucratic environments may not support the use of such intuitive approaches. Instead formal justifications, based on previous (calculative) experience, may be required for the team to be given clearance to take the necessary steps. While such justifications may help the bureaucracy's accountability problem, they may be a serious impediment blocking the team's efforts to take action. Bureaucratic needs may well conflict with the need to act immediately. And in some cases, it may be easier to take a permissible, but ineffective, action than to make the case for action the team thinks will really work.

Another key skill is the ability to reflect on experience in order to analyze failures or upgrade techniques. Wildlife officials need to use the skills of the "reflective practitioner," as Donald Schön (1983) has termed them. Schön argues that in much professional work, intervention is guided by an act–observe–reflect–act cycle rather than a formula. The value of this thoughtful action (or active thought) is that it permits constant checking of assumptions and theory. Reflection in action permits constant upgrading and improvement in tactics. These skills permit the identification of ineffective courses of action and intellectual blind spots. They permit paradigm shifting when it is necessary. Effective teams constantly debrief to determine what went right, what went wrong, and why. Debriefing then becomes an opportunity for reflecting on practice. Teams that do not engage in reflective practice may continue to use the same ineffective techniques, even though they do not work.

Getting a Competent Team

A competent team must be biologically and ecologically knowledgeable, but knowledge itself is not enough. The team must have the practical skills to transform ideas into action. Since every recovery situation is likely to be unique, the mix of skills may vary. Even so, a team should be stocked with people possessing complementary skills, and this means there must be interdisciplinary cooperation. Team members must understand each other's disciplines and respect each other's skills.

Unhappily, teams are often interjurisdictional efforts where bureaucratic domains intersect or overlap. Team members may be political appointees with little interest or expertise in endangered species. These people are likely to prove a barrier to consensus and a frequent source of bureaucratic interference. Rather than seeing themselves as members of a problem-solving group, they may see themselves as the agencies' representatives (Chapter 4). In this situation they may feel they must protect the agency's turf and block any efforts that might diminish it. If such people are minor members of the team, they can cause problems. If they are in charge of the recovery effort, they can stop the team dead in its tracks. Since the average recovery team is likely to include at least one such person, the wise leader must be competent in neutralizing or replacing such members. In some cases, bypassing the person may be possible. More often it may be necessary to accommodate an agenda that includes not only actions in the interest of the species but also actions related to the interests of the agencies represented on the team. Competence in politics is a natural part of preserving fragile alliances.

Interdisciplinary tolerance is difficult to achieve. Getting team members to respect each other is a good first step. The second step is finding out who is competent at what. Being able to assess and blend skills from diverse disciplines is a key skill for team leaders. But team members must be able to make such assessments, too. And they need to be able to handle those with fragile egos as well as those who have good ideas but few social skills. If there is confidence in each other's skills, there will be a willingness to use the team's full intellectual power.

More generally, an effective team is ensured by securing members' commitment, known in the business world as "buy-in." Getting buy-in is essential. And if the members of the team are emotionally committed to saving a species, this will make life easier. Often, however, this kind of commitment—not only to the species, but also to the team's approach—must be reinforced by small symbolic rewards. Voluntary associations are often very good at getting a lot of effort out of people by careful attention to such rewards, and the network builder needs to think about them too. At one Fortune 500 company, for instance, a product developer needed extraordinary help from his suppliers. The product was known in the company as the "Anvil" project, so the developer had a small number of tiny brass anvils made. Whenever a supplier came through with a difficult request, they got one of the anvils. Soon possession of one of these anvils was eagerly sought, and great moving of heaven and earth would take place to secure one. The supply was carefully guarded to prevent excessive giving. The project was so successful that when the next one came along, the CEO of the company suggested that it, too, be "anvilized." This may sound corny, but it worked. There must be many ways that wildlife projects can be "anvilized."

Leadership is essential for an effective team, but the best kind of leadership is coordinate. Coordinate leadership arises when the role of leader temporarily shifts to the person who at that moment knows the answer (Westrum 1992). I have suggested this concept in conjunction with problem-solving in technological areas, such as man/machine systems like air traffic control or production complexes. A vivid example is the problem of landing an aircraft on an aircraft carrier—one of the trickiest technological feats imaginable, yet one that is carried out safely on a routine basis. The key to doing it seems to be that any person in the structure can take charge if they feel they know the right thing to do at that moment (Rochlin et al. 1987). This principle can be applied more generally, however, to situations that require creative problem-solving ability. The person with the correct answer ought to be leading the group. No person is always going to have the right answer, so the key is recognizing which member of the team has the best sense of how to proceed at any given moment. On the old television program *Star Trek*, leadership would shift back and forth between the officers as the situation required different kinds of knowledge. In a well-managed team, leadership can shift back and forth between members of the team as the problem changes.

One important facet of leadership is recognizing the strengths and weaknesses of team members and then assigning jobs to those with the strongest skills for the task. Leaders who know their people well will get more out of them and stir less resentment than those who underestimate or overestimate what their group members can do. A leader must be a coach and encourage high performance. But this demands that the leader know what performance is possible. Good leaders, like orchestra conductors, test their people to see what they can do. Once they have done the testing, they know where they can push and where they must back off.

We have noted that some recovery knowledge is tacit and hard to express. As a consequence, the person who has this intuitive knowledge may not be the team leader or the most forceful person on the team. Since the knowledge is hard to verbalize anyway, what is the chance that the team will use it? How can those on the team match the roles and relevant knowledge? Only if the team's structure is flexible can the necessary role shifting take place.

Here I would like to borrow an analogy from aviation. In what is called "cockpit resource management," pilots are trained to utilize the best information the team possesses, no matter where the information comes from (Helmreich et al. 1986). The idea is to manage the intellectual resources of the flight deck so that good ideas are not suppressed, even if they occur to the "wrong" person. In the past, for instance, accidents have been traced to pilots not paying sufficient attention to the observations and opinions of their copilots (Foushee 1984). This idea of cockpit resource management (CRM) is now starting to be generalized to other fields, such as nuclear power and anesthesia (Howard et al. 1992). What promise might CRM hold for wildlife management? CRM skills are basically interpersonal process skills relevant to the management of information and other intellectual resources. Biologists, even Ph.D. biologists, may not possess these skills. Scientists seldom get training in crisis management as part of their graduate work. Yet these same skills are particularly relevant to team situations that are complicated, urgent, and ambiguous.

I find it interesting to note that many problems on technological teams are solved through telling stories (Brown and Daguid 1991; Zuboff 1988; Schank 1990). Perhaps this is due to the "lateral thinking" involved in storytelling (De Bono 1970). Evidently telling stories calls up relevant scenarios and helps the participants "frame" the situation, as well as providing access to analogies. This may also be the way in which oral corporate memory is accessed. So we might expect that telling stories would be a useful method for solving problems on wildlife teams as well. Telling stories, however, is not a scientifically respectable way to solve problems. Many leaders might not recognize the cognitive value of storytelling and instead encourage linear cause-and-effect thinking. Teams that are competent, however, will recognize the value of stories and encourage an atmosphere in which participants feel comfortable telling them.

The psychology of the competent team is one of openness (Westrum 1988). If a mistake is made, the person who made it ought to be able to discuss it without feeling shame or vulnerability. Recovery work is likely to provide severe disappointments. Often such events involve errors of judgment. If these are the occasion for intense guilt or faultfinding, the team's internal process is likely to become defensive. In such a situation, honest analysis will be blocked out of fear of giving offense or being attacked.

It is the leader's responsibility to maintain the team's internal processes. The leader's main function is not to make correct decisions but rather to create the kind of team and the kind of intellectual environment in which good decisions will be made. This means encouraging wide foraging for key information, avoiding premature closure and groupthink, ensuring that conflicts are resolved, and setting up an effective system of rewards. Technical abilities in a team leader are important, but so too are process skills.

Political Control of Recovery Efforts

One of the preconditions for success in the species recovery process is being able to control recovery efforts. For a variety of reasons, however, this is likely to prove problematic. Sometimes the species may be threatened by a housing development with very powerful sponsors. Wildlife officials may have little direct power, in spite of the law, to stop this development, since the sponsors may have access to the highest political levels (Schneider 1992). Sometimes, in foreign countries, local officials may demand bribes without any real accountability for performance (Begley 1993). Sometimes the protected species may have a high market value, such as the black rhinoceros (*Diceros bicornis*) or the African elephant (*Loxodonta africana*), leading to extensive poaching. Sometimes the species may represent a threat to agriculture or leisure activities, creating a set of human adversaries and possibly nemeses, as with the grizzly bear (*Ursus arctos horribilis*) or the gray wolf (*Canis lupus*). Sometimes the protected species may prey on popular game species, leading to political pressure from hunters.

Protecting the species, therefore, often requires building a coalition among the affected parties, public or private. In Chapter 3, Steven Yaffee demonstrates the enormous complexity that endangered species politics can create. The spotted owl situation involves numerous government agencies, environmental groups, economic interests, and even affected communities. To make any progress in a situation so complicated demands the ability to build effective coalitions among groups who might presume themselves to be adversaries. Building political consensus is not often a skill taught to biologists. Since it is essential, however, for species survival, the system that manages endangered species programs must find a way to teach these important skills.

Find Your Maestros

Now I would like to draw upon my own field of expertise, the social management of high technology. Much of the literature on technology development shows the importance of people with virtuoso talents for running big projects. Arthur Squires has called these people "maestros" (Bowser 1987). A technological maestro combines the following traits:

- A high level of technological virtuosity

- High energy level and span of attention

- High standards

- Ability to ask key questions

- Oriented to hands-on management

If technological projects are run by maestros, the key tasks are accomplished and the work of separate teams is integrated successfully. Without maestros, projects often founder or have key flaws. Tom Peters (1982), a celebrated management consultant, reports on a study carried out by Texas Instruments. Of seventy new product introductions, those that succeeded were generally those that had a "volunteer champion" to lead them. Those without a product champion typically failed. What might wildlife management have to do with maestros? Perhaps endangered species, too, need a "volunteer champion."

But maestro behavior patterns may not fit well into wildlife and land management bureaucracies. After all, maestros tend to be individualists (McLean 1960), and bureaucracies may discourage such people from applying for jobs or put them in positions where they can do little good. Indeed, they may even be weeded out of wildlife agencies through early attrition. Consider, for instance, these comments about NASA culture in the post-*Apollo* era. The speaker, a NASA engineer named Ivy Hooks, discusses why gifted designers and managers were not followed by maestros of comparable quality:

> Today we have no more Fagets, Krafts, and Johnsons. . . . The early NASA people were entrepreneurs in the best sense of the word, by which I mean that they were individualists, creative people, risktakers. . . . In this business you have to take careful calculated risks—something NASA managers are afraid to do nowadays. People like Faget don't find successors because they believe that the cream will rise to the top—just the way they themselves did. They were sure it would happen again. But they were wrong. In the twenty-five years after NASA was founded, it became a bureaucracy, and in a bureaucracy quality doesn't rise. [Cooper 1991:67; see also McCurdy 1993]

Could this also be true of wildlife agencies and land management bureaucracies? Are maestros not bred in the typical agency? Yet maestro traits might be very valuable, as would entrepreneurial skills generally, in protecting or recovering endangered species. Entrepreneurial skills in developing and maintaining the necessary political coalitions would be valuable. Maestros often provide the glue that holds various strands of the project together. Moreover, they focus on maintaining key variables within acceptable limits. In short, the maestro's energy and attention span might be vital to the project's success.

There is another consideration. Maestros breed maestro culture: an emphasis on high standards, high performance, and respect for technical expertise. The presence of maestros attracts other people of talent. They signify an expectation of performance beyond the norm. They turn work into a calling, rather than just a job. They provide incentives to work harder and go the extra mile. This may be important not only for team members but also for outsiders whose cooperation may be essential to success. And remember that maestros can groom future maestros.

"Top Branch": Training the High-Performance Team

Considering the state of the art for recovery teams, two features appear striking: the lack of consensus on what a recovery team ought to be and the need to develop an institutional memory about what has worked and what has not. It is clear that lessons are being learned, but they are not necessarily being coded in a cogent form or transferred from one team to another. So the system as a whole is not learning. The analogy that immediately comes to mind is what happened in the U.S. Navy in regard to air combat just before the Vietnam War.

In the years prior to Vietnam the navy (and the air force as well) had convinced itself that fighter combat in the traditional manner, especially dogfights, was a thing of the past. It was believed that the push-button era had arrived and that air-to-air guided missiles would make training for traditional air combat obsolete. The navy even closed down the Flight Air Gunnery Unit in 1960 (Wilcox 1990). As a result, U.S. fighter pilots were not trained for the kind of combat they experienced in the skies over North Vietnam. Failures in the air eventually led to an investigation overseen by Captain Frank Ault (1989), whose report cited some 242 problems that needed fixing. One result of the report was the creation of the Naval Fighter Weapons School ("Top Gun") at Miramar, California. Top Gun provided a corporate memory and the high-performance training in dogfighting tactics previously lacking. Those who were instructors in the school were maestros; they constantly honed and tested their skills against talented pupils. Top Gun produced a crop of pilots trained in the best techniques available. Air combat in the second phase of Vietnam reflected this improvement in tactics. Victories went up by a factor of six.

Of particular interest to wildlife managers is Captain Ault's observation that his investigation did not so much turn up new solutions as allow the system to use those it already had:

> Interestingly, as the study progressed it became apparent that our contemporaries, either in presenting a problem to us or having been informed of the team's perception of one, generally had some excellent ideas on what should be done. Accordingly, as we approached the end of the effort, there was no need for a session to list the problems and devise the solutions. Almost without exception, *we had the solution (or*

group of solutions) already on hand. Altogether, we visited more than 75 sites; military and industrial, ashore and afloat. It may well have been as near tó an "all hands evolution" as ever achieved in Naval Aviation. As a result, we were virtually assured of a "user friendly" end product. [Ault 1989:37; italics added]

Top Gun allowed maestro culture to flourish in air-to-air combat. What might wildlife management learn from this example? The same concentration of expertise and training might be enormously helpful to the leaders and members of recovery teams. A school might be one way of implementing this program. A once-a-year convocation might be another way that information exchange could take place. A major university could house an institute. Teams could compare notes, exchange information on tactics, perhaps even offer others the chance to assist on difficult problems. Many stories might be shared, and valuable insights along with them.

Lessons

Protecting and recovering endangered species requires management skills and people skills. Traditional knowledge is simply not sufficient for the manager of a recovery team. Endangered species involve special demands and pose special problems. The wildlife manager, in addition to basic scientific knowledge, must also acquire state-of-the-art human relations skills. The wildlife manager must be adept at reflective practice and network building. The manager must be comfortable with ambiguity and competent in formulating problems and then solving them. He or she must be able to secure buy-in, not only of the organization's members, but also of other members of wildlife recovery teams. The manager must recognize the dangers of groupthink and establish procedures to prevent it. Yet neither the traditional training nor the recruitment pattern of wildlife managers prepares them to undertake such management and facilitation. These skills, then, must be acquired on the job.

So much for the skills of the manager. But the system as a whole needs consideration as well. Does it breed, and recognize, its own maestros? Does it determine which teams are the high performers and encourage others to adopt their methods, attitudes, and skills? Would a "Top Gun" institute be useful in teaching such skills and in sharing knowledge generally? Is the wildlife system as a whole a self-reflective, self-organizing, and self-improving system? And if it is not, what process needs to be set in motion to make it that way?

Note

1. Ken Alvarez reminds me that in spite of the value of intuition, conservation biology has now provided quantitative criteria for many of the decisions that the wildlife manager must make. This is one instance in which intuition is often inadequate. The author wishes to thank Ken, who works for the Florida Park Service, for a critical reading and assessment of this chapter.

References

General Sources

Ault, F. 1989. The Ault Report revisited. *The Hook* (Spring):36–39.

Begley, S. 1993. Killed by kindness. *Newsweek*, 12 April, pp. 50–54.

Bowser, H. 1987. Maestros of technology: An interview with Arthur Squires. *American Heritage of Technology and Invention* 3(1):24–30.

Brown, J. S., and P. Duguid. 1991. Organizational learning and communities of practice: Toward a unified view of working, learning, and innovation. *Organization Science* 2:40–57.

Clark, T. W. 1985. Organizing for endangered species recovery. Paper presented at Wildlife Management Directions in the Northwest Through 1990, 2–5 August 1985, NW Section, Wildlife Society, Missoula, Montana.

Clark, T. W., and J. R. Cragun. In press. Organizational and managerial guidelines for endangered species restoration programs. In *Restoring Endangered Species*, M. L. Bowles and C. Whelan (eds.). New York: Cambridge University Press.

Clark, T. W., and R. Westrum. 1987. Paradigms and ferrets. *Social Studies of Science* 17(1):3–34.

———. 1989. High performance teams in wildlife conservation: A species reintroduction and recovery example. *Environmental Management* 13(6): 663–670.

Clark, T. W., R. Crete, and J. Cada. 1989. Designing and managing successful endangered species programs. *Environmental Management* 13(2):159–170.

Cooper, H.S.F. 1991. Annals of space: We don't have to prove ourselves. *New Yorker*, 2 September, p, 41.

De Bono, E. 1970. *Lateral Thinking: Creativity Step by Step*. New York: Harper.

Foushee, H. C. 1984. Dyads and triads at 25,000 feet: Factors affecting group process and aircrew performance. *American Psychologist* 39:885–893.

Helmreich, R. L., H. C. Foushee, R. Benson, and W. Russini. 1986. Cockpit resource management: Exploring the attitude-performance linkage. *Aviation, Space, and Environmental Medicine* 75:1198–1200.

Howard, S. K., D. M. Gaba, K. J. Fish, G. Yang, and F. H. Sarnquist. 1992. Anesthesia crisis resource management training: Teaching anesthesiologists to handle critical incidents. *Aviation, Space, and Environmental Medicine* 63(9):763–770.

Janis, I. 1982. *Groupthink.* Boston: Little, Brown.

Koford, C. B. 1953. The California condor. *National Audubon Society Research Report* 4:1–154.

Maguire, L. A. 1986. Using decision analysis to manage endangered species. *Journal of Environmental Management* 22:245–260.

McLean, W. B. 1960. Management and the creative scientist. *California Management Review* 3(1):9–11.

McCurdy, H. E. 1993. *Inside NASA: High Technology and Organizational Change in the U.S. Space Program.* Baltimore: Johns Hopkins University Press.

Ontiverroz, M. 1985. State says ferret scare unnecessary. *Cody Enterprise,* 14 August, p. 1.

Peters, T. 1982. The rational model has led us astray. *Planning Review* (March):16–22.

Rasmussen, J. 1986. *Information Processing and Human-Machine Interaction: An Approach to Cognitive Engineering.* New York: Elsevier Scientific.

Rochlin, G. I., T. R. Laporte, and K. K. Roberts. 1987. The self-designing high-reliability organization: Aircraft carrier flight operations at sea. *Naval War College Review* (Autumn):76.

Schank, R. C. 1990. *Tell Me A Story: A New Look at Real and Artificial Memory.* New York: Scribner's.

Schneider, K. 1992. U.S. mine inspectors charge interference by agency director. *New York Times,* 22 November, p. 1.

Stein, M. I. 1983. The creative process and the synthesis and dissemination of knowledge. In *Knowledge Structure and Use: Implications for Synthesis and Interpretation,* S. A. Ward and L. J. Reed (eds.). Philadelphia: Temple University Press.

Steiner, G. (ed.). 1965. *The Creative Organization.* Chicago: University of Chicago Press.

Westrum, R. 1988. Organizational and interorganizational thought. Paper delivered at a symposium on Safety Control and Risk Management at the World Bank, Washington, D.C.

———. 1992. *Technologies and Society: The Shaping of People and Things.* Belmont, Calif.: Wadsworth.

Wilcox, R. K. 1990. *Scream of Eagles: The Creation of Top Gun and the U.S. Air Victory in Vietnam.* New York: Wiley.

Yoffe, E. 1992. The silence of the frogs. *New York Times Magazine,* 13 December, p. 40.

Zuboff, S. 1988. *In the Age of the Smart Machine.* New York: Basic Books.

A Short Reading List for the Wildlife Manager

Hackman, R. (ed.). 1989. *Groups That Work and Those That Don't.* San Francisco: Jossey-Bass.

Katzenbach, J. R., and D. K. Smith. 1993. *The Wisdom of Teams: Creating the High Performance Organization.* Boston: Harvard Business School Press.

Schein, E. 1992. *Organizational Culture and Leadership.* 2nd ed. San Francisco: Jossey-Bass.

Schön, D. 1983. *The Reflective Practitioner: How Professionals Think in Action.* New York: Basic Books.

CHAPTER 15

A Professional Perspective
Improving Problem Solving,
Communication, and Effectiveness

Tim W. Clark and Richard P. Reading

Professionals, experts of various kinds, are at the core of endangered species restoration. Experts are usually professionals who have acquired specialized skills and knowledge in a particular subject through formal training, experience, or a combination of the two. Biological experts involved in species conservation include field ecologists, population biologists, geneticists, and captive breeding specialists, to mention just a few. Other experts may be involved in endangered species efforts, as well, such as sociologists, economists, and policy analysts. The work is often demanding. Professionals must demonstrate the very best judgment and performance as they grapple with the challenges of complex, uncertain situations that leave little room for error. Indeed, a mistake can spell extinction for a species.

Professionals, therefore, are constantly seeking new ways to improve their performance. There is always room for new tools in the professional tool kit. Both the tradition and the spirit of professionalism demand this open-ended search. New journal articles, workshops, and day-to-day word of mouth offer new perspectives for professionals and their practice. New ways of thinking about professionals and their work are always active topics of discussion. Finding better ways for professionals to solve problems is of paramount importance in general, and this is especially true in the work of endangered species restoration.

New tools for improvement may come from disciplines not commonly considered part of mainstream conservation. Employing new techniques or insights borrowed from psychology and sociology, for example, can result in improvements by helping professionals understand their role and how they can be most effective. Changing attitudes in order to support species conservation can be very helpful as well. Professional tools that foster effective communication can help. And tools that help professionals transfer their newly learned knowledge to others so that the entire organization can benefit are clearly of value. These and many other tools promise significantly improved professionalism and species recovery.

This chapter reviews several tools for enhanced professionalism. Some of these tools come from psychology, sociology, and organization. All of them are interrelated and mutually supporting. We examine the nature and role of the expert, how experts can improve communication and information sharing, how values, attitudes, and behavior of experts and the public are shaped, and how both experts and their organizations can learn from experience. The discussion proceeds from an examination of the expert's role to a look at alternatives and finally a description of ways to improve learning and effectiveness. We examine these topics here because they are seldom discussed in universities or on the job. Enhancing the professional practices of experts who are responsible for endangered species recovery tasks can be beneficial in a host of ways. It is reasonable to expect, for example, that implementation of the Endangered Species Act (ESA) can be substantially upgraded by improving professional performance.

Experts as Both Solution and Problem

Experts are usually given the task of solving conservation problems and dealing with crises, such as the impending extinction of a species. In this section we look at the role of experts and the benefits of specialized knowledge and problem solving, as well as some of the drawbacks of narrow expertise. In subsequent sections we will look at experts and various models to improve their performance in greater detail.

The cognitive characteristics of experts determine how well they can solve problems, share concepts and communicate knowledge with others, and learn new, improved ways of thinking and problem solving. Ultimately, restoration efforts come down to individuals and the cognitive skills and psychological resources they bring to the task. Success also depends on how well they work together to save endangered species. Strong cognitive skills can make the difference between a program that works and one that fails. The case studies in this book well illustrate the variable performance of experts with different cognitive characteristics. In some programs, the experts' performance contributed to species recovery; in other cases, their performance caused major problems.

To solve extinction problems experts draw on their formal education, their experience, the established models of problem solving they have learned, and traditional ways of relating to their colleagues, administrators, and the public. Sometimes this bank of cumulative expertise is enough to solve the endangered species problem. But at other times the traditional mode of specialized thinking and acting can be a hindrance. When a conservation problem changes rapidly, for example, traditional problem-solving approaches may be ineffective if specialized thought and action cause inflexibility, a lack of creativity, or an inability to communicate with other experts and key figures. And if many different kinds of experts are involved, the challenge of saving species can be especially difficult.

In fact, many different kinds of experts do become involved in endangered species conservation. Some are experts in fieldwork, modeling, captive breeding management, planning, administration, public relations, and coordination. Experts are often grouped by profession (such as conservation biologists) and spend considerable time dealing with others in their profession through collaborative work, oral reports, peer-reviewed articles, and conferences and seminars. To understand how experts perform in general—and in endangered species conservation in particular—it is necessary to examine the nature of expert knowledge and the role of the expert.

Experts usually concentrate on a limited field of learning such as population genetics, reproductive physiology, or habitat ecology. This expertise permits them to acquire rare skills and detailed knowledge, but often at a cost. Experts tend to have facility with just a few areas of knowledge because of their concentrated focus, and they may be unable to communicate their specialized knowledge to others (Kaplan 1990). This shortcoming presents difficulties for those who need to use the expert's knowledge. In endangered species work, decision makers, watchdog organizations, and the public are all interested in expert knowledge and opinion with respect to the species' status, its likely future, and possible conservation measures. But experts commonly have trouble collaborating and sharing this knowledge.

Many people outside of endangered species programs confuse experts with decision makers, though experts are seldom decision makers and vice versa. In fact, nearly all the significant conservation and management decisions are made by agency bureaucrats or politicians. These decision makers may or may not use the expert knowledge available to them. In practice, the knowledge possessed and sought by experts, decision makers, elected officials, and citizens is very different. Experts usually focus on highly detailed knowledge. Decision makers, typically from bureaucratic organizations, usually require somewhat more general knowledge. Elected officials often work with summary knowledge, too. Seldom do they need the expert's detailed information for their decisions. And the general public may only be interested in the most cursory knowledge about a problem and its possible solutions. Thus experts, decision makers, elected officials, and the public value knowledge in quite different ways. Several problems can arise because of these differences. For example, the expert's knowledge is typically portrayed in a form that prevents outsiders, including decision makers, elected officials, and the public, from understanding it. Moreover, the sort of knowledge that interests experts may not interest other people. Such differences can create difficulties when these diverse experts and the outsiders must work together closely.

The way in which experts go about attacking problems and communicating is of special interest for several reasons. First of all, we depend on experts to be superior problem solvers. Second, we count on experts to be flexible and adaptable to new problems and solutions. And third, we rely on experts to police themselves and their professions, constantly upgrading their concepts and methods, following the highest ethical standards. Experts occupy a special role in society. And their standing in society is dependent, to a large degree, on their ability to communicate with other experts, different kinds of experts, and nonexperts. The ability to share information, then, is an essential attribute of effective professional practice.

Communication Barriers

Several barriers make the communication of expert information problematic. In this section, we look at communication and information sharing as an essential tool for successful practice and describe some problems of communication, as well as solutions. These insights into expert communication can significantly upgrade performance. Our discussion relies heavily on the work of Kaplan and Kaplan (1982) and Kaplan (1990).

Conservation experts routinely address highly complex problems that require expertise beyond the ability of any one individual. Because of this demand, as well as legal mandates and great public interest, endangered species recovery efforts often include participants from several state and federal agencies, nongovernmental organizations, and the concerned public. Conservation experts and the public, however, usually have vastly different knowledge, attitudes, and values. Personal agendas, work styles, and methods of relating to others also vary considerably from person to person, expert or not. The challenge is to enable the various participants and members of the public to communicate effectively across disciplines, education levels, and value systems.

Effective communication becomes even more challenging if social science experts take part in recovery efforts, as is recommended by several contributors in this volume and elsewhere (Kellert 1985; Clark and Westrum 1989; Kaplan 1990; Kellert and Clark 1991; Reading et al. 1991; Tear and Forester 1992). Getting all these experts and nonexperts to communicate well is difficult. In some situations, formal collaboration, in the form of alternative dispute resolution or environmental mediation, will be necessary (see Chapter 13). But endangered species conservation has a greater chance of success if such formal dispute resolution mechanisms can be avoided by getting the diverse parties to communicate well and cooperate with one another.

There are a number of obstacles to effective information sharing, both in species recovery efforts and in other situations involving experts. Kaplan and Kaplan (1982) suggest that although people need information and readily seek it, often from the experts, information transfer can be remarkably difficult. Rapid transfer of information is crucial to many endangered species recovery efforts, so it is essential to find ways to make the transfer successful and efficient. In general, three features of information transfer characterize communication. First, there is "leveling," which occurs as information is simplified. Details tend to drop out as information is transferred between people. Second, there is information "sharpening." Sharpening refers to the creation of cognitive landmarks as some details are lost and others are retained and receive greater attention, even exaggeration. Third, there is "rationalization." In an effort to make the information more coherent, people simplify, create landmarks, and make connections. The result of all three features is a more coherent cognitive "map," but with the liabilities of distortion. Although leveling, sharpening, and rationalizing facilitate information transfer, too much can have disastrous cumulative results, especially in endangered species conservation. The key for people interested in good communication is to create a simple, coherent cognitive map for themselves that has a minimum amount of distortion and can be communicated easily. The sharing of a common cognitive map is essential to cooperation. In other words, a common understanding of the conservation problem and its solution should be shared by everyone taking part in a cooperative effort.

Sharing Cognitive Models

Sharing cognitive models of endangered species problems and solutions is vital for teamwork and joint problem solving. Yet people usually hold different cognitive models of the situation. Experts as well as nonexperts hold various mental models of their world—abstract concepts with real-life consequences. According to Kaplan (1990:47–48), a cognitive model "provides a way of understanding how any knowledge domain is acquired, stored and accessed. . . . [It] can be thought of as a structure that stores a considerable amount of information, often acquired in an unsystematic fashion, and accessible in a highly flexible manner." These models are what a person uses to understand the world and take action. Cognitive models may be widely shared or idiosyncratic. In controversial recovery efforts, people with a common interest, often from the same organization, may share a similar cognitive model that differs greatly from those held by competing interests. Effective information exchange requires the ability to share mental models with others. Often people in different disciplines have very different cognitive models and end up talking past each other. To increase the probability of success, collaborative species recovery should strive to develop techniques that improve the chances for effective information exchange.

Sharing cognitive models is perhaps the best means of transferring information. Kaplan (1990) cites a number of factors that can help people share models more effectively. The first principle is simplicity. Endangered species conservation efforts are often wrought with complexity. Nevertheless, to ensure effective information transfer, participants must keep their communication simple. "People prefer environments in which they can make sense of what they perceive and in which they can learn more" (Kaplan and Kaplan 1982:192). People converse best when they use the same temporal and spatial references. Confusing or threatening messages should of course be avoided. Information must be readily available, it must be relevant, and there must be time to find, ingest, and relate that information to the conservation problem at hand. The second and third principles for effective information transfer are concreteness and multiplicity. The more definite the cognitive model being shared, the greater the chances for transfer. Abstract ideas are hard to understand and therefore hard to communicate. Multiplicity refers to the exposure given to thoughts and information. The greater the number of examples and independent sources, as well as the frequency of the message, the greater the probability for information transfer (Chaiken and Stangor 1987; Kaplan 1990; Tessler and Shaffer 1990).

The importance of these three principles led Kaplan and Kaplan (1982) to develop the idea of a "portable cognitive model"—that is, a model which is simple enough to be transferred easily while remaining coherent and useful (see also Kaplan 1990). Portable models must form cogent patterns of interconnected thoughts that are internally consistent and can be related to other models. Still, such models will not be adopted unless there is a need for them. In other words, conservationists must package new problem-solving approaches in easily understandable forms and demonstrate their practical utility.

Attention to several considerations can maximize the chances that cognitive models will be understood by people with differing expertise. Receptivity is the key to the adoption of new models. It depends on social setting, relevance, style of presentation, number of independent sources of information, direct experience, and the strength, salience, clarity, and source of the message. Obviously an expert who is trying to communicate with others has control of only a few of these variables. Recipients of information must be interested in the information being offered and must recognize its utility. Supplanting or even transplanting cognitive models is extremely difficult, however. If people already possess a well-developed cognitive model, it is hard to convince them of the importance of information that does not fit their model. This is especially true if their model is based on knowledge derived from personal experience.

The best method for transferring information may be to provide the necessary information, appropriately packaged, so that recipients can update or build their own new model. The first step in this process is to understand the intended recipients, their cognitive models (present state of knowledge), the way they currently partition knowledge and bound their rationality, and the way they synthesize and understand new information (Simon 1976). This is hard work but necessary. Effective information transfer requires effective collaboration between communicator and recipient. Schön (1983) suggests that effective communication should comprise a "reflective conversation" between an expert and the other party. In reflective conversations, one is both teacher and student; therefore, high levels of feedback and candid recognition of one's own ignorance are required. Such reflective conversations are needed among everyone involved in endangered species conservation.

A number of methods for increasing the effectiveness of information transfer have been suggested by Kaplan and Kaplan (1982). First, limit the amount of information being transferred to no more than three distinct items at a time to avoid overwhelming the recepient (although professionals can usually handle more information if it falls within their field of expertise). Second, develop a vivid presentation style that relies on concrete information and situations. Third, use visual material as much as possible, since it is generally trusted, easily grasped, powerful, and makes for a more enjoyable exchange of information. Fourth, create a sense of mystery and new possibilities. Fifth, keep the message simple, focusing on important issues that are coherent and easily understood. And sixth, relate a positive experience by keeping enthusiasm high, communication clear, and information well organized.

Sharing cognitive models is generally difficult, but the task can be made easier by conveying a simple, concrete message and working to make information transfer an enjoyable experience. As people acquire more specialized knowledge, however, their ability to simplify and communicate that information may diminish. Experts in particular may have trouble transferring information and sharing cognitive models. If this happens in endangered species conservation, recovery efforts can be seriously hampered.

Communicating with Outsiders

Experts are often poor communicators of the knowledge they possess. Yet the costs of poor communication can be dramatic—even the extinction of a species. Training in a particular discipline provides one with a particular way of seeing the world. Schön (1983) argues that the way people view a problem is very much determined by disciplinary biases. Thus problems are often defined according to these biases (Dery 1984, Weiss 1989). For example, declines in the size and distribution of large carnivore populations over the past two hundred years might be viewed by many social scientists as the result of historical and cultural biases toward eliminating predators. But ecologists might see the declines as the direct and indirect result of ecological factors like high mortality and habitat degradation. And organizational theorists might see them as the failure of agencies and organizational systems to protect these species (see Clark et al. 1993).

But professionals are usually unaware that they have professional biases or perceive the world any differently than other people. This lack of self-awareness can be a major problem for those trying to communicate with people outside their discipline—and in endangered species conservation it can be catastrophic. Therefore, transferring information among experts from different disciplines or to nonexperts involves overcoming additional difficulties associated with expertise. Professionals, by definition, possess expertise in a particular discipline or domain. Both affective and perceptual constraints, as we shall see, commonly make it difficult for experts to share cognitive models. Affective constraints include the proclivity of experts to defend their disciplines and discount the opinions of "outsiders." Perceptual constraints include differences in the way professionals from different disciplines view problems and the world. Effective information transfer is further hampered by the different ways in which professionals from different fields communicate (via jargon) and learn. Experts tend to place greater trust in information from their own field of expertise. The highly specialized training of experts can thus prevent them from understanding other perspectives. And yet understanding the cognitive models of other disciplines is the only way to promote interdisciplinary conservation. The challenge is to encourage experts to transcend their own expertise and begin to understand both their own disciplinary biases and the way that other experts and nonexperts see the world.

Moving toward interdisciplinary conservation will require some changes in values, attitudes, and behavior. Even though this is a lot to ask of experts who are unaccustomed to working closely with those from sharply different disciplines, such as sociology and policy sciences, it does promise many advantages.

Changing Values, Attitudes, and Behavior

To increase the chances of restoring a threatened or endangered species, it is often desirable to change the values, attitudes, and behavior of key figures, including the public and other experts, involved in the program. In this section we look at human values, attitudes, and behavior and then describe how they can be changed. Practical examples are given throughout. By increasing the support for restoration efforts, we can do much to promote endangered species conservation.

Changing values, attitudes, and behavior, however, is extremely difficult. Certainly getting biologists to value the role of social scientists as their equals in restoration programs or getting antagonistic groups to value a previously disliked species are desirable goals. As Kaplan and Kaplan (1982:182) have observed, however, many "information providers set out with inflated expectations. Talk of behavior change is intoxicating stuff." Contrary to popular belief, simply providing information to people rarely influences them, for knowledge is only one of several influencing factors. (See Rokeach 1972; Sinden and Worrell 1979; Kaplan and Kaplan 1982; Brown and Manfredo 1987; Reading and Kellert 1993; Kellert in press.) Even with changes in cognitive models, behavior may not change, because behavior is a function of both cognitive models and the environment. Nevertheless, values do change in time as context, knowledge, and experience change.

Change is most easily accomplished when people become aware of internal contradictions among their values. Dissonance reduction theory offers the most widely invoked explanation for these changes. It suggests that people seek to reduce the discomfort they experience due to inconsistent values, attitudes, and behavior by changing their dissonant peripheral values to reflect stronger core values (Rokeach 1979; Williams 1979; Chaiken and Stangor 1987; Tessler and Shaffer 1990). Biologists working on endangered species recovery, for example, have traditionally valued public support, especially local support, but have rarely implemented programs to develop this support (Reading 1993). Several conservation efforts that have failed due to local public animosity have begun to convince biological experts of the importance of developing public support. So biologists are beginning to change their behavior (by implementing education and public relations programs) to make them more consistent with their values (the importance of public support). Still, such changes are only just beginning.

Besides internal contradictions, other things affect the possibility of change. Social context, for example, is important to value, attitude, and behavior changes. As Tessler and Shaffer (1990:493) observe, "making one's own values salient and comparing them to the values of liked or disliked reference groups often results in lasting and desirable

. . . change." Social groups may either promote or thwart change by increasing or decreasing real or perceived dissonance (Rokeach and Grube 1979). Biologists, for example, are often reluctant to enter the policy arena for fear of being labeled biased by their peers (see Chapter 17). Or pressure from local ranchers might induce rural business interests to change from passive support to active opposition of an endangered carnivore reintroduction. Peer pressure can both increase and decrease dissonance; group support may even justify the dissonance. Finally, the group's context may enable members to attribute dissonance to other group members, thus reducing individual accountability and the need for change. Latane's (1981) social impact theory suggests that the degree to which a person is influenced depends on the number, strength, and proximity of sources of change. The greater the number of sources promoting change, the greater the likelihood change will take place.

Changes in values, attitudes, and especially behavior can occur in the absence of strong dissonance, but this process is thought to be less common and often quite slow (Williams 1979). Self-perception theories suggest that even attitudes and behavior that fall within a person's "latitude of acceptability" may be altered for strategic reasons—to please others, for instance, or to improve one's self-image (Bem 1970). In addition, changes may occur through what Chaiken and Stangor (1987) call the "peripheral route," in which marginal changes are made without much consideration.

Chaiken and Stangor (1987) discuss the influence of majority and minority groups on changes in value, attitude, and behavior. Compliance, identification, and internalization motivate change toward agreement with the majority, while information acts as the primary motivator of change toward the minority. Because it requires more careful consideration and is less frequently reinforced by the social environment, change toward minority values is stronger and more persistent. Convincing experts of the importance of working within the policy arena or changing the attitudes of antagonistic groups to make them more supportive of an endangered species recovery program, therefore, may be difficult since both positions represent minority views. But the change, if obtained, is more likely be permanent.

Two major schools of thought explain the way in which people influence the values, attitudes, and behavior of others (Tessler and Shaffer 1990). First, the "cognitive response" perspective suggests that people thoughtfully consider persuasive arguments. Second, the "cognitive miser" perspective argues that people tend to minimize the effort required to assess persuasive arguments. Thus they will thoughtfully consider persuasive arguments only if their customary cues suggest they should do so. Validity is also often assessed using simple decision rules (Chaiken and Stangor 1987). For example, messages from liked (versus disliked) reference groups may be validated automatically. Therefore, biologists should be used to convince others in their field of the need for interdisciplinary approaches to conservation. Similarly, getting local people (farmers, ranchers, loggers) to promote endangered species recovery will probably be quite successful. Recent evidence suggests that both perspectives, cognitive response and cognitive miser, may hold under different circumstances.

Receptivity to persuasion depends on several factors. Chaiken and Stangor (1987) stress the importance of social setting, relevance, and message strength and source. For example, increasing the salience, strength, and clarity of the message supporting endangered species increases the persuasive impact. But for the message to be effective, Tessler and Shaffer (1990) suggest that people must be motivated and capable of receiving and considering persuasive messages. For the endangered species message to be effective, it may require repeated exposures, lack of distractions, a neutral state of mind, extensive prior knowledge, and direct experience. Thus a lot of work is needed to change the values, attitudes, and behavior of experts involved in endangered species recovery programs and to enlist the support of hostile parties.

Just as several factors increase the chances of successful persuasion, several forces inhibit persuasion and make values and attitudes particularly resistant to change. Williams (1979) suggests that the most intransigent values are those that are highly interconnected and congruent with other values, attitudes, and beliefs. A person's central values are supported by strong sanctions and consensus. They are also often associated with high prestige and authority and are symbolically important. In addition, a host of social, cultural, and institutional forces inhibit significant changes in values. As a result, values persist and change is slow, especially in small communities. Thus conservationists working in rural areas should initiate education and public relations as early as possible in order to allow plenty of time for change to occur.

Changing the values, attitudes, and behaviors of people working on species recovery (or those of antagonistic groups) is difficult but not impossible. Often the change requires learning at the individual level. Indeed, individual learning is the chief means by which change occurs. If people can transmit newly learned information to their peers and employers and institutionalize it socially and organizationally, then broad-scale change is in fact possible. Encouraging organizational learning, much less ensuring it, is difficult indeed. Nevertheless, it all begins at the individual level.

Individual and Organizational Learning

We all know about individual learning. After all, we have spent considerable time in schools, classes, workshops, lectures, and meetings of various kinds. These processes result in learning. As a consequence, individual values, attitudes, and behavior are changed. In this section we look at individual learning and organizational learning and their benefits to endangered species conservation. Understanding organizational learning and knowing how to induce it can significantly upgrade species conservation efforts.

Organizational learning is a relatively new concept, dating back only four decades or so. Simply put, organizational learning is "the process whereby organizations understand and manage their experience" (Glynn et al. 1992:3). Many key advances were made in the late 1970s that began to flesh out the concept and foster its application (Argyris and Schön 1978). Recent interest in organizational learning stems from its many practical implications. (See Kofman and Senge 1993; Isaacs 1993; Schein 1993; Ulrich et al. 1993; McGill and Slocum 1993.) If organizations in the private and public sectors can learn quickly, apply lessons, and keep learning, they will be much more successful. But the principles of organizational learning are not easily understood or applied, and they may meet great resistance. This is especially true in large bureaucracies—including the state and federal agencies responsible for endangered species conservation.

The concept of individual learning is familiar to everyone. But the notion of organizational learning is not widely known. For organizations to learn, they must institutionalize new values, new philosophies, new practices. Put another way: an organization learns when it changes its standard operating procedures. Ideally the new procedures permit the organization to become more effective. Feedback from the environment signals success or failure. For learning to be on-going, this cycle of feedback–change in procedures–action–feedback must be endless. But many organizations are poor learners.

In real life most organizations learn slowly, learn the wrong things, or simply do not learn at all. Government bureaucracies seem to be especially resistant to learning (Etheredge 1985; Osborne and Gaebler 1992). Applying organizational learning to endangered species conservation promises significantly improved programs. Programs should be organized and managed in ways that ensure open, quick learning. To be most successful, this means institutionalizing learning mechanisms. For example, the private sector is employing organizational learning in varying degrees to make organizations more effective and more efficient. The concept of organizational learning was introduced to biological conservation by Clark and Westrum (1987), Kellert and Clark (1991), and others, but its value to date has been little recognized. Still, significant improvements in endangered species conservation can come when organizations learn how to learn better.

Many conservationists have learned not only new techniques but entirely new ways of understanding and carrying out their work. (See Schön 1983 for examples of learning in several professions.) The rise of conservation biology in the last few years is an example of improved conservation and represents new ways of thinking about, researching, managing, and saving species and their habitats. But these new approaches have not yet been employed, much less institutionalized, in government agencies. Before these new approaches can be institutionalized, the personnel themselves must learn new ways of solving problems. In most instances, knowledge learned by an individual is not effectively communicated, used, or ultimately institutionalized in the organization. As a result, individual learning occurs but organizational learning does not. If individual learning can be translated into organizational learning, significant improvements in endangered species conservation are undoubtedly possible.

If we expect to improve conservation, therefore, we must examine how organizational learning works (Westrum 1986). Models of organizational learning often divide learning into four components: individual beliefs, individual actions, organizational actions, and environmental responses. (See March and Olsen 1976.) All four are essential. Individual beliefs are important because they are at the heart of organizational learning. If individuals believe they can change organizations and improve conservation, they will try to make improvements.

As they learn improvements, they make efforts to communicate what they have learned to others in their organization. If they are successful, the organization changes its mode of operation. Organizational actions include everything from simply employing a new technique to examining the organization's fundamental "psychology and sociology" and reorganizing its structure and culture. Such major restructuring permits the organization to learn better on an ongoing basis—not only with respect to its own internal operation but also with respect to the external setting. Its external setting obviously responds to its actions. If the organization is sensitive to its environment and behaves adaptively, it is rewarded with success: mission accomplished. In turn, the organization may position itself to learn from the experience and continue learning, acting, learning, in a never-ending process of improved problem solving and adaptation. If all four components—individual beliefs, individual actions, organizational actions, and environmental responses—are actively engaged in improving organizational learning, the organization should be effective at saving endangered species. In practice these components are linked in complex ways. Therefore, organizational learning usually requires careful management by knowledgeable leaders.

Improved organizational performance comes from organizational learning and this, in turn, derives from individual learning. It follows, then, that individuals should be encouraged to learn for themselves and should be rewarded for communicating lessons to the organization as a whole. For any government agency or private organization that is interested in improving the conservation and management of endangered species, it is vital to understand the concept of organizational learning and know how to apply it. Without active, knowledgeable leadership and a management goal of organizational learning, organizations involved in endangered species conservation will fall further and further behind the learning curve and the accelerating extinction crisis.

Lessons

Endangered species conservation involves experts of various types, although biological experts typically dominate recovery programs. Understanding the manner in which these experts seek to solve species endangerment problems is of vital concern, especially if we hope to improve conservation success rates. Traditional practice has carried us a long way in meeting species restoration problems, but, as this chapter suggests, new models of practice could help even more in meeting this vital professional challenge. Four classes of models were examined here that focused on the sociology and role of the professional, improved communication, changing values in order to support species conservation, and improving the transfer of individual knowledge for increased organizational learning.

The range of tools addressed in this chapter holds great promise for professional improvement. Effective communication and information sharing are central to endangered species conservation efforts as experts go about their tasks working closely with other experts and non-experts. In some cases, changes in recovery programs are needed to improve species restoration efforts. This may require the use of new models of professional practice and new means of communication. Broadening participation and increasing the volume and kind of communication in recovery programs—beyond the biological and ecological sciences to include the social sciences—promises to improve the performance of many programs. As participation broadens, however, the difficulty of maintaining effective communication and cooperation increases. And this requires new models of professional problem solving and communication.

How will this change—the use of new tools—come about? Experts must work with agency managers and administrators, elected officials, and the public, as well as other experts, to change and adapt for improved performance. Changing values, attitudes, and behavior is hard, but it is not impossible. For improved conservation to become a reality, there must be improvements in the methods and rate of learning by individuals and organizations alike. This means that individuals involved in endangered species conservation must learn for themselves and then be rewarded for communicating lessons to their peers and employing organizations. For their part, organizations devoted to the conservation and management of endangered species must understand the concept of organizational learning and know how to apply it in ways that foster improved learning by both employees and the organization.

The conceptual tools used by experts, as well as their capacity to communicate key information effectively and their ability to transform their employing organizations, are central to endangered species conservation. As the case studies in this book clearly illustrate, these variables have much to do with a program's success or failure.

Acknowledgments

Peyton Curlee of the Northern Rockies Conservation Cooperative and Rachel Kaplan of the University of Michigan critiqued a draft of this chapter.

References

Argyris, C., and D. A. Schön. 1978. *Organizational Learning: A Theory of Action Perspective*. Reading, Mass.: Addison-Wesley.

Bem, D. J. 1970. *Beliefs, Attitudes, and Human Affairs*. Belmont, Calif.: Brooks/Cole.

Brown, P. J., and M. J. Manfredo. 1987. Social values defined. In *Valuing Wildlife: Economic and Social Perspectives*, B. J. Decker and G. R. Goff (eds.). Boulder: Westview Press.

Chaiken, S., and C. Stangor. 1987. Attitudes and attitude change. *Annual Review of Psychology* 38:575–630.

Clark, T. W., and R. Westrum. 1987. Paradigms and ferrets. *Social Studies of Science* 3:3–33.

———. 1989. High-performance teams in wildlife conservation: A species reintroduction and recovery example. *Environmental Management* 13:663–670.

Clark, T. W., R. P. Reading, and A. P. Curlee. 1993. Conserving threatened carnivores: Developing interdisciplinary, problem-oriented strategies. Unpublished report to Geraldine R. Dodge Foundation. Jackson, Wyo.: Northern Rockies Conservation Cooperative.

Dery, D. 1984. *Problem Definition in Policy Analysis*. Lawrence: University of Kansas Press.

Etheredge, L. S. 1985. *Can Governments Learn?* New York: Pergamon Press.

Glynn, M. A., F. J. Milliken, and T. K. Lant. 1992. Learning about organizational learning theory: An umbrella of organizing processes. Paper presented at Academy of Management Meeting, Las Vegas, Nevada.

Isaacs, W. N. 1993. Taking flight: Dialogue, collective thinking, and organizational learning. *Organizational Dynamics* (Autumn):24–39.

Kaplan, R. 1990. Collaboration from a cognitive perspective: Sharing models across expertise. In *Proceedings of the Twenty-First Annual Conference of the Environmental Design Research Association*, R. I. Selby, K. H. Anthony, J. Choi, and R. Orland (eds.). Oklahoma City: EDRA.

Kaplan, S., and R. Kaplan. 1982. *Cognition and Environment: Functioning in an Uncertain World*. Ann Arbor: Ulrich's.

Kellert, S. R. 1985. Social and perceptual factors in endangered species management. *Journal of Wildlife Management* 49(2):528–536.

———. In press. Public attitudes toward bears and their conservation. In *Ninth International Bear Conference*, J. J. Clear, C. Servheen, and L. J. Lyon (eds.).

Kellert, S. R., and T. W. Clark. 1991. The theory and application of a wildlife policy framework. In *Public Policy and Wildlife Conservation*, W. R. Mangun and S. S. Nagel (eds.). New York: Greenwood Press.

Kofman, R., and P. M. Senge. 1993. Communities of commitment: The heart of learning organizational. *Organizational Dynamics* (Autumn):5–23.

Latane, B. 1981. The psychology of social impact. *American Psychology* 36:343–356.

March, J. G., and J. P. Olsen. 1976. Organizational learning and the ambiguity of the past. In *Ambiguity and Choice in Organizations*, J. G. March and J. P. Olsen (eds.). Bergen, Norway: Universitetsforlaget.

McGill, M. E., and J. W. Slocum, Jr. 1993. Unlearning the organization. *Organizational Dynamics* (Autumn):67–79.

Osborne, D., and T. Gaebler. 1992. *Reinventing Government: How the Entrepreneurial Spirit Is Transforming the Public Sector.* New York: Penguin Books.

Reading, R. P. 1993. Toward an endangered species reintroduction paradigm: A case study of the black-footed ferret. Ph.D. dissertation, Yale University.

Reading, R. P., and S. R. Kellert. 1993. Attitudes toward a proposed black-footed ferret (*Mustela nigripes*) reintroduction. *Conservation Biology* 7:569–580.

Reading, R. P., T. W. Clark, and S. R. Kellert. 1991. Toward an endangered species reintroduction paradigm. *Endangered Species Update* 8(11):1–4.

Rokeach, M. 1972. *Beliefs, Attitudes, and Values: A Theory of Organization and Change.* San Francisco: Jossey-Bass.

———. 1979. Introduction. In *Understanding Human Values: Individual and Societal*, M. Rokeach, (ed.). New York: Free Press.

Rokeach, M., and J. W. Grube. 1979. Can values be manipulated arbitrarily? In *Understanding Human Values: Individual and Societal*, M. Rokeach (ed.). New York: Free Press.

Schein, E. H. 1993. On dialogue, culture, and organizational learning. *Organizational Dynamics* (Autumn):40–51.

Schön, D. A. 1983. *The Reflective Practitioner: How Professionals Think in Action.* New York: Basic Books.

Simon, H. 1976. *Administrative Behavior.* 3rd ed. New York: Free Press.

Sinden, J., and A. Worrell. 1979. *Unpriced Values: Decisions Without Market Values.* New York: Wiley.

Tear, T. H., and D. Forester. 1992. Role of social theory in reintroduction planning: A case study of the Arabian oryx in Oman. *Society and Natural Resources* 5:359–374.

Tessler, A., and D. R. Shaffer. 1990. Attitudes and attitude change. *Annual Review of Psychology* 41:479–523.

Ulrich, D., M. A. Von Glinow, and T. Jick. 1993. High-impact learning: Building and diffusing learning capability. *Organizational Dynamics* (Autumn):52–66.

Weiss, J. A. 1989. The powers of problem definition: The case of government paperwork. *Policy Science* 22:97–121.

Westrum, R. 1986. Management strategies and information failures. Paper delivered at NATO Advanced Research Workshop on Failure Analysis of Information Systems, August 1986, Bad Winsheim, Germany.

Williams, R. M., Jr. 1979. Change and stability in values and value systems: A sociological perspective. In *Understanding Human Values: Individual and Societal*, M. Rokeach (ed.). New York: Free Press.

16

A Sociological Perspective
Valuational, Socioeconomic, and Organizational Factors

Stephen R. Kellert

This chapter explores the importance of valuational, socioeconomic, and organizational factors in the implementation of the U.S. Endangered Species Act (ESA). It is my contention that the success of most, if not all, endangered species programs depends greatly on systematic consideration of various human dimensions rather than just assessing biological and technical elements. These social variables are, unfortunately, often ignored, viewed as only marginally important, or inadequately considered in most efforts to protect and recover species in jeopardy of extinction (Clark 1992; Reading 1993). Although the reasons for this situation are many, the major causes include the biological and technical bias of most wildlife professionals, the difficulty of understanding human behavior and institutions, and the political risks associated with attempts to manage socioeconomic and cultural factors.

Background: The Analytical Perspective

A conceptual framework described elsewhere (Kellert and Clark 1991) is used here to elucidate the interactive importance of biophysical, valuational, socioeconomic, and organizational factors in endangered species policy. This analytical framework (Figure 16-1) views the management of endangered species as the product of four basic forces expressed primarily through the competitive interactions of varying constituencies and changing over time depending on the stage of the decision-making process (Brewer and deLeon 1983).

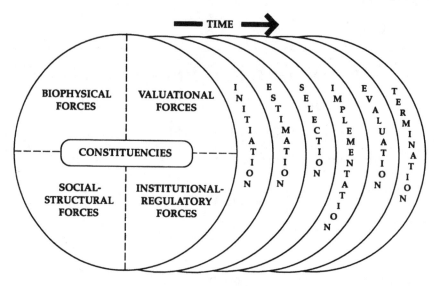

FIGURE 16-1.   Conceptual framework for analyzing wildlife conservation issues.

In this chapter the framework is used to describe a number of critical valuational, socioeconomic, and organizational factors affecting endangered species conservation and recovery—illustrated here by the historic plight of the great whales and the gray wolf (*Canis lupus*). The framework is also used to offer suggestions for improving the consideration of nonbiological dimensions in endangered species policy. Concluding remarks emphasize the urgency of a more comprehensive, interdisciplinary approach to species conservation if the current extinction crisis is to be understood and addressed.

Before beginning, however, it would be well to note that an analysis of this kind can be highly negative and somewhat unfair—a characteristic affliction of the academic critic free from the technical constraints and political realities of the professional. Many endangered species programs are depicted here as unsuccessful and ineffective, echoing the judgment of many recent assessments (Yaffee 1982; Kellert and Clark 1991; Kohm 1991; Rohlf 1991; Houck 1993). On the other hand, no one can deny that the protection of endangered species is among the most difficult tasks facing the conservation field today. Many valiant efforts have been made in the face of enormous obstacles to reverse the seemingly inexorable path toward extinction faced by many species, and notable successes have in fact occurred. Still, the overall record of endangered species management is one of significant shortcomings and many failures (GAO 1988, 1992). This analysis should be viewed, then, in the spirit of constructive criticism, and recognition of the courageous efforts of those who are struggling to push against and even reverse the tide of history.

Nonbiological Impediments

Various valuational, socioeconomic, and organizational factors represent serious impediments to the protection and recovery of endangered species. Eleven factors are specifically noted here, illustrated by examples drawn from experience with the great whales and gray wolf.

Valuational Dimensions

Many endangered species programs tend to ignore the importance of human beliefs and motivations as factors profoundly influencing species recovery. Valuational factors are often viewed as vague, insubstantial, and unworthy of scientific exploration. As a consequence, the fundamental significance of human values and perceptions in determining behavior is often insufficiently considered or treated in an intuitive and biased fashion.

A major problem in many endangered species programs is the tendency to ignore the importance of attitudinal differences as an obstacle to the recovery of species. Studies in Minnesota and Michigan of the gray wolf, for example, have identified deeply ingrained differences in basic attitudes among varying urban, rural, hunter, trapper, and farming groups as a major barrier to species recovery (Kellert 1987, 1991a). Analogous differences were found among the general public, commercial fishermen, and sealers in their attitudes toward great whales and other marine mammals in Canada (Kellert 1991b). These attitudinal variations often affect the chances for successful endangered species recovery. Yet they are seldom systematically considered, assessed, or understood in most endangered species programs (Kohm 1991; Reading 1993). A consistent problem has been the underestimation of the significance of these variations among critical social groups essential to species recovery. A related failure has been the limited use of this information to clarify values among opposing groups, to educate varying constituencies, and to work toward the resolution of conflicts (see Chapter 3). These value differences are often regarded as being minor political considerations, as possessing limited relevance, or as being unworthy of scientific consideration. This narrow disciplinary emphasis can frequently result in naive and inadequate endangered species program formulations and recovery efforts.

A second problem has been the failure of most resource managers to appreciate the attitudinal biases and ideological orientations of the management agencies themselves and the relationship of these orientations to the bias of major client groups. Various indicators suggest that the two principal agencies mandated to manage endangered wolves and whales—the U.S. Fish and Wildlife Service (FWS) and the National Marine Fisheries Service (NMFS)—have strong utilitarian, consumptive-use, and commodity-oriented value orientations (Clarke and McCool 1985). These attitudes have been nurtured by a long association and funding dependence on varying consumptive-use groups, such as hunters and sportfishermen at FWS and commercial fishing interests at NMFS. Indeed, these agencies have often functioned in what has been called a "secondary service capacity," inclined to serve the traditional interests of their primary clientele. The need for agencies to function as regulatory bodies when attempting to protect endangered species has often conflicted with their traditional activities, particularly when restrictions are placed on consumptive-use groups with strong utilitarian values. In these situations, both FWS and NMFS have been reluctant to shift from a service to a regulatory orientation—an administrative bias that has frequently resulted in the agencies being perceived as "uncomfortable" with species preservation, nonconsumptive use, and the assertion of federal control over endangered wildlife (Yaffee 1982). A related problem has been a lack of agency understanding of the values of emergent environmentalist groups (Tober 1989). This combined tendency—deferring to the interests of primary consumptive-use clients while lacking an affinity for nontraditional wildlife interests—has led some to charge that the FWS and NMFS have been "captured" and has led as well to wasteful adversarial conflict with wildlife protection groups.

A third valuational problem has been the tendency of many endangered species programs to overrate the importance of public knowledge as a critical determinant of program success (Reading and Kellert 1993). A naive response to opposing value differences among competing constituencies has been the presumption that supplying more information will resolve these differences, presumably in the direction desired by the agencies. The intransigence of underlying value differences has rarely been appreciated (Shaw 1977), however; nor has the related tendency of most groups to use additional information to reinforce and rationalize their prevailing perspectives (Reading and Kellert 1993). This situation may be illustrated by the results of a study of wolf recovery in Michigan, where opponents and supporters of wolf restoration held widely varying attitudes toward this species, although each side possessed similar knowledge. Correspondingly, groups inclined to exploit and dominate whales (sealers, fishermen, Newfoundland residents) had knowledge equivalent to those with a strong protectionist ethic toward marine mammals (young adults, the highly educated, urban residents, members of environmental organizations). These studies (Kellert 1991a, 1991b) reveal that proponents and opponents of endangered species conservation typically use knowledge not to change their basic values but to reinforce their established attitudes toward species recovery. Many endangered species programs erroneously assume that providing additional facts will shift public attitudes in the direction the agency favors. This stress on the cognitive rather than the affective dimensions of human perception has often contributed to program failure (Kellert 1983).

Socioeconomic Dimensions

A variety of sociological, economic, and political factors often seriously affect endangered species programs. These forces are profoundly present at the local and community level and may be manifest both in formal institutions and in more subtle relationships. Most endangered species programs make little effort to explore the importance of these socioeconomic factors. As a consequence, the programs are often viewed by local communities as "insensitive" and in conflict with traditional use and behavior patterns.

The first socioeconomic problem centers on property rights. Many endangered species pass back and forth across varying political and landownership boundaries (Kellert 1986a). This mobility often results in ambiguous management responsibilities and creates problems of assigning culpability and imposing sanctions on harmful behavior (Bean 1983). The ability of most great whale species to traverse long distances across jurisdictional boundaries is especially illustrative of this problem (Kellert 1986b). As creatures of the open oceans, these species have often been exploited as resources of the global commons—rendering management and control difficult, despite formal agreements such as the International Whaling Convention. The extension of national jurisdictions from 3 miles to 200 miles along the continental shelf, however, has resulted in vastly improved cetacean protection measures. Wolves have similarly been the victims of unclear and conflicting property rights (Peek et al. 1991). Not only does this species occupy large territorial ranges crossing state, public, and private property boundaries, but it also has the tendency to inflict injury on livestock, typically viewed as private property even when grazing on public lands.

A related issue has been the problem of effective intergovernmental relations. In the case of wolves, this difficulty has been manifest in the frequent conflict between federal and state governments. Although the gray wolf is an endangered species capable of crossing state borders, the FWS has typically deferred to state governments in the management of this species (Goldman-Carter 1983). The prevailing convention has been to uphold the state's rights over its "residential" wildlife, frequently provoking lawsuits by contending nongovernmental organizations. In the case of the great whales, relations between most of the Western nations and Japan and Norway have been contentious at best, causing great uncertainty over the future management of these species. The point is that complicated property rights issues are often involved in the management, protection, and restoration of endangered wildlife. Yet these factors are rarely considered or systematically explored as essential components of a successful recovery program. The consequence has been confusion regarding the underlying basis for intergroup conflicts, considerable uncertainty about the assignment of management responsibility, and an inclination on the part of consumptive-use groups to oppose endangered species protection out of fear of losing traditional property and resource rights on both public and private lands.

A second socioeconomic factor has been the problem of conflicts between the perceived economic needs of resource-dependent groups and the protection of endangered wildlife. Endangered species programs are often opposed by groups with a primary dependence on the extraction of natural resources (Kellert 1986a). These groups frequently assume that protection efforts will limit their economic activities or block their traditional access to resources. In the wolf and whale studies cited, despite the sharp difference in species and ecosystems, almost identical attitudes were observed among widely varying groups such as livestock producers, loggers, commercial fishermen, and sealers. In each case, endangered species conservation was viewed as antithetical to the interests of these groups and a significant threat to their way of life; often it was opposed for symbolic reasons. These attitudes are often so strongly held that they lead to intense "entrepreneurial" politics—that is, industry groups invoke a variety of economic and political strategies designed to capture or maintain control over the regulatory agencies (Wilson 1980). Very often management agencies are unaware of these efforts and unprepared to deal with subtle measures exercised by pressure groups to influence their behavior.

A third socioeconomic problem is the tendency of management agencies to overestimate the power of economic incentives as a means of encouraging the restoration of endangered species. Examples can again be cited in the case of the great whales and the gray wolf. Regarding the great whales, it has long been suggested that economic interests devoted to the exploitation of these animals have the greatest incentive for ensuring their survival (Clark 1973). Although this notion has a certain compelling logic, the history of the great whale "fishery" has been a repeated story of nonsustainable exploitation (Norris 1978). One reason for this situation has been inadequate knowledge of the breeding biology and population ecology of these species. More important, however, has been the tendency of whalers to harvest the resource to the maximum extent possible and then reinvest the surplus profits in other areas of monetary return (Scharff 1980). This behavior often occurs, as in the case of the great whales, when large profits are obtained, the species' reproductive potential is limited, and considerable capital has been invested in the technology needed for exploitation. In the case of the gray wolf, it has been argued that enhanced support for species recovery would be more likely, especially among rural groups, if the animal had greater economic value. This argument has been used to advocate the resumption of a limited season for harvesting wolf pelts. Research has revealed very little support for this contention (Kellert 1987). Moreover, the creation of legitimate markets for depleted species frequently leads to increased illegal poaching of the targeted animal (Geist 1988). The point is not to argue the complicated issue of sustainable exploitation of wild species, but to stress that endangered species programs must be careful in exploring various economic strategies advanced to enhance recovery.

Organizational Factors

Understanding the structure, functioning, and dynamics of complex organizations is one of the most serious challenges facing the wildlife management field today. Indeed, inadequate organizational performance, particularly among government agencies, may be among the foremost reasons for the ineffectiveness and failure of many endangered species programs. In recent years, a number of excellent studies have begun to identify the many subtle and complex organizational factors that can subvert the protection and recovery of endangered species. (See, for example, Yaffee 1982, 1994; Clark and Westrum 1987; Clark and Harvey 1988; Reading 1993.) These studies reflect an emerging realization of the need to obtain a far better understanding of organizational behavior as a basis for greatly improving the performance of endangered species programs (Clark and Cragun 1991).

The first organizational impediment has been the narrow professional training of most wildlife managers (Clark and Harvey 1988, Clark 1993). The professional's education has almost exclusively stressed the biological and technical sciences, often omitting exposure to economics, sociology, anthropology, psychology, law, and organizational behavior. The professional needs the insights of environmental history and ethics as well (Chapters 15 and 17). Management of the great whales and the gray wolf would have benefited greatly from an interdisciplinary examination of the complex social factors responsible for their endangerment and essential for their recovery. Yet most marine mammal and furbearer biologists have only limited training (or interest) in this type of integrated, multidimensional knowledge. Indeed, wildlife professionals often choose these career paths in the hope of dealing more with wildlife than people and their institutions.

A second organizational problem has been the temporal and spatial narrowness of most endangered species programs. The problem of time has been expressed in a proclivity to view endangerment from a short-term perspective—neglecting the necessity of understanding the historical context, recognizing the importance of planning for recovery over a long period, and considering the cumulative impacts of diverse human activities (Grumbine 1992). A common spatial problem has been the tendency to view endangered species conservation as largely the result of local factors and to ignore its links to ecosystem and landscape-level forces. An adequate spatial perspective should recognize that the recovery of endangered species requires a landscape-level understanding and the consideration of factors both directly and indirectly involved in species dynamics. A narrow temporal and spatial perspective in the case of the great whales has led to insufficient consideration of species' varying territorial requirements, the importance of stationary stocks, and the ecological role played by cetaceans in their habitats (Norris 1978). Analogous problems have occurred with wolves. The contentious management history of both taxa attests to the need to consider the concerns of diverse local, national, and even international groups who have a strong stake in the conservation and recovery of these species.

A third organizational problem has been conflicting institutional goals and objectives. (See Clark and Kellert 1988; Clark and Westrum 1989; Reading 1993.) This conflict has frequently arisen among the lead government agencies: the FWS and NMFS. As previously noted, both organizations have well-established traditions of functioning as service agencies primarily concerned with the needs of consumptive-use and commodity-oriented groups. This narrow utilitarian orientation, often reinforced by a financial dependence on the consumptive user, frequently conflicts with the regulatory objectives of protecting endangered wildlife—particularly if it results in controls on the activities of traditional client groups (Yaffee 1982).

A recent analysis by Houck (1993) has shown that in attempting to avoid conflicts between endangered species protection and traditional consumptive-use activities, as well as conflicts with varying economic interests, the ESA has come to function more as a discretionary than a prohibitive regulatory policy. The agencies' desire to maximize organizational flexibility has allowed them to minimize many listing and recovery actions considered either highly controversial or antagonistic to the interests of traditional clients. In the process, the ostensible goal of protecting endangered species has often been supplanted by the objective of maintaining agency prerogatives and the support of its long-term constituents (Coggins and Russell 1985).

A fourth organizational problem is the government agencies' increasingly obsessive concern with maintaining bureaucratic control over personnel and resources. While these bureaucratic considerations are not in themselves problematic, they can become dysfunctional to species recovery when they become, in effect, ends in themselves (Clark and Harvey 1988; Reading 1993). This has sometimes occurred when the agencies are especially fearful of changing socioeconomic and political circumstances, such as new endangered species management mandates and the emergence of public interest groups. Traditional agencies, in these circumstances, are often inclined to focus their energies on controlling personnel and funding sources, even to the extent of allowing these activities to take precedence over endangered species recovery objectives. To accomplish these control goals, the bureaucracies may resort to restricting the flow of information, stressing performance standards based on rigid rules of conduct, and jealously guarding traditional powers and prerogatives (Westrum and Samaha 1984; Etheredge 1985).

These organizational maintenance activities can become the primary concern of agencies, even to the point of subverting other goals such as the conservation and recovery of endangered species. Although it is difficult to predict when such "goal inversion" may occur, studies of the black-footed ferret (Clark and Harvey 1988), spotted owl (Yaffee 1994), grizzly bear (Grumbine 1992), and other imperilled wildlife suggest it may not be uncommon. Goal inversion has certainly occurred in the case of both the gray wolf (Lopez 1978; Goldman-Carter 1983; Houck 1993) and the great whales (Kellert 1986b).

A fifth organizational problem has been the development of rigid organizational structures (Argyris and Schön 1978). This type of inflexible decision-making environment has been noted among agencies with a strong traditional bias, such as the FWS and NMFS, where the inclination is to resist criticism and ignore "outside" interests. A related problem has been an intolerance of novel approaches to endangered species management—particularly when these innovative strategies are offered by nontraditional interest groups. Such expressions of organizational insecurity can lead to efforts designed to achieve monopolistic control over recovery programs, especially when new interest groups advocate innovative approaches (Clark et al. 1989).

These organizational problems are often associated with the lack of creative, risk-taking leadership. In this regard, both the FWS and NMFS have been characterized as preferring a low-profile, timid leadership style where good administration is often equated with the absence of controversy. This authority pattern can be perpetuated by concentrating resource professionals at the bottom of the decision-making hierarchy; career-minded individuals collect at middle management levels; persons deferential to political interests are clustered at the top (Wilson 1980). These decision-making tendencies have occurred in the case of the gray wolf, where efforts to thwart this species' reintroduction and defer to commodity and states' rights interests have often been found among the agency leadership. Timid leadership has occurred in the case of the great whales, as well, resulting in infrequent attempts to exercise powerful aspects of the Marine Mammal Protection Act to exact compliance among commercial fishing interests and foreign nations whose activities conflict with cetacean recovery efforts.

Government agencies often justify their timid behavior by emphasizing narrow biological criteria and rendering unduly optimistic assessments of the chances for species recovery (Clark 1993). Agencies can resist novel information by rejecting new ecological data or denying the relevance of unconventional knowledge (Clark and Westrum 1987). The presumption of having all the necessary data helps them to deny the possibility of scientific uncertainty in the management of endangered species. This tendency usually involves undue optimism and a focus on strictly biological data.

All of these organizational impediments have been encountered in the management of the great whales and the gray wolf. Perhaps the most egregious example in the case of the great whales has been the use of maximum sustainable yield to determine harvest levels. In more recent times, inflexible presumptions have been reflected in the reluctance to consider the ecological importance of whales in various marine ecosystems or the importance of radically changing public values as a factor affecting the management of these species. Regarding wolves, similar blinders have been encountered regarding the species' social behavior, its interactions with prey populations, and the actual amount of encountered depredation on livestock. For both taxa, the management agencies have consistently overrated the state of biological knowledge and downplayed the relevance of important socioeconomic and valuational information.

Lessons

Eleven factors—three valuational, three socioeconomic, and five organizational—have been cited here as frequent impediments to effective endangered species conservation and recovery efforts. Taken by itself, each of these factors is typically of limited significance. What makes them so problematic is their concurrent, cumulative, and synergistic impact in many endangered species situations. While large-scale human development, the exploitation of species, and the introduction of nonnative organisms represent the primary causes for species endangerment, deficiencies in professional behavior may constitute the major reason for our limited success in the recovery of most endangered wildlife. It is, of course, far easier to criticize than to offer remedies. The advantage of being an academic is that one can often avoid the difficulties of program implementation, especially when the challenge is to improve the efficiency and effectiveness of complex organizations. Nevertheless, in the hope of providing constructive suggestions, the following recommendations are offered for enhancing endangered species programs.

Lesson 1. Wildlife managers need to become far more serious about the idea that "wildlife management is really people management." Professional degree programs should be better designed to incorporate an interdisciplinary understanding recognizing the need for integrating the natural, social, and policy sciences and to expose students to environmental history and ethics. Increased emphasis should be placed on studying the diverse values people attach to living resources and their conservation. Professional training should also consider how economic and property structures affect the utilization of biotic resources in different cultural and biogeographical circumstances. Awareness of legal and regulatory process is important, too, as well as the study of complex organizations, both governmental and nongovernmental. Above all, the assumption that knowledge of the human dimensions is "soft," political, and inferior to the "hard" sciences must be dispelled as an insidious bias undermining the objective of balanced professional training.

Lesson 2. A comprehensive, interdisciplinary approach should be incorporated into the design of endangered species programs as well. One promising means for accomplishing this objective is to create professional teams that represent diverse disciplines and a variety of value orientations. As these management teams would incorporate varying epistemological perspectives, they would be more likely to respond to the interests of competing groups.

Lesson 3. The effectiveness of interdisciplinary management teams would be enhanced by concentrating their efforts at the local level and granting them a high degree of autonomy and flexibility. Operating in close proximity to the problem fosters opportunities for matching professional expertise with diverse concerns among competing constituencies. It is also important to guard against co-optation by local interest groups, which may occur when professionals work closely with locals and live in small communities.

Lesson 4.   Autonomy and flexibility can help to maximize the possibilities for responding to the uncertainties of endangered species problems. Many imponderables are encountered in attempting to recover endangered wildlife, and management teams function best when they can be adaptive in responding to new information and evolving situations. But if there is to be considerable independence, there must also be careful evaluation and review to ensure the effective pursuit of program objectives.

Lesson 5.   Endangered species programs are more likely to succeed if there is a systematic effort to incorporate a knowledge of human values and socioeconomic systems. This information can be obtained through rigorous surveys, focused interviews, community appraisals, participant observation, and secondary data. It is also important to assess the complex structure of both formal and informal property and authority relations.

Lesson 6.   Agency leaders should strive to link science and policy and avoid undue emphasis on maintaining the bureaucracy. This type of leadership is difficult to achieve under the best of circumstances and is particularly challenging in complex endangered species recovery efforts. One means of achieving these objectives might be to adopt the Marine Mammal Commission's successful approach (Kellert 1986b). The Marine Mammal Commission (MMC) was created by the Marine Mammal Protection Act and is restricted to policy, science, and oversight responsibilities. All program implementation tasks are concentrated with the management agencies, the NMFS and FWS. The separation of policy from management functions has allowed the MMC to distance itself from bureaucratic, personnel and political pressures and focus its energies on the connection between scientific knowledge, policy options, and management performance. The creation of a similarly constituted "Endangered Species Commission" might provide the basis for more effective leadership focusing on the link between policy and science goals—as distinguished from management and bureaucratic objectives.

Lesson 7.   It is also important to increase the visibility and accountability of endangered species programs. One means of achieving this goal is to provide greater nongovernmental access to information about the progress of species recovery efforts through periodic, open, and systematic program review. A related approach might allow nonagency groups to participate in the planning and review process for endangered species recovery.

Lesson 8. A frequent cause of inadequate organizational performance is the absence of competitive pressure. This tendency is highly characteristic of government-controlled programs. It might be helpful to experiment with the creation of nongovernmental recovery teams that work in cooperative but independent association with agency groups. Another idea might be to subcontract recovery efforts to nongovernmental groups, particularly if government programs are consistently ineffective.

Lesson 9. Considerable conflict is associated with competing constituencies in many controversial endangered species programs. This war of values could be diminished by greater reliance on dispute resolution techniques. The situation might also be improved by developing more innovative public education and awareness programs intended to clarify competing values, provide information on recovery program objectives, and enhance public knowledge of imperiled wildlife.

Lesson 10. It is important to cultivate an appreciation among recovery teams and agency leadership of endangered species conservation as an ongoing process rather than a single event. The ambitious goal of enhancing intergroup cooperation and consensus around the concept of endangered species recovery necessitates long-term and fundamental change. The recovery of a species may represent but one stage in the extended effort to restore adequate ecosystem functioning and health.

Lesson 11. It has become apparent that basic shifts in agency culture and tradition at the FWS and NMFS may be necessary before the reforms suggested here can be implemented. Until there are fundamental shifts in agency perspective and organization, major improvement in endangered species conservation and recovery remains doubtful. Recommended alterations in agency culture include the following:

- Change the primary biological focus of the FWS and NMFS from game and commodity species to all wildlife and the ecosystems that support them. This will necessitate a major change in emphasis from large vertebrates to lower vertebrates and especially invertebrate life, as well as from the habitats of single species to ecosystem and landscape relationships.

• Broaden the focus of these agencies to include the equitable consideration of all values associated with wildlife—particularly ecological, naturalistic, aesthetic, and ethical values—as well as utilitarian benefits. Agencies must achieve a better understanding of the diversity of attitudes toward wildlife in our highly pluralistic society.

• Persuade these agencies to train and employ professionals with divergent backgrounds in the natural, social, and policy sciences. Professionals with varying demographic backgrounds should be placed in a variety of management and leadership positions.

• Expand the commitment of these agencies to aspire to be the biological arm of government rather than functioning as low-profile, secondary service agencies mainly concerned with the needs of traditional client groups. This means broadening the funding basis for public wildlife management through an expanded user fee system less dependent on taxation of consumptive users.

• Increase the willingness of these agencies to place national and even global conservation above the needs of parochial interests and states' rights. This will necessitate a far more ambitious system of national parks, forests, wildlife refuges, and national marine sanctuaries and a commitment not to defer to the demands of traditional interest groups or to state efforts to control wildlife.

The U.S. Endangered Species Act, as originally conceived, may be regarded as a model of rational policy formulation (Yaffee 1982; Kellert and Clark 1991). The ESA is remarkably comprehensive; it is unequivocal in its commitment to protect endangered wildlife; and it places strong emphasis on scientific management. Among its many forceful provisions are a definition of wildlife including all lifeforms, a mandate based on a wide range of public values, a scientific listing procedure, a methodology for designating habitat critical to a species' survival, prohibitions against the direct and indirect taking of endangered wildlife, restrictions against potentially harmful actions of federal agencies, a recognition of the global character of the endangered species problem, and harsh penalties against violators (Chapter 2). Various legal challenges have further strengthened the ESA's commitment to protect and recover endangered wildlife (Bean 1983).

Despite these extraordinary characteristics, considerable evidence calls into question the ESA's effectiveness as a means of conserving and restoring endangered wildlife. Based on various numerical performance criteria—the number of species listed, the pace of the listing process, the number of candidate species awaiting listing, the amount of critical habitat designated, the number of recovery plans proposed and approved, the number of species delisted for reasons of recovery, the number of listed and candidate species gone extinct, the frequency of interagency consultations, the number of restrictions on governmental or private actions—one can conclude that the ESA has not been especially successful (GAO 1988, 1992; Kellert and Clark 1991). Given the scope of the current global extinction crisis, an even more discouraging conclusion may be reached—that the ESA borders on failure as a policy mandate for protecting and recovering endangered wildlife.

The good news is that our society has never been more committed to the importance of protecting the nation's endangered wildlife. We are making progress in beginning to alter the inertial rush of an enormous freight train that has for centuries careened downhill in a single, devastating direction. We are just beginning to assert the will and develop the knowledge necessary to halt the ominous drift toward biological catastrophe. A necessary first step may require nothing less than major organizational reform of our federal and state agencies mandated to manage and protect endangered wildlife. If the analysis presented here is correct, basic changes may be needed in the operation and performance of various government agencies and their leadership.

This recommendation may seem utopian. Yet the current biological crisis demands bold and innovative action. We can no longer wait for the presumed progress of incremental reform. If the agencies rise to the enormous challenge of today's extinction crisis, future generations will laud them for their courage, wisdom, and insight. If they fail, they may be regarded as being as much to blame for the twentieth century's biological impoverishment as any of the more obvious causes of endangerment and extinction.

Acknowledgment

The author very much appreciates the helpful editorial suggestions of Professor John Wargo of the Yale University School of Forestry and Environmental Studies.

References

Argyris, C., and D. Schön. 1978. *Organizational Learning*. Reading, Mass.: Addison-Wesley.

Bean, M. 1983. *The Evolution of National Wildlife Law*. New York: Praeger.

Brewer, G., and P. deLeon. 1983. *The Foundations of Policy Analysis*. Homewood, Ill.: Dorsey Press.

Clark, C. 1973. The economics of overexploitation. *Science* 181:630–634.

Clark, T. W. 1992. Practicing natural resource management with a policy orientation. *Environmental Management* 16:423–433.

———. 1993. Creating and using knowledge for species and ecosystem conservation: Science, organizations, and policy. *Perspectives in Biology and Medicine* 36:497–525.

Clark, T. W., and J. Cragun. 1991. Organization and management of endangered species programs. *Endangered Species Update* 8:1–4.

Clark, T. W., and A. Harvey. 1988. Implementing endangered species recovery policy: Learning as we go? *Endangered Species Update* 5:35–42.

Clark, T. W., and S. Kellert. 1988. Toward a policy paradigm of the wildlife sciences. *Renewable Resources Journal* 6:7–16.

Clark, T. W., and R. Westrum. 1987. Paradigms and ferrets. *Social Studies of Science* 17:3–33.

———. 1989. High-performance teams in wildlife conservation: A species reintroduction and recovery example. *Environmental Management* 13:663–670.

Clark, T., R. Crete, and J. Cada. 1989. Designing and managing successful endangered species recovery programs. *Environmental Management* 13:159–170.

Clarke, J., and D. McCool. 1985. *Staking Out the Terrain: Power Differentials Among Natural Resource Agencies*. Albany: State University of New York Press.

Coggins, G., and I. Russell. 1985. Beyond shooting snail darters in pork barrels: Endangered species use in America. *Georgetown Law Journal* 70:1433–1525.

Etheredge. L. 1985. *Can Governments Learn?* New York: Pergamon.

Geist, V. 1988. How markets in wildlife meat and parts, and the sale of hunting and fishing privileges, jeopardize wildlife conservation. *Conservation Biology* 2:15–26.

General Accounting Office (GAO). 1988. *Endangered Species: Management Improvements Could Enhance Recovery Program.* GAO/RCED-89-5. Washington, DC: Resources, Community, and Economic Development Division.

———. 1992. *Endangered Species Act: Types and Numbers of Implementation Actions.* GAO/RCED-92-131BR. Washington, DC: Resources, Community, and Economic Development Division.

Goldman-Carter, J. 1983. Federal conservation of threatened species by administrative discretion or by legislative standard? *Boston College Environmental Affairs Law Review* 11:63–104.

Grumbine, R. 1992. *Ghost Bears: Exploring the Biodiversity Crisis.* Washington, DC: Island Press.

Houck, O. 1993. The Endangered Species Act and its implementation by the U.S. Departments of Interior and Commerce. *University of Colorado Law Review* 64:278–369.

Kellert, S. 1983. Affective, evaluative and cognitive perceptions of animals. In *Behavior and the Natural Environment,* I. Altman and J. Wohlwill (eds.). New York: Plenum.

———. 1985. Socioeconomic factors in endangered species management *Journal of Wildlife Management* 49:528–536.

———. 1986a. Social and perceptual factors in the preservation of species. In *The Preservation of Species,* B. G. Norton (ed.). Princeton: Princeton University Press.

———. 1986b. Marine mammals, endangered species, and intergovernmental relations. In *Ocean Resources and U.S. Governmental Relations in the 1980s,* M. Silva (ed.). Boulder: Westview Press.

———. 1987. The public and the timber wolf in Minnesota. *Anthrozoos* 1:100–109.

———. 1991a. Public views of wolf restoration in Michigan. *Transactions of the North American Wildlife and Natural Resources Conference* 56:152–161.

———. 1991b. *Canadian Perceptions of Marine Mammal Management and Conservation in the Northwest Atlantic.* Technical Report 91-04. Guelph, Ont.: International Marine Mammal Association.

Kellert, S., and P. Brown. 1985. Human dimensions information in wildlife management, policy, and planning. *Leisure Planning* 7:269–280.

Kellert, S., and T. Clark. 1991. The theory and application of a wildlife policy framework. In *Public Policy and Wildlife Conservation,* W. R. Mangun and S. Nagel (eds.). New York: Greenwood Press.

Kohm, K. A. (ed.). 1991. *Balancing on the Brink of Extinction: The Endangered Species Act and Lessons for the Future.* Washington, DC: Island Press.

Lopez, B. 1978. *Of Wolves and Men.* New York: Academic Press.

Norris, K. 1978. Marine mammals and man. In *Wildlife and America*, H. P. Brokaw (ed.). Washington, DC: Government Printing Office.

Peek, J., D. E. Brown, S. R. Kellert, L. D. Mech, J. H. Shaw, and V. Van Ballenberghe. 1991. Restoration of wolves in North America. *Wildlife Society Technical Review* 91(1):1–20.

Reading, R. P. 1993. Toward an endangered species paradigm: A case study of the black-footed ferret. Ph.D. dissertation, Yale University.

Reading, R. P., and S. Kellert. 1993. Attitudes toward a proposed black-footed ferret (*Mustela nigripes*) reintroduction. *Conservation Biology* 7:569–580.

Rohlf, D. J. 1991. Six biological lessons why the Endangered Species Act doesn't work—and what to do about it. *Conservation Biology* 5:273–282.

Scharff, J. 1980. Ethical issues in whale and small cetacean management. *Environmental Ethics* 3:241–279.

Shaw, W. W. 1977. A survey of hunting opponents. *Wildlife Society Bulletin* 5:383–396.

Tober, J. A. 1989. *Wildlife and the Public Interest: Nonprofit Organizations and Federal Wildlife Policy*. New York: Praeger.

Westrum, R., and K. Samaha. 1984. *Complex Organizations*. Englewood Cliffs, N.J.: Prentice-Hall.

Wilson, J. Q. 1980. *The Politics of Regulation*. New York: Harper & Row.

Yaffee, S. 1982. *Prohibitive Policy: Implementing the Federal Endangered Species Act*. Cambridge: MIT Press.

———. 1994. *The Wisdom of the Spotted Owl: Policy Lessons for a New Century*. Washington, DC: Island Press.

17

A Policy Sciences Perspective
Improving Implementation

Garry D. Brewer and Tim W. Clark

Implementation is a key step in the public policy process, but it often receives inadequate attention. All the case studies in this book are about policy implementation, specifically of the Endangered Species Act (ESA). Individually and collectively, they illustrate the importance and complexities of policy implementation as a social process. Regardless of what the formal ESA policy may call for, the real ESA policy is made in the implementation process. This is because the act's implementers often modify the formal policy through their actions. As a result, policy researchers distinguish between the formal and real policy (Brewer and deLeon 1983). The case studies in this book describe real policy. How then should scientists, managers, and administrators participate in real implementation processes? Who possesses the right knowledge and skills in endangered species conservation—wildlife biologists, agency bureaucrats, or nongovernmental conservationists? Can a broad knowledge of the policy process and the implementation stage help professionals implement ESA policy more effectively, regardless of their work setting? What must professionals know about the implementation stage to maximize the chances that ESA policy will be effective?

We will try to answer these and other questions central to improving endangered species policy implementation. We begin by presenting the the traditional scientific response to species restoration. We then present the policy sciences perspective, focus on several aspects of the implementation stage, and examine some options for improving implementation. We conclude by recommending that conservation biologists, agency participants, and others involved in ESA implementation develop and apply an explicit policy orientation. Our experience in endangered species conservation, natural resource management, and other aspects of implementation has shown that a policy orientation can result in improved professional responses, better management of programs, and more successful species conservation.

The Traditional Response

The traditional response to endangered species restoration and conservation is built largely on rational, disciplined, scientific processes. Endangered species restoration is often seen as a biological task requiring application of the biological sciences by the many people involved in conservation programs. To these people, this view seems only natural. After all, they say, the problem is about a biological species threatened with extinction—therefore the problem is biological and the solution is biological. This narrow view makes a lot of assumptions and, unfortunately, it is misleading.

The traditional approach to species restoration is deeply ingrained in current *modus operandi.* Over the last few hundred years, procedural matters have been invented, refined, and utilized to such a degree that science's epistemological foundation and methodology are accepted without reflection (Barrett 1978; Clark 1993). Science is eminently successful in framing problems and finding answers to questions within a certain realm—that is, situations that can be controlled. Science tests or develops theory by formulating, experimenting, confirming, refuting, reformulating, and refining with the ultimate end of understanding natural phenomena. Truth, both as an objective of science and as a concept, is embodied in scientific theory. Scientific theories—observations about particular sets of circumstances—can be organized such that predictions about the future are possible. Their success, however, is in large part due to the care with which scientists select the problems they examine.

But for the many problems today that fall outside the bounds that science has established, "the rational bias, tight discipline, and quantitative procedures erected to support the scientific edifice provide little help" (Brewer and deLeon 1983:3). Species restoration is one such problem. Science is necessary for conservation but not sufficient, a fact that may surprise some readers. There is much more to species restoration than the biological sciences. Understanding endangered species conservation as a policy implementation task embodies what is really involved in the work.

For practical problems, such as endangered species extinction, that cannot be entirely selected by biologists, managers, or administrators and demands a policy response, the traditional scientific approach may not suffice. No theory is capable of predicting whether ESA implementation will be successful, for example. The evidence both for and against ESA's success is considerable, but it is not as reliable as the evidence about most phenomena in the natural sciences and it cannot foretell outcomes or explain all the events and processes involved in species restoration. Nevertheless, outcomes do occur and events and processes are quite real. These "soft" variables have hard consequences, and they call for some means to understand and deal with them (Westrum 1988).

To understand how traditional science can contribute most to species conservation, it is essential to understand its limitations. One key area where traditional science falls short is understanding human values. Human values come into play in all aspects of species restoration. Values, according to Bem (1970) and Rokeach (1979), are preferred modes of conduct or end states of existence. Values have affective, cognitive, and directional components that guide us in making judgments, expressing preferences, making choices, and in our behavior (Reading 1993). Values figure prominently in all aspects of the human enterprise, including science. Yet traditional science has been unable to predict how values operate in complex policy processes, even though it has provided much insight. Apart from values, many other aspects of complex social and policy processes are not readily amenable to traditional science. It is vital to understand where and when science can help or inhibit policy improvements, including implementation in general and ESA implementation in particular.

The Policy Sciences Perspective

The cases cited in this book well illustrate the complexity and subtlety of policy processes, especially those in the implementation phase. A policy sciences view of the public decision process is somewhat different from that of many biologists, agency managers, and nongovernmental conservationists in ESA implementation.

Understanding Other Viewpoints

The ESA policy processes and implementation are subject to different interpretations. A review of almost any program shows that various participants have different views of what happened. In the din of conflicting voices, who really knows what happened or what is to be done? Brewer and deLeon (1983) and Simon (1985) note that the way people perceive reality is itself "unreal," as it is limited by intellect and colored by personality and experience. In other words, people involved in endangered species restoration will see, comprehend, and value events differently. When one person says that a program faltered for this or that reason, for example, that description only partly captures the reality felt by the full complement of participants and observers. So whose perception is correct or even "better"? Brewer and deLeon (1983:2) advise that programs should be operated "to be sensitive to the differences [in people's perceptions] by including them in a composite, evolving portrait of the problem and what needs to be done to cope with and overcome it." The value of this approach is that it introduces broad knowledge for democratic decision making and promotes a consensus-building process that may be essential to a program's success.

The species restoration problem is itself complex and ever-changing. Different parties, even within the same organization, will perceive the problem differently and thus favor different means to overcome it. Potential solutions advanced by one party may not be immediately clear to others. In some instances, solutions may depend on imagination for their development and implementation, a characteristic that all parties may not share. Solutions themselves may be problematic because they require change and action. In other words, being aware of a problem and trying to solve it can affect both the field of view and the very nature of the problem itself. The dynamic nature of problems and solutions, as conceived and acted upon by various participants, must be understood in all species restoration efforts if ESA is to be implemented well.

The Policy Sciences Response

The policy sciences were developed to deal with this kind of situation (Lasswell 1971). Professional policy scientists seek to understand and describe situations, outcomes, events, and processes in their real-life context. They also offer options for action to people with authority or those with the desire and ability to make a difference. The complexity of these technical and social situations—and the fact that these professionals do not get to choose their problems (endangered species problems are thrust upon them)—put policy scientists in a different situation compared to traditional scientists. It is often impossible to find the perfect solution in such situations. Technical matters, social complexity, human values and perceptions, uncertainty—all conspire to make that ideal unattainable (Reading 1993). Any solution, general or specific, must reflect these and many more requirements, including political feasibility, without which there can be no solution (Brewer 1981).

The policy sciences perspective views endangered species conservation as a social problem. The extinction crisis and its solutions mean very different things to decision makers, to biological scientists, and to the public. Moreover, many elements of this problem do not remain static: neither the problem, nor its context, nor possible solutions. They are all evolving based on their own dynamic and in response to human efforts to investigate and control them. What should be done in such circumstances? It is tempting to rely on traditional science and its many outstanding successes in seeking an optimal solution to the extinction crisis. In fact, though, the conventional scientific disciplines treat information in a highly selective manner and deal largely with what-has-already-occurred. In contrast, policy scientists focus on what-will-be. How, then, can the scientific method be employed in finding solutions to current and future social problems? It can be used in many ways.

Policy scientists are, of course, not without their own values and preferences, which may clash with those of the clients for whom the analysis is done. In most situations, there is a convergence between the values held by the policy scientist and the employer. When extreme divergence occurs, policy scientists should give top priority to the value of human dignity. This value overrides all other considerations. In practical terms, this may mean that the policy professional should resign. This choice is unlike that confronting scientific specialists, for their "client" is the discipline or, more broadly, science itself (Brownowski 1965).

Therefore, the policy sciences approach recognizes that human values are at the very center of policy processes, including endangered species implementation. The problems that policy scientists try to solve are created by society, not by the theoretical interests of a scientific discipline (such as conservation science). The extinction problem was defined by society and a response was codified in the ESA as amended. Because the extinction problem is a social problem, the policy scientist must consider numerous human perspectives to find solutions.

Foundations of the Policy Sciences

Policy is a term with many definitions. Tober (1989:142) offers this one: "Policy is a proposed course of action of a person, group or government within a given environment providing obstacles and opportunities which the policy was proposed to utilize and overcome in an effort to reach a goal or realize an objective or purpose." It is clear, then, that policy is subtle and complex. Tober (1989:142) goes on to say that "policy is necessarily an abstraction, therefore, to be approached through aggregative or summing analytical procedures. It is a pattern of behavior rather than separate discrete acts which constitute policy."

Real policy must be deduced as much from the collective actions of the policy participants as from a formal articulation of specific goals and procedures (Kellert and Clark 1991). The case studies in this book focus on real ESA policy. Policy development, often misconceived by traditional scientists, is not an attempt to replace political decisions with rational, objective decisions based on applied economics, operations research, or some other disciplinary perspective (Brunner and Ascher 1992; Brunner 1982). The policy sciences are not a simple modification of the standard approaches in common use today. They require a fundamental change in outlook. The challenge posed by policy research is not equivalent to that of conventional scientific work. The two approaches are distinct; they are different enterprises with different objectives.

The policy sciences' philosophical underpinnings can be found in the writings of John Dewey and other pragmatists in the early 1900s and more recently in the writings of Harold D. Lasswell and others (see Quade 1970; Lasswell 1971; Muth et al. 1990). Their approach and methods took substance in the 1920s and 1930s at the University of Chicago. These pioneers were interested in understanding patterns in human decision and policymaking in a fundamental sense. The policy sciences are both theory and applied guidelines that resulted from the decades-long investigations of these first policy scientists. The policy sciences center on problem-oriented, multimethod, comprehensive, and human-centered inquiry leading to purposeful action; they derive from the cumulative work of many scholars, researchers, practitioners, and analysts over the last half century. In the words of Brewer and deLeon (1983:9), the policy sciences "defined an approach concerned with knowledge of the decision or policy process and knowledge in that process. The policy sciences join and integrate theory (knowledge of) and practice (knowledge in) to improve them both for human benefit."

All the conventional disciplines are potentially policy-relevant, but none—not philosophy, history, economics, psychology, sociology, political sciences, or the physical, chemical, or biological sciences—has the same frame of reference as the policy sciences. The term "policy sciences" was chosen deliberately (Reynolds 1975): policy refers to the most important decisions of a society; science refers to the acquisition and use of reliable knowledge. Policy science is "an invitation to anyone concerned with both policy and science to share the distinctive frame of reference" (Brewer and deLeon 1983:9).

Operational Principles

Just as it is unlikely that a person can chart a successful travel route across the landscape without a map, it is also unlikely that one can navigate public policy processes, including ESA implementation, without a map. Maps orient us and ascertain location, movement, and direction. To be most useful, they must portray all the significant parts of the "landscape"—in this case, all parts of the endangered species policy process—and they do not remain static. Social and biological sciences theory as well as practical experience assist in policy-relevant mapmaking, but neither alone is sufficient. In Chapter 4, for example, Reading and Miller discuss a case where essential knowledge about the restoration task was ignored or dismissed by decision makers to the harm of the species and the overall conservation effort. Policy scientists would conclude that officials were using an inadequate map. To address the complexities of endangered species conservation, theory must be connected appropriately and realistically with on-the-ground application. Improved modes of thinking and mapmaking are necessary in such cases. Human situations of great complexity and uncertainty provide the fundamental reason for the development of the policy sciences as a distinctive approach to improve thinking, mapmaking, and action.

To locate a problem with respect to its form, content, status, and age is to begin to solve it. A new problem is not the same as an old problem. When a species is first thought to be endangered, for example, an initial and tentative recognition of a problem occurs. More must be learned and an interpretation made about what it might mean. After these initial information-building activities, a decision may be made about an appropriate response. Then it must be implemented. Work is undertaken, time passes, and the problem changes—for better or worse. We learn more about it through active engagement with it. What are its consequences? Did the conservation program succeed or did it make matters worse? Depending on the answer, it may be necessary to bring the effort to a halt or at least to change course or try another approach. Knowing where a problem is located in the policy process is a vital procedural and substantive matter.

A key activity in all this is to learn how to organize, compare, and accumulate knowledge about the overall policy process itself and its manifestations. Methods of doing so are not fixed. Indeed, they vary widely and their effectiveness is limited by the experience, judgment, and skill of the people involved and by the nature and context of the problem itself. If knowledge about the process were limited to a single disciplinary perspective, sufficient coverage of a conservation problem might not be possible. It is vital to understand that in complex policy matters, no single person or organization is responsible for the entire process. Many parties are involved, each holding a somewhat different perspective that may obstruct or promote smooth operations. Each party may see the problem differently. A number of the endangered species cases examined in this book describe this multiple reality and the consequent competition among groups, sometime fierce, to gain dominance or impose one group's reality on others. To understand this process and participate effectively, a constantly updated map of this dynamic is necessary.

Methods

The policy sciences offer a set of principles by which to contribute to such complex processes: problem orientation, contextuality, and multiple methods (Lasswell 1971). Clark (1992) has described these principles in the context of natural resource management. First, problem orientation requires that five tasks be conducted for a problem to be understood and addressed. Brunner (1991) has described these tasks as standpoints that must be developed:

1. Historic standpoint: description of trends in the problem, including analysis of both context and process

2. Scientific standpoint: explanation of trends

3. Projective standpoint: projection of trends

4. Normative standpoint: evaluation and projection of trends

5. Operational standpoint: intervention, evaluation, and selection of alternatives that might solve the problem

Second is understanding the context of the problem and "understanding the relationship between the parts and the whole" (Brewer and deLeon 1983:13). This principle requires knowledge about the past, the present, and what is likely to happen in the future. Contextuality requires a comprehensive conceptual framework to direct a person's attention to the details as well as to an evolving appreciation of the whole. We must know who the participants are, what their perspectives are, whether the situation is organized or not, what strategies they seek, what values they are promoting, and what outcomes they prefer (Lasswell 1971).

Third, a wide variety of methods can be used in problem orientation and contextual mapping. A host of biological and social science methods are called for in endangered species conservation, including surveys of the species' status, surveys of public values, and surveys of how well the program is working, among many others. (See Manheim 1977 and Romesburg 1981.) Not only are content methods needed to get a fix on substantive issues, but procedural methods are needed to see how problems come to be understood and how their contexts are viewed and dealt with. Ideally this array of methods will result in a comprehensive and comprehensible understanding of the problem.

The policy sciences offer guidance to anyone interested in understanding complex human problems, including endangered species conservation. The beginner's skill in applying the policy sciences will be limited, of course, but with time and experience, significant insights and achievements are possible. The policy sciences are not a panacea, however, nor are they foolproof. Applying the policy sciences is hard work. But surely it is worth the effort. Significant decisions are being made daily in endangered species conservation— many of them based on incomplete problem orientation, inadequate contextual mapping, and limited methods. The policy sciences offer a means to avoid oversights, eliminate problem-solving weaknesses, and identify goals worth striving for. They offer a significant opportunity to help us understand the basic problems confronting, in this case, humanity's growing desire to preserve biological diversity.

The Policy Process Model

Understanding a complex endangered species case requires an appreciation that the effort is part of a larger policy process. And this includes recognition by the public that the loss of species is a problem for the nation. Proposals are advanced, debated in the media and in scientific circles, and eventually considered by Congress. Congress, in turn, formulates a national response to the problem and authorizes a course of action. Government agencies are then given responsibility for implementing Congress's policy prescriptions. Numerous interests participate in implementation, wield their influence, and offer their evaluations about how well the policy is being applied. Often changes, fine-tuning, or even alternative ways to implement the policy are suggested by nongovernmental interests. If the policy as implemented is successful in solving the endangered species problem as defined, then it can be terminated or changed to solve some new problem.

This simplified description shows that from beginning to end the endangered species policy process follows more or less predictable steps. As shown in Figure 16-1, essentially six steps can be recognized (Brewer 1973):

1. *Initiation:* recognition of a problem; creative thinking about it; preliminary investigation of concepts and claims.

2. *Estimation:* scientific study of the problem, likely impacts, and outcomes; normative assessments; development of outlines of a programmatic response.

3. *Selection:* focused debate on the issues; choice about a program to solve the problem.

4. *Implementation:* development and application of a specific program.

5. *Evaluation:* comparison of estimated performance of the program with what was actually attained; reconciliation of the differences.

6. *Termination:* stopping the program or changing it to solve a new problem.

Despite its oversimplification, this policy model is useful both conceptually and practically. It is easy to learn and apply.

The policy sciences perspective provides a constructive standpoint for species restoration and opens up new understanding and insights into the process—especially to those who approach the task only from the traditional biological, disciplinary, or agency viewpoints. It is also useful to those who may tend to oversimplify contexts or demonize their opponents. The policy sciences are intellectually and practically well grounded in philosophy, science, and human affairs. They have been developed and applied in a wide variety of circumstances in recent decades. Their operational principles—problem orientation, contextuality, and multiple methods—can be enormous assets to improving policy processes and implementation in endangered species conservation. The policy sciences model, whatever its variations, identifies several phases through which nearly all policies and programs go. What happens in the early phases greatly affects what happens later.

Policy Implementation

Implementation is the stage where actions are done to or for endangered species. Despite its importance, it is frequently overlooked as an explicit step deserving close attention. Without proper implementation, however, the ESA is of little value. The significance of the implementation phase in policy processes became fully appreciated in the 1970s during an intense period of policy research. (See Pressman and Wildavsky 1973; Bardach 1977; Fesler 1980; Nakamura and Smallwood 1980.) Today there is a substantial literature on implementation that includes case studies and conceptual frameworks linking levels of policy and programs, organization theory, and analysis. This body of literature, although it offers considerable insight into effective implementation, has hardly been applied to work on endangered species. In fact, this book was undertaken to enlarge the body of literature on the ESA's implementation as a basis for learning and improvement.

Policy Roles and Implementation

Professional conservationists seldom see themselves as policy implementers who need to learn about implementation theory. Yaffee (1982:69), however, concludes from his study of ESA implementation that it is "not much different from implementing other types of policy." Knowledge of implementation in general could surely aid people involved in species restoration.

Many people mistakenly distinguish between the roles of policymaker and policy implementer, believing that the two are separate and different. Clark (1992) has examined the activities of natural resource managers, conservationists, and wildlife biologists as policy implementers in the overall natural resource policy and management process. Conventional wisdom in these professions holds that policy is formulated by "policymakers"—legislators, the courts, and top-level administrators—and that most biologists and managers have nothing to do with it. These people see themselves as simply fulfilling technical roles without being intimately involved in the policy process (Gruber 1987). Such a view is not only incorrect. It also limits improvements in endangered species conservation.

Frontline biologists and managers do in fact implement—something very different from being only a "biologist" or a "resource manager." These people, whether they understand it or not, are involved in streams of decisions and actions that collectively determine what happens to wildlife. They affect implementation by exercising their considerable discretion and by providing information to policymakers. In failing to see themselves as implementers, they are merely reflecting their profession's norms. These people possess specialized language and thought that impose their own categories on the world of experience and serve as a lens through which they view themselves and the outside world and act upon that view. And like all such frameworks of thought and action, this one—through its basic design, its fundamental propositions, and its intellectual, sociological, and political arrangements—hides alternative possibilities. Enforced too strongly, such a view can block critical insight and change.

The process of policy implementation is largely determined by government bureaucracies. In endangered species conservation, agency implementers create real ESA policy by their actions. They have great discretion in applying the ESA, and they use it to fit the policy to their bureaucracy's ideology and needs. Yaffee's (1982) study of ESA implementation notes that government experts use a "mix of science, art, and politics" in making decisions, that "their individual attitudes, values, and professional norms weigh significantly in the process," that the process hides "enormous amounts of administrative discretion," and that ESA policy is "heavily influenced by the political context in which implementation takes place." These observations reveal biologists, managers, and agency administrators to be policy implementers, with their own values and political interests, whether they see themselves that way or not.

Implementation is a complex process in which people's aspirations are turned into specific actions through government programs. Thus the structure and conduct of programs, as well as organization theory and design, receive considerable attention in analysis of policy implementation. The cases in this book describe a range of responses to the endangered species problem. Although each species and its problems are unique and the government's programs contain certain local elements, it is easy to see the common conditions that are needed for good policy implementation: Goals and objects must be specified; they must be specified in operational terms; programs must be designed; responsibility for their conduct must be assigned; and resources must be allocated for their operation (Brewer and deLeon 1983). This simple sequence, however, is far more demanding than it might seem—as the many cases in this book amply illustrate.

Forces Shaping Implementation

Several causes of implementation failure must be understood if they are to be avoided. In some cases, the implementation task is given to agencies ill suited or even opposed to the policy (see Chapter 4). Successful implementation depends not only on commitment to the policy's aims—in this case the ESA—but also the organization's ability to carry them out. Other failures result from inefficient, inept, corrupt, or deceitful practices that, individually or collectively, can result in muddled programs or, in extreme cases, failed efforts and extinction of a species.

Three additional causes involve the context of implementation, the enabling legislation, and the nature of the implementing agencies. It is vital to appreciate the real-life context in which implementing decisions are made—including who participates in implementation. The cases examined in this book contain many diverse parties, each with a different role to play. Bureaucratic agencies play a central role in ESA policy implementation, including the U.S. Fish and Wildlife Service, National Marine Fisheries Service, and many more at the federal level. At the state level, the central players are game and fish departments, wildlife management agencies, and departments of natural resources, as well as other bureaucracies. Bureaucratic rivalry clearly plays a role in ESA implementation. The judicial system is important, too, since implementation is seldom free from conflict. ESA implementation actions have gone to the courts in several precedent-setting cases, and litigation will continue to be an important dimension. There are also nongovernmental interest groups, sometimes numerous and powerful: industry, conservation groups, and many more participate in most endangered species restoration efforts. Thus the context is a matter of paramount consequence.

The policy selection/implementation connection should not be overlooked either. For example, the ESA policy selected must match the implementation capabilities of the organizations assigned to carry it out. Chapters 2 and 3 examine this connection in some detail. Programs to implement ESA policy rarely start with a clean slate.

Usually existing federal and state programs are assigned responsibility for implementation, and these programs bring accumulated baggage that may be overwhelming—especially when a multiplicity of federal and state agencies are given responsibility for implementation. Many structural and procedural matters need attention. Even if ESA were a perfect prescription, its success would depend on the implementing agencies and their common problem of bureaucratic inertia. Old values, fixed operating procedures, organizational history, and many other factors figure prominently into prospects for successful implementation. In such bureaucratic settings, the likelihood of effective implementation is dim. One strategic response to such circumstances is to shift implementation responsibility away from the government and into the quasi-public or private sector (such as The Nature Conservancy or other not-for-profit organizations).

The nature of the implementing agencies is vital as well. Some agencies seem incapable of implementing policy well or managing programs successfully. The increased demands made on government with fewer resources can figure into a program's faltering or failing. There are two common responses to weak government implementation. First is centralization of information and authority to extend the reach of government. Second is decentralization and delegation to reduce the central government's ability to control implementation. There are other reasons to centralize or decentralize, but both strategies are evident in the ESA case material in this book. Neither is a cure-all; both strategies suffer from problems well illustrated in the cases. Finally, because ESA implementation falls largely to bureaucrats, it should not be forgotten that bureaucrats are human and suffer from many weaknesses—confusion, lack of energy, muddleheadedness, self-interest, and much more. Even if the ESA were the most carefully written prescription, it would still be open to interpretation and misrepresentation. Bureaucrats are not renowned for leadership or creative problem solving.

Several other factors that influence implementation must also be kept in mind: the source of the policy, clarity of the policy, support for the policy, complexity of the administration, incentives for implementers, and resource allocation (Brewer and deLeon 1983). Each of these factors is affected by others in the context. Fesler (1980) has noted that implementation success is related to able and committed leadership, clear objectives, organizational capabilities, and initial success, while implementation failure is related to inertia, ill-defined clientele, inadequate scope of effort, weak organizational capabilities, inadequate funds, and loss of leadership. Although they are not ordinarily analyzed in these terms, variables accounting for both success and failure appear in many of the endangered species cases.

Practical Applications

A better understanding of several paradigms of the social and biological sciences could improve ESA implementation by showing us how implementation works or fails in real political contexts. Much can be gained by reexamining policy implementation through the lens of several disciplinary perspectives.

Multi-Disciplinary Viewpoints

To improve our understanding of both the process and practice of implementation, we must understand it from multiple perspectives simultaneously. *Valuation,* for example, recognizes that people ascribe values to wildlife, and the variables that shape people's values figure significantly in efforts to conserve endangered species. Comprehensive study of the valuational aspects of species restoration is essential if we are to understand the attitudes held by people influencing ESA implementation, by people involved in the actual programs, and by people affected by the programs' outcomes (Reading 1993). This knowledge, however, is often lacking in endangered species restoration efforts (Kellert 1985).

Power, as discussed by Reading (1993), is a major variable in almost all human endeavors. Weber (1968:53) says that power is "the probability that one actor within a social relationship will be in a position to carry out his will despite resistance, regardless of the basis on which this probability rests." Power is a characteristic of both individuals and organizations and derives from authority and control of resources (Lindblom 1980; Wilson 1980). Knowing about the sources and uses of power—and using it appropriately—are essential in every program.

Scientific management, based on an engineering approach, was espoused early in this century as a way to operate organizations efficiently. Bureaucracies were set up for this kind of efficiency and treated as closed, mechanical systems (Perrow 1986). As it was assumed that both the system and the individual (the "mechanical bureaucrat") sought optimal behavior, rules, regulations, and routines were used to control behavior. Much of the behavior of today's bureaucracies implementing ESA policy is based on this theory, although there is growing evidence that many implementation weaknesses flow from these assumptions (Wilson 1980).

Innovation perspective, a reaction to bureaucratic management, recognizes that people are not machines and their behavior cannot easily be regulated or routinized (Daft 1983). Since bureaucracy prescribes rules, roles, and regulations for all actions, innovation is the only means to combat its ineffectiveness and inefficiency. The weaknesses of bureaucracy are widely known, but correcting them is difficult as innovations are often blocked or stifled (Wilson 1980). Many different organizational alternatives based on cooperation and participation have been proposed. But replacing deeply ingrained bureaucratic practices with a different set of perspectives is seldom easy and usually causes pain.

Psychological perspectives of implementation include people's personalities, interests, and training, all of which have significant effects on implementation (Daft 1983). Designing incentives that accommodate human psychology is essential for a functioning program. Differences in people's early or late adoption of new approaches is related to their psychological predisposition as innovators, imitators, or resisters. These predispositions may be directly mirrored at organizational or professional levels. Differences exist, too, in how people adopt organizational norms or professional ethos.

Small-group theory is another area vital to understanding current ESA implementation and expediting improvements. Small groups play key roles in endangered species conservation as recovery teams, field teams, working groups, and core decision groups (Chapter 14). Small, dynamic groups are often able to diagnose causes of problems and plan for improvement (Varney 1989). There are several major concerns in small-group dynamics: starting up, defining members' roles and responsibilities, setting individual and group goals, solving problems and making decisions, and leading effectively. Clark and Westrum (1989) describe small-group structure and dynamics in terms of high-performance teams in wildlife conservation. As important as small groups are, the outcome of implementation involves other organizational considerations as well.

Organization theory is important in understanding policy implementation. Organizational studies are common in the published literature of public and business administration and policy research (Wilson 1980; Daft 1983; Perrow 1986). Organization theory can provide numerous insights into how and why policy implementation takes the forms it does, and it can offer options for organizing and managing endangered species restoration efforts more usefully. Clark et al. (1989) have suggested ways to design and manage endangered species recovery programs, for example, and Clark and Cragun (1994) offer organizational and managerial guidelines for such complex tasks. Two key variables in organizational systems are structure and culture; both can be managed to improve implementation.

Communication theory and *cybernetics* offer additional insights into implementation. Research into networks and feedbacks of formal and informal communication has demonstrated their central role in implementation—for example, by identifying who supplies key knowledge (technical, ideological, political) about the implementation task (Brewer and deLeon 1983). Cybernetic control theory is helpful in understanding systems-level issues; the concepts of lead time, loads, and lags in system responsiveness are essential to understanding policy implementation. Complex implementation systems with lengthy chains of command and communication and dispersed authority are slow to act.

Biology, of course, is vital to ESA implementation. At one level, species restoration is about the biological sciences, and recovery efforts have traditionally focused on the biological aspects of the problem. Autecology (life history characteristics, habitat requirements, behavior), population ecology (demographics, genetics, social structure, scarcity/abundance, dispersal), community ecology (predator/prey relations, competition, biotic and abiotic interactions), and habitat considerations (quality, quantity, productivity, resilience, insularization) are among the many variables essential to conserve species (Chapter 12). This kind of knowledge directly affects implementation.

Implementation, then, is complex in theory and in practice; none of the disciplinary perspectives gives complete understanding. The best prospects for good implementation come from weaving all these perspectives into a whole picture.

Implementation as Social Process

ESA implementation is a social process. Participants, their perspectives, values, strategies, and desired outcomes are key reference issues in this social process. Like much policy implementation, the ESA implementation effort can be very complex. Many things demand consideration.

First, it is clear that policies and programs that closely accord with the values, needs, and power of implementers are most likely to succeed. Thus it is important to find a match between the requirements of ESA policy and its potential implementers. The case studies in this book demonstrate more than one mismatch.

Second, two paradigms are especially useful to synthesize the various disciplinary perspectives of endangered species conservation into a coherent whole: the policy sciences and organization theory. The policy sciences perspective was described earlier. Organization theory takes into account the nature of the problem itself, its size and complexity, its context, the difficulty of solutions, the temporal and other constraints, and how best to organize and manage the program to solve the problem (Wilson 1980; Daft 1983; Perrow 1986). Small, simple, and flexible organizational responses are more likely to succeed than big, complex, and rigid bureaucratic responses (Vroom and Jago 1988).

Third, given the complexity of the implementation process, it is a wonder that anyone's plan for action ever gets used. The process should be thought of as a social learning experience in which participation is maximized, evaluations are frequent and open (both inside and outside the implementation effort), and flexibility is the watchword. Whatever the implementation mechanism, it should be designed as simply as possible.

Fourth, improving implementation will require fundamental changes in bureaucratic thinking about its standard "solution" to complex endangered species policy implementation. Those with power in the executive, legislative, and judicial branches should consider that endangered species conservation might best be accomplished though quasi-governmental or nongovernmental mechanisms. Enlarging the base of participation in restoration programs beyond the current domination of government bureaucracies is warranted on several grounds. Considerable knowledge, experience, and skill exist outside government. Broadened participation in decision making and implementation meets the overriding goal of enhancing democratic values. The ubiquitous weaknesses of government-dominated ESA implementation force one to consider alternative strategies.

In the final analysis, policy implementation is a social process, not a scientific one. The cases in this book illustrate many manifestations and contexts of the complex social process of endangered species conservation. Unless we understand this fact and conceive a useful model of the social process, effective participation and organization may be impossible.

Developing and Using a Policy Orientation

For the many field biologists, agency administrators, and other participants at the heart of the current effort to restore endangered species, there is no time to learn the professional skills of a policy scientist, despite their obvious advantages. Two solutions to this problem come to mind. First, the perspectives of the policy sciences can be worked into ESA implementation on a case-by-case basis by bringing professional policy scientists into the process. Second, the many professionals deeply immersed in endangered species conservation could develop a policy orientation on their own. Such an enlightened orientation would complement rigorous science so that scientific professionalism need not be sacrificed in the process.

Unfortunately, few university programs or in-service government training programs offer professionals the opportunity to develop such an orientation. Through such programs, professionals involved in endangered species conservation would largely retain their traditional norms but would come to understand policy processes better and would come to view themselves as policy implementers in many situations. An introduction to the policy sciences, a policy model, and detailed consideration of the implementation phase have already been offered here. The case for conservation managers and biologists developing and applying a policy orientation themselves has been presented by Clark (1992, 1993), Clark et al. (1992), and others. It is reasonable to expect that such an orientation would help conservation biologists and others become more effective in achieving their conservation aims.

Few would deny that the ultimate causes of species endangerment are human: valuational, economic, social, and political. For conservation biologists and managers to be successful in ESA implementation, they must understand the processes that drive environmental degradation and be proficient at providing remedial strategies and tactics. As Pool (1990:673) has noted: "Clearly, there is a whole new set of considerations for a researcher whose work has consequences for public policy, even though his basic role—getting answers—is unchanged." Orr (1990:9) asks: "How should those calling themselves conservation biologists deal with politics and the question of management in their research, writing and teaching?" And Norton (1988:238) questions whether conservation biologists have an obligation to "participate with the public in a debate regarding the very nature of ecological health, even while trying to protect it."

Can professionals in the biological sciences actually play a useful role in ESA implementation without sacrificing their effectiveness and credibility as scientists? We believe the answer is yes. Clark and Kellert (1988:7) observe that if the field of conservation science "is to contribute fully and adequately to the critical societal decisions affecting the future abundance and well-being of our nation's flora and fauna, then it seems essential that young . . . professionals be sufficiently educated in the complexities, subtleties and techniques of the policy process."

Lessons

Although implementation is often overlooked as a definite stage in the policy process, specific knowledge is needed to implement policy well. It is essential, for example, to understand the multiple values and perspectives at play in this complex social and policy process. But the traditional scientific responses to endangered species conservation—such as that of wildlife biology—are ill suited to deal with the complexity and intricacies of implementation.

The policy sciences approach was developed to deal with exactly this kind of situation. Policy scientists utilize a set of operational principles—including contextual mapping, problem orientation, and use of multiple methods—to aid in practical problem solving. Indeed, the policy sciences provide an explicit model to guide ESA implementers. Many factors shape implementation, including commitment and organizational capability. Bureaucracies often seem particularly ill equipped to carry out ESA implementation effectively. Multiple perspectives are simultaneously needed to understand and implement ESA policy. In sum, then, an explicit policy orientation is necessary if the professional is to become fully effective in the implementation game.

Acknowledgments

Denise Casey and Steve Primm critically reviewed a draft of this chapter.

References

Ascher, W. 1986. The evolution of the policy sciences: Understanding the rise and avoiding the fall. *Journal of Policy Analysis and Management* 5:365–373.

Bardach, E. 1977. *The Implementation Game*. Cambridge: MIT Press.

Barrett, W. 1978. *The Illusion of Technique: A Search for Meaning in a Technological Civilization*. New York: Doubleday/Anchor.

Bem, D. J. 1970. *Beliefs, Attitudes, and Human Affairs*. Belmont, Calif.: Brooks/Cole.

Brewer, G. D. 1973. Experimentation and the policy process. In *25th Annual Report of the Rand Corporation*. Santa Monica, Calif.: Rand.

———. 1981. Where the twain meet: Reconciling science and politics in analysis. *Policy Sciences* 13:269–279.

Brewer, G. D., and P. deLeon. 1983. *The Foundations of Policy Analysis*. Homewood, Ill.: Dorsey Press.

Brownowski, J. 1965. *Science and Human Values*. New York: Harper & Row.

Brunner, R. D. 1982. The policy sciences as science. *Policy Science* 15:115–135.

———. 1991. The policy sciences as a policy problem. *Policy Sciences* 24:65–98.

Brunner, R. D., and W. Ascher. 1992. Science and social responsibility. *Policy Sciences* 25:295–331.

Clark, T. W. 1992. Practicing natural resource management with a policy orientation. *Environmental Management* 16:423–433.

———. 1993. Creating and using knowledge for species and ecosystem conservation: Science, organizations, and policy. *Perspectives in Biology and Medicine* 36:497–525.

Clark, T. W., and J. R. Cragun. 1994. Organizational and managerial guidelines for endangered species restoration programs. In *Restoring Endangered Species*, M. L. Bowles and C. Whelan (eds.). Cambridge: Cambridge University Press.

Clark, T. W., and S. R. Kellert. 1988. Toward a policy paradigm of the wildlife sciences. *Renewable Resources Journal* 6:7–16.

Clark, T. W., and R. Westrum. 1989. High-performance teams in wildlife conservation: A species reintroduction and recovery example. *Environmental Management* 13:663–670.

Clark, T. W., R. Crete, and J. Cada. 1989. Designing and managing successful endangered species recovery programs. *Environmental Management* 13:159–170.

Clark, T. W., P. Schuyler, T. Donnay, P. Curlee, T. Sullivan, M. Cymerys, L. Sheeline, R. Reading, R. Wallace, T. Kennedy, Jr., A. Marcer-Battle, and Y. DeFretes. 1992. Conserving biodiversity in the real world: Professional practice using a policy orientation. *Endangered Species Update* 9(5/6):12–16.

Daft, R. L. 1983. *Organization Theory and Design*. St Paul: West.

Fesler, J. W. 1980. Implementation: Success and failure. In *Public Administration: Theory and Practice*, J.W. Fesler (ed.). Englewood Cliffs, N.J.: Prentice-Hall.

Gruber, J. E. 1987. *Controlling Bureaucracies: Dilemmas in Democratic Governance.* Berkeley: University of California Press.

Kellert, S. R. 1985. Social and perceptual factors in endangered species management. *Journal of Wildlife Management* 49:528–536.

Kellert, S. R., and T. W. Clark. 1991. The theory and application of a wildlife policy framework. In *Public Policy and Wildlife Conservation,* W. R. Mangun and S. Nagel (eds.). New York: Greenwood Press.

Lasswell, H. D. 1970. The emerging concept of the policy sciences. *Policy Sciences* 1:3–14.

———. 1971. *A Pre-view of the Policy Sciences.* New York: American Elsevier.

Lindblom, C. E. 1980. *The Policy-Making Process.* Englewood Cliffs, N.J.: Prentice-Hall.

Manheim, H. L. 1977. *Sociological Research: Philosophy and Methods.* Homewood, Ill.: Dorsey Press.

Muth, R., M. M. Finley, and M. F. Muth. 1990. *Harold D. Lasswell: An Annotated Bibliography.* New Haven: New Haven Press.

Nakamura, R. T., and R. Smallwood. 1980. *The Politics of Policy Implementation.* New York: St. Martin's Press.

Norton, B. G. 1988. What is a conservation biologist? *Conservaton Biology* 2: 237–238.

Orr, D. 1990. The question of management. *Conservation Biology* 5:10–15.

Perrow, C. 1986. *Complex Organizations: A Critical Essay.* New York: McGraw-Hill.

Pool, R. 1990. Struggling to do science for society. *Science* 248:672–673.

Pressman, J. L., and A. Wildavsky. 1973. *Implementation.* Berkeley: University of California Press.

Quade, E. S. 1970. Why policy sciences? *Policy Sciences* 1:1–2.

Reading, R. P. 1993. Toward an endangered species reintroduction paradigm: A case study of the black-footed ferret. Ph.D. dissertation, Yale University.

Reynolds, J. F. 1975. Policy science: A conceptual and methodological analysis. *Policy Sciences* 6:1–27.

Rokeach, M. (ed.). 1979. *Understanding Human Values: Individual and Societal.* New York: Free Press.

Romesburg, H. C. 1981. Wildlife science: Gaining reliable knowledge. *Journal of Wildlife Management* 45:293–313.

Simon, H. A. 1985. Human nature in politics: The dialogue of psychology with political science. *American Political Science Review* 79:293–304.

Tober, J. A. 1989. *Wildlife and the Public Interest: Nonprofit Organizations and Federal Wildlife Policy.* New York: Praeger.

Tobin, R. J. 1990. *The Expendable Future: U.S. Politics and the Protection of Biological Diversity.* Durham, N.C.: Duke University Press.

Varney, G. H. 1989. *Building Productive Teams: An Action Guide and Resource Book.* San Francisco: Jossey-Bass.

Vroom, V. H., and A. G. Jago. 1988. *The New Leadership: Managing Participation in Organizations.* Englewood Cliffs, N.J.: Prentice-Hall.

Weber, M. 1968. *Economy and Society: An Outline of Interpretive Sociology.* Vol. 1. New York: Bedminster Press.

Westrum, R. 1988. Organizational and interorganizational thought. Paper delivered at World Bank Conference on Safety Control and Risk Management, October 1988, Washington, DC.

Wilson, J. Q. 1980. *The Politics of Regulation.* New York: Harper.

Yaffee, S. L. 1982. *Prohibitive Policy: Implementing the Endangered Species Act.* Cambridge: MIT Press.

PART IV

Lessons

Lesson-drawing is practical; it is concerned with
making policy prescriptions that can be put into
effect. Lessons are not learned in order to pass
examinations; they are tools for action. . . . Lessons
can be positive, leading to prescriptions about
what ought to be done.
RICHARD ROSE

18

Synthesis

Tim W. Clark, Richard P. Reading, and Alice L. Clarke

There is a deeply rooted national commitment to preserving healthy populations of plant and animal species in the United States. This social goal has been apparent in activities from grassroots conservation groups to international organizations, in the operation of government agencies directed to manage the nation's living resources, in the growing awareness and changing behavior of U.S. citizens, in popular culture and media coverage, and in many other facets of American life. The nation's commitment to maintaining biological diversity has been codified in several important pieces of legislation—above all, in the 1973 Endangered Species Act (ESA) as amended. But the twenty-year record of conservation efforts since the act's passage—including the cases detailed in this book—is not one of successful implementation of the law or, presumably, public will. Indeed, the ESA's history has included many controversies, battles, and species losses but few success stories. Currently, as we approach yet another reauthorization of the ESA, there is considerable debate about how endangered species protection and recovery policies should best be prescribed by law. Much less attention, however, has been given to how the law should be implemented, particularly with regard to two variables that are critical to success: professional and organizational performance.

Despite dedication and hard work by many people in and out of government and vast expenditures of time and money, the ESA's implementation has been weak. The weak performance is fairly well documented (see Chapter 1) but poorly understood. As a result, it has been difficult to build consensus on how to improve the ESA and its implementation. Still, some restoration programs clearly work better than others. We believe this is because of the kinds of professionals involved, the way these experts go about their work, and the organizational arrangements they employ. As noted in Chapter 1, professions

and organizations are the repositories of much of the knowledge society requires to solve its problems; thus they lie at the very heart of
ESA implementation. Yet they have received little systematic attention to date. This is especially unfortunate because there is much potential for improving these elements to match the demands of the ESA
implementation task. Contributors to this book have documented
both the profound need and the numerous opportunities that exist to
improve professional and organizational responses to the endangered
species crisis.

The preceding chapters have cited a number of practical lessons
that are often overlooked but must be heeded if these improvements
are to become reality. Many of the lessons derive from the experiences
of participants in ongoing programs—including some of the highest-
profile programs in the United States, programs that receive some of
the greatest attention, resources, and talent this nation has to offer.
This firsthand, expert experience of the way the ESA is actually implemented across a broad range of species and circumstances provides
a practical basis for suggesting improvements. Added to these firsthand experiences are the views of a broad range of conservation and
social scientists, including conservation biologists, conflict and dispute
resolution managers, organization theorists, sociologists, and policy
scientists, most of whom have been directly involved in endangered
species issues for years. Collectively, the authors in this book have
over two hundred years' experience working on endangered species
conservation, and their cumulative knowledge constitutes an invaluable source of ideas for improving ESA implementation. And besides
the contributors to this book, there are, of course, many other highly
qualified people whose enormous experience, knowledge, and insights are waiting to be tapped. With this brief background, let us turn
to a discussion of the key lessons offered in the preceding chapters and
some possible directions for the future.

"Meta-Lessons"

In this chapter we distill the dozens of useful lessons offered
throughout the book. As many lessons from different chapters are
quite complementary, we have grouped similar lessons together to
create a set of "meta-lessons" that capture the most important guidelines for improvement. By summarizing, we do not wish to diminish
the importance of the individual lessons. They have been hard-won

from field experience or scholarly pursuit. We hope, instead, to illustrate the similarities of implementation problems and to demonstrate that the task of improving ESA implementation is possible by attending to a few significant variables. Some of these lessons are already familiar to professionals in the conservation community and even to the public; others are not so obvious.

Meta-Lesson 1: Endangered species conservation is a multifaceted task of interacting biological, professional, sociological, organizational, economic, political, and policy dimensions. Regardless of the biological status of the species and its habitat, the ultimate causes of most species' endangerment lie in human values that are manifest in varying social, economic, and political institutions and activities. All of these complex "ultimate causes," as well as the biological features of the conservation task, must be integrated into a holistic understanding of the problem that should then receive the interdisciplinary focus of the conservation community. Attempting to restore species by ignoring everything but the species' biology invites failure.

Viewing the endangered species crisis from such a holistic perspective demands an interdisciplinary approach. A diverse group of qualified professionals can bring different knowledge, professional styles, and problem-solving skills to bear on an endangered species problem. Nearly all of the book's contributors call for, or support, such an integrated, multifaceted approach. Brewer and Clark (Chapter 17) suggest that to make such efforts a reality requires mechanisms that foster interdisciplinary problem solving, and they introduce methods to analyze the endangered species problem at all scales of time, space, and complexity. Chapter 16 discusses a range of valuational, economic, and organizational variables requiring attention in endangered species cases. Several contributors cite the need to integrate the biological and social sciences. They call for simultaneous, integrated, problem-oriented analyses instead of discipline-based, sequential analyses in which various kinds of biologists, and perhaps a few social scientists, work independently and employ traditional problem-solving approaches. The conventional arrangement offers little basis for mutual ongoing education, integration, or synthesis of the parts into a whole picture of the problem, much less the solution. Instead the participants (and advisers) should represent a wide variety of disciplines, and mechanisms should be established and vigorously supported to bring these experts together to work as a unified, interdisciplinary team. Finding the means to do problem-oriented, interdisciplinary work will not be easy. It will require innovative approaches—possibly including changes in legislation, new organizational arrangements, creative leadership, additional funding, and special training programs as suggested by Westrum in Chapter 14.

This lesson calls for a fundamental change in our perception of the endangered species problem and effective solutions. Several contributors call for all participants to acquire a "policy orientation" to their work—that is, an understanding of the social, political, and organizational process surrounding the endangered species recovery effort and the participant's own role in that process. A policy orientation complements an interdisciplinary, problem-oriented view of endangered species challenges. In real life, species endangerment cannot be separated from context. They are inextricably intertwined. Chapter 3 illustrates the utility of taking an integrated, interdisciplinary view, and several other chapters describe the complex connection between conservation problems and their social contexts. Many obstacles to species restoration are rooted in the valuational, economic, or political dimensions of the situation.

Meta-Lesson 2: There is no substitute for sound professional training well grounded in state-of-the-art theories and techniques. Endangered species restoration efforts are usually risky, urgent, and highly complex. Programs should therefore rely on state-of-the-art principles and practices to improve the chances for successful recovery. This, in turn, necessitates employing professionals, be they biologists or social scientists, who are knowledgeable about the species and the problems it faces, well grounded in their disciplines, up to date in both concepts and method for conserving and recovering endangered populations, and, as discussed in Meta-Lesson 1, willing and able to work in an interdisciplinary setting. Finding and training such professionals will be no easy task.

Lack of reliable knowledge about the species and the problem it faces can hamper efforts to develop effective conservation programs and convince decision makers of the urgency of providing additional support, as illustrated in Chapter 11. Jackson (Chapter 7) admonishes those involved in endangered species recovery programs to gain reliable knowledge about the species and its plight as a basis for management actions and credibility. This enterprise should include an honest assessment of uncertainty coupled with an independent outside review of the reliability of the current state of knowledge.

Every contributor to this volume has emphasized the necessity of updated professional training, even interdisciplinary training. Conserving and restoring endangered species usually requires specialized training. But traditional education and practice in game management do not prepare wildlife biologists for managing small, endangered populations. Instead, they must be trained in the concepts of conservation science. Wildlife biologists and social scientists alike must maintain their familiarity with current technical knowledge and methods and be able to assess the utility and rigor of theoretical advances. Snyder (Chapter 8) suggests that doing so may require an increased reliance on professionals from outside the agencies. Minta and Kareiva (Chapter 12) call for a stronger connection between science and policy. They review current advances in conservation science, discussing the utility of these ecological theories and techniques with respect to endangered species management and policy. The chances for successful species recovery will surely increase if professionals not only understand but can apply the most recent and rigorous theories and techniques for conservation.

Meta-Lesson 3: High-performance teams should be employed to carry out the work of species restoration. Restoration teams, whatever form they take, should be established as soon as possible in the recovery effort and supported continuously until the species is fully recovered and delisted. Recovery teams, task forces, working groups, and advisory boards are often used in endangered species conservation, but they seldom are the kind of team that is actually needed to meet the restoration challenge. Special groups should be structured, staffed, empowered, and encouraged to act as flexible, high-performance teams oriented toward the goal of species recovery. Several contributors discuss the attributes of high-performance teams and offer suggestions for developing such teams.

The parent organizations that establish high-performance teams must support them financially, politically, morally, and organizationally and buffer them from outside pressure. Although the recovery effort may involve a combination of groups, a single team should ultimately be responsible for the work. The team should function as a "human envelope" around the species; if the envelope fails, the species is likely to go extinct. Teams are often the best means to accomplish the complex, uncertain, and urgent work of species restoration, and teams, given their interactive nature, may learn from experience. Even though coordinators can be very helpful, no person is an adequate substitute for a team of qualified specialists. Even after delisting, the team should be called on periodically to evaluate the species' status and the adequacy of ongoing management activities.

In designing and operating successful teams, it is important to understand why some teams have failed: intention, incompetence, ignorance, and ill fortune are several reasons. Teams should be composed of people with a wide variety of skills, expertise, and experience, rather than simple political representation of the major interests. They must show keen observation, use reliable knowledge, generate numerous ideas and options, and manifest good leadership. They must be organized flexibly and managed to encourage open participation by all members. They must have stable, independent funding and authority, clear and frequent contact with top administrators and key decision makers, open communication, a competent, task-oriented staff, an incentive system to get the work done quickly, and short chains of command. Teams should have visibility and accountability; they must be competent in dealing with the social as well as the biological dimensions of the task; and, insofar as possible, they must have control over the political processes that directly shape recovery efforts. Chapter 16 suggests that nongovernmental teams be set up to compete with the agencies or in lieu of government teams: they may well be more flexible, more diverse, more successful, and less costly.

Meta-Lesson 4: *Government agencies must be made more effective in dealing with endangered species recovery; in some cases, this means an overhaul of the entire government system that is involved.* Current recovery efforts are dominated by federal and state governments, and every chapter of this book offers lessons for improving government or even reinventing it. If the high-performance teams recommended in Meta-Lesson 3 are to maximize their effectiveness, they must be backed by supportive organizations that are themselves well organized. Yet a consistent theme throughout is that well-managed programs are the exception rather than the rule. Several chapters discuss problems with traditional agency organization and management. The authors condemn rigid bureaucratic arrangements and call for open, flexible, task-oriented efforts. Most biologists and agency bureaucrats will be unable to bring the essential organization and management skills to the recovery program. Instead, organization and management specialists are probably needed.

Making government programs more effective will involve a range of considerations and eventual changes. Chains of command should be kept short. Field-level decisions should be made by field-level people instead of micromanaged from the top down by administrators. Agencies should strive to do a better job of analyzing a progam's context, including the diversity of human values involved, and improving policy and decision-making processes in endangered species restoration efforts. Yaffee (Chapter 3) suggests that the agencies should promote integration rather than fragmentation of knowledge, authority, and effort, promote long-term rather than short-term views of the restoration challenge, and promote cooperation rather than competition among the varing interests. Furthermore, agencies should be more open. They should actively solicit new information and support for endangered species conservation by working openly with individuals and groups outside the government—including regular and ongoing review by outside appraisers. Chapter 2 recommends limiting agency discretion by changing the ESA's language; Chapter 5 recommends changing agency job performance standards so that incentives favor saving species. Agencies must remember that their legitimacy is based on their technical expertise and the kinds of information they bring to public decision making; every effort, therefore, should be made to ensure that the expertise and information are of the highest quality. Before these specific changes can be applied with any chance of success, some contributors believe there must be major changes in agency cultures and structures. In other words, government must be reinvented to save endangered species. Groves (Chapter 10), however, suggests that since changing the agencies will be too difficult, participants in recovery programs should learn to maximize their effectiveness working within government bureaucracies.

As the organizational dimension of endangered species work is invisible, its importance is grossly underappreciated. The contributions to this book suggest that great advances in endangered species recovery may be possible through improving the organization and management of recovery programs. Because government so strongly dominates endangered species conservation, improving performance at the program and team levels may offer the best hope of saving species.

Meta-Lesson 5: Clearly written prescriptions and guidelines in the ESA, founded on modern conservation biology and social sciences, themselves will facilitate implementation by professionals and organizations. The direction of endangered species conservation is ultimately authorized by the ESA. The goals and provisions of the act must rest soundly on modern conservation biology, permit rapid inclusion of new knowledge, and actually work to make implementation successful. Most of the case studies cite the need to clarify the language of the ESA during its reauthorization in order to support implementation efforts.

Specific improvements to the ESA suggested by the authors in this book would address both biological and nonbiological concerns. Greater protection for ecosystems, plants, and invertebrates, for example, as well as modified listing and taking requirements, should be addressed. Additionally, the ESA should be amended to limit government agency discretion and minimize the dominant role of agencies. Mattson and Craighead (Chapter 5) urge changes in the ESA to clarify time frames and levels of confidence for defining recovery, as well as whether the burden of proof should fall on demonstrating that a proposed action does or does not harm the species. J. Alan Clark (Chapter 2) suggests that the standard of judicial review and the basis for legal standing to sue the government under the ESA should be changed to make it easier for outside parties to sue. A multidirectional approach to endangered species conservation—including changes in legislation, agency behavior, program structure, and more, involving both governmental and nongovernmental participation—should be authorized and encouraged. And, of course, adequate resources must be made available.

Without a sound policy prescription, implementation is virtually impossible. Currently the ESA is open to broad agency interpretation and political manipulation. Congressional representatives must understand real-life implementation problems—and so too should the conservation groups pushing for improved endangered species conservation. Apart from calling for the ESA to clarify time frames for conservation goals and degrees of assurance for species recovery, Chapter 5 recommends that all participants in endangered species programs should familiarize themselves with the mandates of the ESA and the means available to achieve the goals.

Meta-Lesson 6: Organizations and professionals must initiate conservation actions earlier, set recovery goals high enough to ensure long-term species conservation, and continuously monitor overall progress. Nearly every Chapter calls for earlier initiation of conservation efforts, clearer goals, and more and better evaluation. Currently, populations of many species are permitted to fall to nearly irreversible lows before they are recognized as deserving special remedial attention. In such dire circumstances, even the most vigorous programs and best-intentioned and well-qualified professionals may not be able to reverse the trends. As a species continues to decline and approaches extinction, management options narrow, costs rise sharply, and the sense of urgency grows nerve-rackingly high. Fear of failure can become paralyzing; flexibility for experimentation approaches nil. As a result, the context of the recovery program deteriorates to a politically charged and conflict-laden mess with little room for maneuvering. Simply starting conservation *before* a species is severely endangered would alleviate much of the pressure, keep more options open, and reduce the costs.

Once a recovery program is under way, many adjustments in approaches and methods must be made according to success or failure on the ground. In other words, ongoing feedback (adaptive management) is essential for good programs—and this, in turn, depends on the program's having clear goals against which performance can be measured. Nearly all the chapters call for upgraded appraisal, feedback, and learning. As noted earlier, the ESA should be changed to alter the standard of judicial review and the basis for bringing suits against legally responsible authorities; this too is a form of appraisal and feedback. Chapter 4 calls for impartial periodic review of the whole endangered species conservation process and each species recovery program—which, in turn, will likely require an interdisciplinary national recovery evaluation team with formal advisory and oversight responsibility. Nonagency experts should staff the appraisal team, which should be a permanent oversight committee to examine progress and suggest improvements. Chapter 7 concludes that an "evaluator organization" should be created as one element of the entire national recovery effort. And Chapter 11 cites regular and thorough program review as essential for conservation.

Early action, coupled with clear recovery objectives and continuous program monitoring and evaluation, allows participants to develop more effective programs and thereby increase the chances for successful species recovery. This is one of the strongest lessons from all the contributors.

Meta-Lesson 7: Learning should be enhanced at both professional and organizational levels, and mechanisms to promote learning should be institutionalized. We now have more than twenty years' experience in ESA implementation and a wealth of accumulated individual, organizational, and policy knowledge that needs to be shared with everyone interested in more successful conservation. This knowledge can provide a basis for learning and consensus. Yet many of the chapters in this book describe programs where little learning has taken place. Obstacles to learning include rigid professional norms, compartmentalized disciplinary boundaries, and narrow organizational allegiances. In many cases, evidence suggests that professionals do learn useful lessons but have great difficulty translating them into organizational learning. Simply put: The individual lessons of hindsight have not been translated into the organizational lessons of foresight. Individual learning is much more common than organizational learning, and bureaucracies are notorious for limited learning. Moreover, the various models of professionalism described in Chaper 15 promote insight to different degrees, and insight is essential as a basis for individual learning and improvement.

As mentioned in several chapters, there are two kinds of organizational learning: simple (or single loop) and complex (or double loop). Simple learning takes place in all endangered species programs. When a program upgrades a census method or otherwise improves its collection and use of information, and thus its performance, simple learning has occurred. But complex learning—as when program personnel examine and improve the operating norms, structures, and decision-making procedures of the program and organization—appears to be very rare. In double-loop learning, the progam essentially "rewires itself" for improved performance.

Good program appraisal, from outside and from within, can improve both simple and complex learning. Many mechanisms are available to accomplish this; "Top Gun" and "maestro" schools (Chapter 14) and improved models of professionalism are good examples of things that can be done now. Decision analyses, scenario writing, decision seminars, groupthink analyses, and case study analyses could also be used to promote learning (Chapter 17). Program leaders, managers, and participants could learn "how to learn" more effectively through in-service programs that use several of these techniques. Organizations both in and out of government could hire better-educated and more experienced people, employ them in positions appropriate to their backgrounds, and allow them to exercise their professional judgment. This means that new employees must be knowledgeable in policy implementation processes and organizational systems management, as well as traditional knowledge areas.

Ultimately, university curricula in wildlife and natural resource management, conservation biology, and related programs must take the lead in educating young professionals—not just workers—in the conservation biology, social sciences, and especially the interdisciplinary approaches required. Ideally, such education would take place before professionals enter the workplace.

Meta-Lesson 8: Conservation leaders must be encouraged, trained, and empowered. Endangered species conservation needs leaders who are well versed in on-the-ground ESA implementation and who have broad training and experience to understand the many facets of the restoration challenge. Because of the interdisciplinary nature of conservation efforts, leaders must demonstrate not only technical skills but also the process skills that promote interdisciplinary teamwork. Backhouse and colleagues (Chapter 11) suggest that strong leadership at all levels in the program, from special teams to the highest levels of decision making, is necessary to project strategic vision, mobilize resources, make decisions, and inspire workers.

Like the "maestros" described in Chapter 14, good leaders must possess a high level of technical virtuosity, a high energy level, and a broad span of attention. They must set high standards, ask key questions about the task continually, and be hands-on managers. They must not be rigid agency bureaucrats—a style of leadership that spells failure for endangered species programs. Leaders must promote free discussion and the open flow of information. They must protect vital communication channels that encourage early reporting of both problems and solutions. Yaffee (Chapter 3) calls for better agency leadership in building consensus, developing broad support for programs, and working with diverse people and groups. Many of the program weaknesses cited in Chapter 4—bureaucratic obsession with control, ineffective working groups, poor use of science and experimentation—relate to patterns of weak leadership. Chapter 11 calls for improved leadership at all levels, especially among senior managers, middle managers, and recovery teams, and illustrates why high rates of turnover in leadership should be avoided. Support and direction from higher levels of leadership are very helpful but sometimes quite difficult to secure. Snyder (Chapter 8) calls for agency leaders to stay open to information. Kellert (Chapter 16) calls for top agency leaders to link science and policy and avoid the pitfalls of bureaucratic over-control.

Developing leaders with the vast array of skills recommended here would increase the success rate significantly. Many leadership traits, however, are innate or derive from years of practical experience. Such traits are not easily taught, but other more basic skills can be learned. Leadership training for these skills could be a focus of agency in-service training and professional development efforts for endangered species programs, as well as part of a university education.

Where Do We Go from Here?

This book is an initial attempt to improve endangered species conservation by learning from professionals with years of experience either implementing recovery programs or developing the theoretical knowledge needed to facilitate them. It examines only part of the endangered species problem—professional and organizational performance. As such, the volume focuses on the professional and organizational implementation problem and goes a long way toward defining it usefully but remains far from reflecting a complete solution. Many other opportunities exist to explore and learn further from past cases of ESA implementation. The benefits to endangered species conservation would be great.

Efforts similar to this one, but examining a broader array of taxa, programs, experiences, and disciplines, are necessary and we encourage work in this area. Moreover, there is a great need for more well documented case studies and for advances in both single and interdisciplinary approaches to endangered species conservation. Because most participants in endangered species restoration programs are not trained to critically evaluate, or even notice, professional and organizational implementation problems, this effort will probably require special forums to encourage such analyses and special training to help participants understand the problems they face and extract useful lessons for others. Collaboration with social scientists would greatly facilitate such a process. As the body of case material and theory grows, we will be able to experiment with new approaches to professional and organizational implementation and thereby continually improve our strategies for the conservation and recovery of endangered species.

To transform such grandiose hopes into reality will require far more than the support of a few participants from a few endangered species recovery programs and universities. What the contributors to this book here called for necessitates a fundamental change in the way the major institutions (government agencies, universities, and conservation organizations) conduct themselves. It requires honest self-appraisal, a concerted and deliberate move from single-disciplinary to truly interdisciplinary efforts, and a willingness to hire and train personnel to work as an effective part of an interdisciplinary team. The approach we advocate is forward-looking, constructive, and supportive of our nation's intent to protect its endangered species.

We hope that with this volume we will begin a dialogue within the conservation community at large. The eight "meta-lessons" define a standard that can transform weak performance to a strong show of success measured in the most tangible of terms: species survival. We understand that even those who agree in general with our analysis may disagree with particulars of our findings and interpretation. And that means they have something to contribute to the discussion, which we hope will continue beyond the pages of this book. Perhaps the Society for Conservation Biology, the Ecological Society of America, and the relevant social science societies could join forces in a series of conferences to stimulate further discussion of the problems relating to endangered species policy implementation—working to include the wide interdisciplinary focus we have identified here as essential to future improvements. Perhaps the journals of these societies could devote a special issue to improvements in ESA implementation. The specific form that the dialogue takes is, at this point, less important than the fact that the dialogue must continue.

We encourage the traditional resources management societies, many of whose members are now responsible for endangered species recovery, to play a pivotal role in directing the future of professional training. We call upon The Wildlife Society and The Society of American Foresters to help us evaluate how the changing demands on our professionals—both in terms of new areas of disciplinary training (such as the biology of small populations) and interdisciplinary training (recognition of the social, economic, and political aspects of species recovery)—can best be met within university curricula and on-the-job-training. These professional societies, as well as others, have a long history of involvement in university professional programs. Their expertise, when focused specifically to address problems of improving ESA implementation through recognition of the professional and organizational weaknesses identified in this book, could provide the momentum needed to update conservation training across this country.

Similarly, government agencies and nongovernmental conservation groups, as well as universities, can help us to reflect on our past experiences and discover a more productive future for ESA policy implementation. High-level decision makers within government might best contribute by revamping current incentive systems at the professional and organizational levels to reward performance leading to effective and efficient species recovery and maintenance of biological diversity. Nongovernmental organizations might help by identifying key figures to serve on local, state, and national recovery teams and by continuing to bridge the conversation gap between government, universities, and the public. Universities might help by improving training in disciplinary areas and interdisciplinary problem-solving approaches. Universities could offer some of the retraining programs needed by returning professionals. Again, the specifics are less important than the fact that these important steps must be taken. We must do more than manipulate endangered species' biology if we hope to save them. We must address our own performance as well.

Too many recovery efforts fall prey to their own narrow and inadequate problem analysis, crippling organizational arrangements, and limited learning approaches. The complexity of the endangered species conservation task is often little appreciated. Sometimes it is even overlooked. Weak science (or the delayed use of science and other reliable knowledge) is also all too common. Little complex learning about program performance ever occurs. These are only a few recurring weaknesses that limit the efficacy of conservation efforts. As a result, species and habitats continue to erode or disappear. Yet the recurring weaknesses in the professional and organizational aspects of endangered species programs are largely preventable and correctable.

The lessons from this book can be applied on both large and small scales. The government departments and agencies currently responsible for endangered species recovery could shoulder the responsibility for overhauling national policy for recovery teams. At a more modest level, they could institute in-service training and professional development programs in problem solving, decision making, leadership, conservation biology, interdisciplinary teamwork, and more. These same improvements could be effected at regional or local levels in order to capitalize on the extensive field experience and incorporate the interdisciplinary expertise of sociologists, organization theorists, and policy scientists. Universities, conservation organizations, state game and fish departments, and other participants could encourage, institute, and participate in these efforts.

The lessons are clear. And the opportunity to apply them is now.

There is too much at stake not to learn and continue refining our approach to endangered species restoration as we gain more knowledge. Indeed, the continuing search for improved conservation policy and better implementation is the principal challenge for people committed to the conservation of biodiversity.

Contributors

Ken Alvarez has been a biologist with the Florida Park Service since 1970. He currently supervises a staff of three biologists in a regional headquarters for state parks in southwestern Florida. He served on a U.S. Fish and Wildlife Service recovery team for the Florida panther from 1976 to 1981. In 1983 he was appointed by the governor to the Florida Panther Technical Advisory Council, a body established by the state legislature to advise the Florida Game and Fresh Water Fish Commission. Out of these experiences came a book, *Twilight of the Panther: Biology, Bureaucracy and Failure in an Endangered Species Program*, published in 1993.

Gary N. Backhouse is a principal planner with the Wildlife Management Branch of the Victoria Department of Conservation and Natural Resources. He received his B.S. from La Trobe University in 1976. He has a background in policy and project planning and management, and his current work involves threatened species policy and planning. He has been involved with the eastern barred bandicoot recovery program since late 1991 and has been program leader since late 1992. He has coauthored a book on freshwater fishes and recently completed a natural history text on wild orchids of Victoria. He has published over twenty papers on a variety of topics including freshwater fish, feral animals, and biology and conservation of birds and mammals.

Garry D. Brewer is professor of resource policy and management, business administration, and public policy studies as well as dean of the School of Natural Resources and Environment, University of Michigan. Between 1974 and 1991, he was a member of the faculty of Yale University holding the Frederick K. Weyerhaeuser (1984–1990) and Edwin W. Davis (1990–1991) chairs. He is author, coauthor, or editor of nine books and over 175 other professional publications on a wide range of substantive topics, including contributions on organizational complexity and behavior, computer applications to social and national security problems, political and economic development, forecasting and strategic planning, and several forms of environmental management and resource policy matters. Brewer has been editor of the journals *Policy Sciences* and *Simulation & Games* and has served on the editorial boards of seven other professional journals, including the *Journal of Conflict Resolution* and *Public Administration Review*. Other professional activities include membership on the boards or executive committees of the Woods Hole Oceanographic Institution, the Organization for Tropical Studies, and the Yosemite National Institute. He also serves the National Academy of Science as a member of its Board on

Environmental Studies and Toxicology, its Committee on the Outer Continental Shelf, and as chair of a panel on socioeconomics related to oil and gas operations in the outer continental shelf.

J. ALAN CLARK received his J.D. from the University of Michigan Law School in 1992 and his M.S. in natural resource policy from the School of Natural Resources and Environment at the University of Michigan in 1993. Clark has worked on endangered species issues with the National Wildlife Federation and the Environmental Defense Fund. On a research fellowship in 1992, he worked on air policy at the Environmental Protection Agency. While on an international travel fellowship during 1993, he researched legal and policy issues surrounding introduced species, endangered species, and biodiversity in Australia and New Zealand. He also spent several months working with the Royal Forest and Bird Protection Society of New Zealand. Clark has published several popular articles on endangered species and government policy in both the United States and New Zealand. Currently he is practicing environmental law with the Seattle firm of Preston, Thorgrimson, Shidler, Gates, and Ellis.

TIM W. CLARK is professor adjunct in the School of Forestry and Environmental Studies at Yale University and board president of the Northern Rockies Conservation Cooperative. He received his Ph.D. from the University of Wisconsin–Madison in 1973. His interests include conservation biology, organization theory and management, and the policy sciences. He has written over 170 papers and several books and monographs, including *Mammals of Wyoming* (coauthor, 1987), *Conservation Biology of the Black-Footed Ferret* (1989), and *Tales of the Grizzly* (coauthor, 1993). His edited books include *Conservation and Management of Small Populations* (coeditor, 1983). He has received various awards, including the Outstanding Contribution Award from the U.S. Fish and Wildlife Service and the Presidential Award from the Chicago Zoological Society. Clark has served as a member of the Policy Review Committee on Research and Resource Management in the U.S. National Park Service. He is also a member of three species survival commissions of the IUCN–World Conservation Union. He has worked for over twenty years on endangered species conservation in the United States, Australia, Inner Mongolia, and Indonesia.

ALICE L. CLARKE has worked for the Minnesota Department of Natural Resources and the U.S. Forest Service. She has served as editor of *Endangered Species Update* and as a technical consultant for a children's book on endangered species. She received her Ph.D. from the School of Natural Resources and Environment, University of Michigan, in 1993. While at the University of Michigan, she was an active participant in the Population and Environment Dynamics Program, whose members, representing a wide variety of disciplines, share a common interest in an interdisciplinary approach to issues of global change. Her dissertation explored the behavioral ecology of human dispersal, and her work has been published in traditional biological journals (*Animal Behaviour*) as well as social science journals (*Population and Development Review, Human Nature*). Clarke is currently an NSF–NATO postdoctoral fellow at the Norwegian Institute for Nature Research, University of Trondheim, where she is investigating the potential contribution of the theoretical field of behavioral ecology to critical problems of conservation biology.

JOHN J. CRAIGHEAD is currently professor emeritus of zoology and forestry at the University of Montana and chairman of the board of the Craighead Wildlife–Wildlands Institute. From 1952 to 1977 he was leader of the Montana Cooperative Wildlife Research Unit. He is recognized for his definitive research on the biology of the grizzly bear, which has spanned over thirty years. He and his brother, F. C. Craighead, Jr., pioneered radio-tracking and satellite biotelemetry techniques for large mammals and were leaders in originating the National Wild and Scenic Rivers System. He has coauthored four books with F. C. Craighead, Jr. and has also authored or coauthored five monographs and more than eighty scientific papers. The monograph *A Definitive System for Analysis of Grizzly Bear Habitat and Other Wilderness Resources* (1982) gives methodologies for describing and evaluating vegetation complexes and landforms using computer technology and LandSat multispectral imagery. Craighead has received numerous awards, including the National Geographic Society's John Oliver La Gorce gold medal for ecological accomplishments (1979). Following his conviction that scientists have a responsibility to explain to the public the results of scientific research, he has devoted almost as much time to public education as research. This has been accomplished through numerous popular articles, motion pictures, wildlife photography, television documentaries, seminars, and lectures.

JAMES E. CROWFOOT was formerly on the faculty of the University of Michigan's School of Natural Resources and Environment, where he taught courses on management of environmental organizations and was chair of the Resource Policy and Behavior Concentration. He is now President of Antioch College in Ohio. He is also director of the Pew Scholars Program in Conservation and Environment and a member of the university's Program on Conflict Management Alternatives. His recent publications include *Environmental Disputes: Community Involvement in Conflict Resolution* (coauthored with Julia Wondolleck) and chapters titled "Conservation Leadership in Academia," "Towards an Interactive Process for Siting National Parks in Developing Countries," and "Multicultural Organization: What Is Needed in the USDA Forest Service." Crowfoot received his Ph.D. in organizational psychology from the University of Michigan in 1972 and has earlier degrees in physics and theology. From 1983 to 1990 he was dean of the School of Natural Resources and Environment.

CRAIG R. GROVES is director of the Western Heritage Task Force for The Nature Conservancy in Boulder, Colorado. Prior to this position he was employed as a staff biologist in the Nongame and Endangered Wildlife Program of the Idaho Department of Fish and Game and as coordinator of Idaho's Natural Heritage program. He received his M.S. in zoology from Idaho State University in 1981. His primary interests are conservation biology (particularly the ecology of threatened, endangered, and sensitive species), neotropical migrants, information management, and gap analysis. He has received Regional Forester awards from two regions of the Forest Service for his work with sensitive species and a Special Recognition Award from the Idaho Fish and Game Commission for work with endangered, threatened, and sensitive wildlife and plants. Groves is the author of numerous technical and popular papers on nongame species. He has worked for over ten years on endangered species conservation in the western United States and the Caribbean.

JEROME A. JACKSON is professor of biological sciences at Mississippi State University. He has worked with the red-cockaded woodpecker for more than twenty-five years; served as recovery team leader and principal author of the first recovery plan for the species; participated as an expert witness in several legal actions involving the species (both for and against federal agencies); worked with several federal and state agencies, private industries, and conservation groups regarding the species; and worked throughout the range of the species. He is a fellow of the American Ornithologists Union and the Explorers Club and has received numerous awards for his research and teaching. He has published more than 150 articles in professional journals, nearly a hundred in popular magazines, and is the author of a forthcoming book about red-cockaded woodpeckers. Jackson is also a member of the U.S. Fish and Wildlife Service Advisory Committee for the ivory-billed woodpecker and has searched for that species and studied its habitat needs and decline in the United States and Cuba. He is cohost of the weekly television feature "Mississippi Outdoors."

PETER M. KAREIVA is professor of zoology at the University of Washington. His research emphasizes the application of simple models to environmental problems ranging from conservation to agriculture and risk analysis. He has critically examined the uncertainty of viability analyses for spotted owls and desert tortoises, and he is generally concerned with spatial statistics as a tool for melding extensive geographical data with population models. His empirical work concerns the interplay of dispersal and habitat heterogeneity in governing population dynamics. Kareiva serves on the editorial boards of *Ecology, Oecologia, Theoretical Population Biology, Molecular Ecology,* and *Comments on Theoretical Biology.*

STEPHEN R. KELLERT is professor of social ecology at the Yale University School of Forestry and Environmental Studies. He has published over a hundred articles, books, and monographs, mainly on natural resource policy, human dimensions in wildlife management, and the conservation of biological diversity. He has conducted extensive research on human values relating to nature and its conservation and protection. His most recent books include *The Biophilia Hypothesis* (1993 with E. O. Wilson) and *Ecology, Economics, Ethics: The Broken Circle* (1991 with F. H. Bormann).

DAVID J. MATTSON is currently a research scientist with the University of Idaho Cooperative Park Studies Unit and is pursuing a Ph.D. in fish and wildlife resources. He previously worked eleven years as a wildlife biologist for the Interagency Grizzly Bear Study Team in the Yellowstone ecosystem and held primary responsibility for investigating grizzly bear habitat relationships, including their relations with humans. During his Yellowstone tenure, he was also engaged in helping managers develop grizzly bear conservation strategies. He has written over fifty papers and technical reports, most related to grizzly bear habitat relationships and conservation. He has also been consulted by researchers and managers from Canada, Russia, Norway, and France on bear research and management programs, reintroductions, ecosystem management, and cumulative effects assessments. In addition to his graduate studies that emphasize the analysis of habitat selection by Yellowstone grizzlies, Mattson is exploring the implications of behavioral structuring and gene flow to population viability, and he is interested in natural resource policy implementation.

BRIAN J. MILLER has a postdoctoral position with the National Autonomous University of Mexico (UNAM) and is a research associate with the Smithsonian Institution. He received a Ph.D. from the University of Wyoming on black-footed ferret behavior in December 1988 and then received a Smithsonian Postdoctoral Fellowship Award to study reintroduction techniques for that species. He has published thirty scientific articles, written a monograph for the *Ethology* series, coedited a book on prairie dog management for black-footed ferret reintroduction sites, and is cowriting a book on black-footed ferret biology and conservation. While working on ferrets, he participated in founding a conservation education course on the Navajo Nation. Miller is presently working with UNAM to initiate a protected area in the Chihuahuan desert that would include the largest remaining black-tailed prairie dog complex in North America.

STEVEN C. MINTA is assistant professor in the Environmental Studies Board, University of California at Santa Cruz. His dissertation focused on carnivore ecology. From 1988 to 1993 he was affiliated with the University of Montana and NGOs, such as the Northern Rockies Conservation Cooperative, and pursued his interests in population monitoring, predation, and spatial analysis. He has tried to bring the tools of academic research to bear upon practical problems of land use and species management (black-footed ferret, wolf, fisher, grizzly bear, ungulates). In 1993, Minta was funded by the National Council for Air and Stream Improvement, Inc., as a visiting researcher at the University of Washington, Seattle. There he worked on large-scale monitoring of forest predators as well as other landscape-level issues and methods, work that he now continues at U.C. Santa Cruz. He has published in *Ecology*, *Ecological Applications*, *Oecologia*, and *Journal of Mammalogy*.

RICHARD P. READING is a wildlife biologist and endangered species specialist with the U.S. Bureau of Land Management in Malta, Montana, and a research associate with the Northern Rockies Conservation Cooperative. He received his Ph.D. in wildlife ecology from Yale University in 1993. In his dissertation on endangered species reintroduction paradigms, he took an interdisciplinary approach focusing on the biological/technical, valuational, and organizational aspects of endangered species reintroductions. His publications in technical bulletins and journals, such as *Conservation Biology, Endangered Species Update, Environmental Management,* and *Society and Natural Resources,* reflect his interest in developing an interdisciplinary approach to endangered species conservation. He is currently coauthoring a book on black-footed ferret conservation. Reading is a member of the Montana Black-Footed Ferret Working Group and the Reintroduction Specialist Group of the IUCN–World Conservation Union's Species Survival Commission.

NOEL F. R. SNYDER is currently director of parrot programs for Wildlife Preservation Trust International, although he has worked in the Endangered Wildlife Program of the U.S. Fish and Wildlife Service for most of his career. From 1972 through 1976, he was field leader of the Puerto Rican parrot recovery program; from 1978 to 1980, he led research efforts on the Everglade kite in Florida; from 1980 to 1986, he led conservation and research efforts for the California condor. His current efforts center on leading a cooperative reintroduction program for thick-billed parrots in Arizona. His most notable books and monographs include *The Parrots of Luquillo, Natural History and Conservation of the Puerto Rican Parrot* (coauthored with J. W. Wiley and C. B. Kepler, 1987), *Biology and Conservation of the California Condor* (coauthored with H. Snyder, 1989), and *New World Parrots in Crisis: Solutions from Conservation Biology* (coedited with S. Beissinger, 1992). Snyder's awards include the William Brewster Award of the American Ornithologists Union in 1989, a Distinguished Achievement Award from the Society for Conservation Biology in 1989, and the Conservation Award of the Zoological Society of San Diego in 1992.

RICHARD L. WALLACE is a doctoral student in the School of Forestry and Environmental Studies at Yale University, where he is studying the organization and implementation of federal programs under the Endangered Species Act and the Marine Mammal Protection Act. Before attending Yale, he was special assistant to the executive director of the Marine Mammal Commission. There he participated in the development, review, and evaluation of government programs for the research and management of marine mammals and their habitats, assisted in the production of the commission's annual report, and compiled *The Marine Mammal Commission Compendium of Selected Treaties, International Agreements, and Other Relevant Documents on Marine Resources, Wildlife, and the Environment*, the definitive collection of international law in the environmental field. Wallace continues to work for the commission on a contractual basis. He has also written or contributed to published papers on subjects as diverse as the application of E. O. Wilson's biophilia hypothesis to human interactions with the marine environment and the role of a policy orientation in the conservation of biological diversity.

RON WESTRUM is a social scientist who specializes in the study of highly creative organizations. Educated at Harvard (B.A.) and the University of Chicago (Ph.D. in sociology), he is professor of sociology and interdisciplinary technology at Eastern Michigan University. He is a member of the Human Factors Society, the American Association for the Advancement of Science, and the Institute for Management Science, as well as several other professional organizations. Westrum is also president of Aeroconcept, a management consulting firm. His publications include two books, *Complex Organizations* (1984) and *Technologies and Society: The Shaping of People and Things* (1991), and numerous articles. He is the editor of the newsletter *Social Psychology of Science*, which reviews cognitive issues in science and technology. Westrum is currently writing a history of the Sidewinder missile.

JULIA M. WONDOLLECK is on the faculty of the School of Natural Resources and Environment at the University of Michigan, where she teaches courses in environmental conflict management and negotiation skills for resource managers. She is the author of *Public Lands Conflict and Resolution: Managing National Forest Disputes* (1988) and coauthor (with James Crowfoot) of *Environmental Disputes: Community Involvement in Conflict Resolution* (1990). Wondolleck received her Ph.D. in 1983 from the Massachusetts Institute of Technology in environmental policy and planning. Her research and writing since that time have focused on the causes and consequences of environmental disputes, specifically in the management of public lands and the conservation of endangered species, and on finding better ways of managing these conflicts.

STEVEN L. YAFFEE is a faculty member in the School of Natural Resources and Environment at the University of Michigan, where he teaches courses in natural resource policy and administration, negotiation skills, environmental history, and biodiversity and public policy. His research focuses on understanding and improving public decision-making processes and exploring the behavior of administrative agencies and interest groups involved in implementing public policies. He has worked for more than fifteen years on federal endangered species policy and is the author of *Prohibitive Policy: Implementing the Federal Endangered Species Act* (1982) and *The Wisdom of the Spotted Owl: Policy Lessons for a New Century* (1994). Yaffee received his Ph.D. in 1979 from the Massachusetts Institute of Technology in environmental policy and planning and has earlier degrees in natural resources. He has taught at MIT and the Kennedy School of Government at Harvard and has been a researcher at the Oak Ridge National Laboratory and the Conservation Foundation/World Wildlife Fund.

Index

Adaptive management, 240–41, 288, 294, 295
Administrative Procedures Act, 64
Agencies:
federal, *see* Federal agencies
lessons, 422–23
see also specific agencies
Agua Caliente Indian Tribe, 316
Alternative dispute resolution, *see* Conflict management perspective: alternative dispute resolution
Alvarez, Ken, 205–24, 433
American Ornithologists Union, 164
Andean condors, 192
Argyris, C., 146, 148
Army, U.S., 330–31
Section 7 consultations, 166–67
Army Corps of Engineers, 163, 320
Artificial cavity excavation techniques, 161, 162
Attitudes of professionals, changing, 359–63
Ault, Captain Frank, 345
Australian eastern barred bandicoot recovery program, 251–68
captive breeding, 255, 257, 267
diet of, 253
evaluation of conservation effort in 1991, 256–62
geographic distribution, 252–53
history of, 251
lessons of, 263–68
on defining the problem, 264
gain reliable knowledge early on, 263–64
on leadership, 265
organization and management must match the problem faced, 265–66
program review, 266–68
life cycle, 253
management of wild population, 254–56
near extinction of Victoria's population, 254
population viability analysis, 254, 261
public education, 255
recovery teams, 255, 259
reorganization and progress, 262–63
threats facing, 253–54
weaknesses in original program, 256–62
in communication, 260
comprehensive management, absence of, 260–61

in expertise and breadth of view, 258
inadequate definition of the problem, 256–57
lack of reliable knowledge, 256–57
in leadership, 258–59
ongoing evaluation, lack of, 261–62

Babbitt, Bruce, 294–95
Backhouse, Gary N., 251–68, 433
Bad luck and failure of recovery efforts, 329–37
Bass, Oron P. (Sonny), 207
Baudy, Robert, 207, 214, 215, 216–17
Bean, M. J., 20, 25, 28–29, 31, 36, 39, 314
Beatley, T., 314
Behavior of professionals, changing, 359–63
Bem, D. J., 393
Berg, W. E., 91
Big Cypress National Reserve, 207, 215
Biodiversity, legislation to conserve, 66
Biological sciences, 407
see also Conservation science perspective
Black-Footed Ferret Advisory Team (BFAT), 80, 83
Black-footed ferret recovery program, 37, 73–95, 186, 332
captive breeding, 76, 85
conflicts in, 78–79
description of black-footed ferret, 75
history of, 75–76
lessons, 94–95
prairie dogs and, 75, 76
professional and organizational performance, 77–94
measuring program success, 91–94
program control, 84–87, 380
program structure, 77–79
science, role of, 87–91, 92
working groups, 80–84
reintroduction, 76, 79, 86, 87, 89–90
Bonytail, 318
Boyce, M. S., 287–88
Brewer, Garry D., 391–410, 419, 433–34
Brown, G., 114
Brunner, R. D., 398
Bureaucracies, 78, 194–95, 219, 220, 224, 235, 237, 241, 242, 260, 265–66, 328, 336, 408
intuitive thinking and, 338
maestros and, 344
methods for avoiding, 383